T0335568

VOLUME ONE HUNDRED AND EIGHT

ADVANCES IN
COMPUTERS

VOLUME ONE HUNDRED AND EIGHT

ADVANCES IN
COMPUTERS

Edited by

ATIF M. MEMON
College Park, MD,
United States

ACADEMIC PRESS

An imprint of Elsevier

Academic Press is an imprint of Elsevier
50 Hampshire Street, 5th Floor, Cambridge, MA 02139, United States
525 B Street, Suite 1800, San Diego, CA 92101-4495, United States
The Boulevard, Langford Lane, Kidlington, Oxford OX5 1GB, United Kingdom
125 London Wall, London, EC2Y 5AS, United Kingdom

First edition 2018

Notices

Knowledge and best practice in this field are constantly changing. As new research and
experience broaden our understanding, changes in research methods, professional practices,
or medical treatment may become necessary.

Practitioners and researchers must always rely on their own experience and knowledge in
evaluating and using any information, methods, compounds, or experiments described
herein. In using such information or methods they should be mindful of their own safety and
the safety of others, including parties for whom they have a professional responsibility.

To the fullest extent of the law, neither the Publisher nor the authors, contributors, or editors,
assume any liability for any injury and/or damage to persons or property as a matter of
products liability, negligence or otherwise, or from any use or operation of any methods,
products, instructions, or ideas contained in the material herein.

ISBN: 978-0-12-815119-8
ISSN: 0065-2458

For information on all Academic Press publications
visit our website at https://www.elsevier.com/books-and-journals

Working together
to grow libraries in
developing countries

www.elsevier.com • www.bookaid.org

Publisher: Zoe Kruze
Acquisition Editor: Zoe Kruze
Editorial Project Manager: Shellie Bryant
Production Project Manager: James Selvam
Cover Designer: Christian J. Bilbow

Typeset by SPi Global, India

CONTENTS

PREFACE

This volume of *Advances in Computers* is the 108th in this series. This series, which has been continuously published since 1960, presents in each volume four to seven chapters describing new developments in software, hardware, or uses of computers. For each volume, I invite leaders in their respective fields of computing to contribute a chapter about recent advances.

Volume 108 focuses on five topics. In Chapter 1, entitled "Model-Based Testing for Internet of Things Systems," authors Ahmad et al. posit that the Internet of Things (IoT) has become a means of innovation and transformation. Applications extend to a large number of domains, such as smart cities, smart homes, and healthcare. The Gartner Group estimates an increase up to 21 billion connected things by 2020. The large span of "things" introduces problematic aspects, such as interoperability due to the heterogeneity of communication protocols and the lack of a globally accepted standard. The large span of usages introduces problems regarding secure deployments and scalability of the network over large-scale infrastructures. The chapter describes the challenges for IoT testing, includes a state-of-the-art testing of IoT systems using models, and presents a Model-Based-Testing-As-A-Service approach to respond to its challenges through demonstrations with real use cases.

In Chapter 2, entitled "Advances in Software Model Checking," authors Siddiqui et al. observe that society is becoming increasingly dependent on software which results in an increasing cost of software malfunction. At the same time, software is getting increasingly complex and testing and verification are becoming harder. Software model checking is a set of techniques to automatically check properties in a model of the software. The properties can be written in specialized languages or be embedded in software in the form of exceptions or assertions. The model can be explicitly provided in a specification language, can be derived from the software system, or the software system itself can be used as a model. Software model checkers check the given properties in a large number of states of the model. If a model checker is able to verify a property on all model states, it is proven that the property holds and the model checker works like a theorem prover. If a model checker is unable to verify a property on all model states, the model checker is still an efficient automated testing technique. This chapter discusses advances in software model checking and focuses on techniques that use the software as its model and embedded exceptions or assertions as the properties to be verified.

Such techniques are most likely to be widespread in the software industry in the coming decade due to their minimal overhead on the programmer and due to recent advances in research making these techniques scale.

Chapter 3, "Emerging Software Testing Technologies," by Lonetti et al., provides a comprehensive overview of emerging software testing technologies. Beyond the basic concepts of software testing, the chapter addresses prominent test case generation approaches and focuses on more relevant challenges of testing activity as well as its role in recent development processes. An emphasis is also given to testing solutions tailored to the specific needs of emerging application domains. Software testing encompasses a variety of activities along the software development process and may consume a large part of the effort required for producing software. It represents a key aspect to assess the adequate functional and nonfunctional software behavior aiming to prevent and remedy malfunctions. The increasing complexity and heterogeneity of software pose many challenges to the development of testing strategies and tools.

In Chapter 4, "Optimizing the Symbolic Execution of Evolving Rhapsody Statecharts," authors Khalil and Dingel present two optimization techniques to direct successive runs of symbolic execution toward the impacted parts of an evolving state machine model using memoization (MSE) and dependency analysis (DSE), respectively. Model-driven engineering (MDE) is an iterative and incremental software development process. Supporting the analysis and the verification of software systems developed following the MDE paradigm requires to adopt incrementally when carrying out these crucial tasks in a more optimized way. Communicating state machines are one of the various formalisms used in MDE tools to model and describe the behavior of distributed, concurrent, and real-time reactive systems (e.g., automotive and avionics systems). Modeling the overall behavior of such systems is carried out in a modular way and on different levels of abstraction (i.e., it starts with modeling the behavior of the individual objects in the system first and then modeling the interaction between these objects). Similarly, analyzing and verifying the correctness of the developed models to ensure their quality and their integrity is performed on two main levels. The intralevel is used to analyze the correctness of the individual models in isolation of the others, while the interlevel is used to analyze the overall interoperability of those that are communicating with each other. The evaluation results of both techniques showed significant reduction in some cases compared with the standard symbolic execution technique.

Chapter 5, "A Tutorial on Software Obfuscation," by Banescu et al., discusses the important problem of protecting a digital asset once it leaves the cyber trust boundary of its creator; this is a challenging security problem. The creator is an entity which can range from a single person to an entire organization. The trust boundary of an entity is represented by all the (virtual or physical) machines controlled by that entity. Digital assets range from media content to code and include items such as music, movies, computer games, and premium software features. The business model of the creator implies sending digital assets to end-users—such that they can be consumed—in exchange for some form of compensation. A security threat in this context is represented by malicious end-users, who attack the confidentiality or integrity of digital assets, in detriment to digital asset creators and/or other end-users. Software obfuscation transformations have been proposed to protect digital assets against malicious end-users, also called Man-At-The-End (MATE) attackers. Obfuscation transforms a program into a functionally equivalent program which is harder for MATE to attack. However, obfuscation can be used for both benign and malicious purposes. Malware developers rely on obfuscation techniques to circumvent detection mechanisms and to prevent malware analysts from understanding the logic implemented by the malware. This chapter presents a tutorial of the most popular existing software obfuscation transformations and mentions published attacks against each transformation. The chapter presents a snapshot of the field of software obfuscation and indicates possible new research directions.

I hope that you find these articles of interest. If you have any suggestions of topics for future chapters, or if you wish to be considered as an author for a chapter, I can be reached at atif@cs.umd.edu.

PROF. ATIF M. MEMON, PH.D,
College Park, MD, United States

CHAPTER ONE

Model-Based Testing for Internet of Things Systems

Abbas Ahmad*,‡, Fabrice Bouquet*, Elizabeta Fourneret†, Bruno Legeard*,†

*Institut FEMTO-ST – UMR CNRS 6174, Besançon, France
†Smartesting Solutions & Services – 18, Besançon, France
‡Easy Global Market, Valbonne, France

Contents

Advances in Computers, Volume 108
ISSN 0065-2458
https://doi.org/10.1016/bs.adcom.2017.11.002

1

Abstract

The Internet of Things (IoT) is nowadays globally a mean of innovation and transformation for many companies. Applications extend to a large number of domains, such as smart cities, smart homes, and health care. The Gartner Group estimates an increase up to 21 billion connected things by 2020. The large span of "things" introduces problematic aspects, such as interoperability due to the heterogeneity of communication protocols and the lack of a globally accepted standard. The large span of usages introduces problems regarding secure deployments and scalability of the network over large-scale infrastructures. This chapter describes the challenges for the IoT testing, includes state-of-the-art testing of IoT systems using models, and presents a model-based testing as a service approach to respond to its challenges through demonstrations with real use cases involving two of the most accepted standards worldwide: FIWARE and oneM2M.

1. INTRODUCTION

The Internet of Things (IoT) interconnects people, objects, and complex systems. While this is as vast as it sounds, spanning all industries, enterprises, and consumers, one of the challenges facing the IoT is the enablement of seamless interoperability between each connection. The massive scale of recent Distributed Denial of Service (DDoS) attacks (October 2016) on DYNs servers [1] that brought down many popular online services in the United States, gives us just a glimpse of what is possible when attackers can leverage up to 100,000 unsecured IoT devices as malicious endpoints. Therefore, ensuring security is a key challenge.

Systematic and automated testing of the IoT based on models is a way of ensuring their security. Nevertheless, traditional testing methods, whether they are based on models or not, require adaptation due to the IoT heterogeneous communication protocol, complex architecture, and insecure usage context, the IoT must be tested in their real use case environment: service-based and large-scale deployments.

A model-based testing as a service (MBTAAS) approach is proposed to respond to these challenges where the IoT are tested in their natural environment: large-scale and service-oriented. The MBTAAS approaches goal is to ensure conformance and security over proposed solutions (APIs, devices, etc.). The models of the approach mimic the intended behaviors, thus the services are seen as Black Boxes for testing.

Models can be described in different ways, depending on the discipline. They can be described by use of diagrams (UML, BPMN, etc.), tables, text, or other types of notations, such as the ETSI TPLan. They might be expressed in a mathematical formalism or informally. Although computing equipment and software play an important role for the application of modeling in science and engineering, modeling is not as ubiquitous in software engineering as it is in other engineering disciplines. However, formal methods are one discipline where modeling has been applied in the areas of safety and security-critical software systems. Test phases are in the last stages of system development. They are therefore strongly constrained by delivery periods. In general, they suffer from the accumulated delays during the overall project.

Currently, the verification and validation process for the IoT having no systematic or automatic process is often underestimated and outsourced to reduce costs. All these factors affect the quality of current products. Testing is a difficult and time-consuming activity. Indeed, it is not possible to test everything because the quantity of test cases to be applied could be potentially infinite for the majority of modern systems. The difficulty of testing lies in the selection of relevant test cases for each phase of validation. It also lies in the absence of a true reflection of test optimization guided by the maintenance of the final quality while minimizing costs.

We present hereafter a technology which is used ever more in the IoT domain to avoid bugs, to improve large-scale quality, interoperability, and reduce security threats: model-based testing (MBT). MBT allows the automatic generation of software test procedures, using models of system requirements and behavior. Although this type of testing requires more upfront effort in building the model, it offers substantial advantages over traditional testing methods [2]:

- Rules are specified once.
- Project maintenance is lower. There is no need to write new tests for each new feature. Once a model is obtained, it is easier to generate and regenerate test cases than it is with hand-crafted test cases.
- The design is fluid. When a new feature is added, a new action is added to the model to run in combination with existing actions. A simple change can automatically ripple through the entire suite of test cases.
- Design more and code less. This makes the testing process generic with respect to the development languages used to implement an application, for instance, for different operating systems (OS).
- High coverage. Tests continue to find bugs, not just regressions due to changes in the code path or dependencies.
- Model authoring is independent of implementation and actual testing, so that these activities can be performed by different members of a team concurrently.

- Unit testing is elementary and it is not sufficient to check the functionalities, thus MBT is complementary to unit testing.
- Ensures that the system is behaving in the same sequence of actions.
- MBT adoption in the industry as an integrated part of the testing process has largely increased.
- Mature and efficient tools exist to support MBT.

Section 2 presents the specific key challenges of testing IoT system. Section 3 gives an insight on the state of the art, in general, about MBT and existing MBT tools to select solutions adapted to the IoT testing approach. Section 4 presents MBT for IoT conformance testing. Section 5 presents MBT for IoT security testing. Section 6 presents online MBT for IoT robustness testing. Section 7 depicts an exploration of the MBTAAS approach. We finally provide our lessons learnt and discuss the different approaches in Section 8, before providing concluding remarks in Section 9.

2. CHALLENGES OF TESTING IoT SYSTEMS

Gartner states that more than 6.4 billion IoT devices were in use in 2016, and that number will grow to more than 20 billion by 2020. Testing these devices, which range from refrigerators that automatically place orders to the supermarket to self-driving cars, will be one of the biggest challenges encountered by the device manufacturers and integrators in the coming years. Effective testing is critical and defining the best approach implies dissecting the different IoT layers involved.

2.1 IoT Layers

There are four major layers in an IoT system, as illustrated in Fig. 1. From bottom to top, the first is the *sensor layer*. Then, there is the *gateway and network layer*. On the third layer, the *management service layer* and finally as a fourth layer, the *application layer*.

The *sensor layer* uses and produces data. Thus, the *IoT devices*, at this bottom layer, can behave differently whether their purpose is to generate data (a *data producer*), or to use produced data (a *data consumers*). Note that it is significantly different from the *application layer* where data are also used and produced. For example, the complex event processing (CEP) application is a data consumer application within the application layer. Its purpose is to read input data and produce output data depending on a defined rule. An example of a rule defined in a CEP application: *if the data of two temperature sensors in*

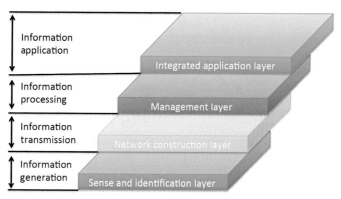

Fig. 1 IoT layers.

the same room exceed a certain value, then an alert is triggered. The produced data are in this case the generated alert.

The *service connectivity and network layers* are represented by the sensors, actuators, tags (which include radio frequency IDs (RFID) and barcodes), and other types of data producers/consumers. At the gateway and network layer, wide area networks, mobile communication networks, Wi-Fi, Ethernet, gateway control, etc. are considered. Then, in the management service layer, device modeling configuration and management is a major focus. Data flow management and security control need to be provided at the management service layer. Finally, the overall application layer is where the applications dedicated to energy, environment, health care, transportation, supply chain, retail, people tracking, surveillance, and many other endless applications are located.

2.2 Key Challenges

The different layers of the IoT bring a new level of complexity to performance monitoring and conformance testing in terms of scalability, interoperability, and security. Testers must adapt to new platforms and techniques to ensure that they can address the challenges of testing the IoT devices and applications to deliver the best experience to the end user. With large numbers of devices involved when talking about the IoT, scalability is defined as one of the most important key challenges. The mass of different devices that the IoT introduces, each with different protocols, brings along all the perks of a heterogeneous environment: the interoperability challenge. With many different devices and the amount of generated data in the IoT infrastructure, it is difficult to keep track of the privacy and confidentiality of the produced data. Security is a major challenge for the success of the IoT expansion, and we will address this key challenge for testing the IoT systems in the next sections.

2.2.1 Scalability

Connected IoT devices rely on fast communication. Consequently, network status can have a significant effect on device performance. Smart devices often experience problems with network infrastructure, such as overburdened Wi-Fi channels, unreliable network hardware, and slow or inconsistent Internet connections. The IoT devices and applications must be tested across these different conditions to ensure that they respond correctly without losing data.

2.2.2 Interoperability

Each IoT device has its own hardware and relies on software to drive it. Application software will also integrate with the IoT devices, issuing commands to the device and analyzing data gathered by the device. Connecting "things" as devices requires one to overcome a set of problems arising in the different layers of the communication model. Using device data or responding to a devices request requires an IoT deployment to interact with a heterogeneous and distributed environment. Indeed, devices are most likely to be running several protocols (such as HTTP, MQTT, COAP), through multiple wireless technologies. Devices have many particularities and it is not feasible to provide a testing solution where one size fits all. Some devices are resource constrained and cannot use full standard protocol stacks because they cannot transmit information too frequently due to battery drainage; they are not always reachable due to the wireless connection based on low duty cycles, their communication protocols are IoT-specific, lack an integrated approach, and use different data encoding languages. A global IoT platform deployment is difficult to foresee as a direct result of these limitations. Developers face complex scenarios where merging the information is a real challenge.

2.2.3 Security

Securing an entire "classical" infrastructure is a challenge in itself, but the IoT demands an even larger security approach to keep endpoints and networks protected against more sophisticated cybercrime techniques and tools, such as sniffer attacks, DoS attacks, compromised-key attacks, password-based attacks, and man-in-the-middle (MITM) attacks.

Security certification is needed to ensure that a product satisfies the required security requirements, which can be both proprietary requirements (i.e., defined by a company for their specific products) and market requirements (i.e., defined in procurement specifications or market standards). The process for certification of a product is generally summarized in four phases:

1. Application. A company applies a product for evaluation to obtain certification.
2. An evaluation is performed to obtain certification. The evaluation can be mostly done in three ways:
   (a) The evaluation can be conducted internally to support self-certification.
   (b) The evaluation can be performed by a testing company, who legally belongs to the product company.
   (c) It can be a third-party certification where the company asks a third-party company to perform the evaluation of its product.
3. In the case of an internal company or a third-party company evaluation, the evaluation company provides a decision on the evaluation.
4. Surveillance. This is a periodic check on the product to ensure that the certification is still valid or it requires a new certification.

In the market requirements case, the requirements are also defined to support security interoperability. For example, to ensure that two products are able to mutually authenticate or to exchange secure messages. Security certification is needed to ensure that products are secure against specific security attacks or that they have specific security properties, such as:

- *Authentication*: Data sender and receiver can always be verified.
- *Resistance to replay attacks*: Intermediate nodes can store a data packet and replay it at a later stage. Thus, mechanisms are needed to detect duplicate or replayed messages.
- *Resistance to dictionary attacks*: An intermediate node can store some data packets and decipher them by performing a dictionary attack if the key used to cipher them is a dictionary word.
- *Resistance to DoS attacks*: Several nodes can access the server at the same time to collapse it. For this reason, the server must have DoS protection or a fast recovery after this attack.
- *Integrity*: Received data are not tampered with during transmission; if this does not happen, then any change can be detected.
- *Confidentiality*: Transmitted data can be read only by the communication endpoints.

- *Resistance to MITM attacks*: The attacker secretly relays and possibly alters the communication.
- *Authorization*: Services should be accessible to users who have the right to access them.
- *Availability*: The communication endpoint can always be reached and cannot be made inaccessible.
- *Fault tolerance*: The overall service can be delivered even when a number of atomic services are faulty.
- *Anonymization*: If proper countermeasures are not taken, even users employing pseudonyms could be tracked by their network locator.

Nevertheless, in the IoT domain, initiative exists for security certification, such as the alliance for IoT innovation (AIOTI) through its standardization and policy working groups that complement the existing certifications, such as the common criteria (CC). However, enabling technologies that help to formally address the specific testing challenges and that enable the IoT certification today are lacking [3]. In the next section, we present MBT approaches with several paradigms, usages, and tools that can behave as well as an enabler for the IoT certification. Then, we will show how the key challenges can be tackled with the MBT approach.

3. STATE OF THE ART

The key challenges in testing for scalability, interoperability, and security of the IoT have proven to be problematic when trying to resolve them with traditional testing methods, which is a strong roadblock for wide adoption of this innovative technology. This section aims to describe how the MBT has addressed this problem though IoT testing.

MBT designates any type of testing based on or involving models [4]. Models represent the system under test (SUT), its environment, or the test itself, which directly supports test analysis, planning, control, implementation, execution, and reporting activities. According to the World Quality Report 2016 [5], the future testing technologies require agile development operations for more intelligence-led testing to meet speed, quality, and cost imperatives. Hence, MBT, as a promising solution, aims to formalize and automate as many activities related to testing as possible and thereby increase both the efficiency and effectiveness of testing. Following the progress of model-based engineering [6] technologies, MBT has increasingly attracted research attention. Many MBT tools are developed to support the practice and utilization of MBT technologies in real cases. The MBT tools provide functions that cover three MBT aspects, i.e., generation of test cases, generation of test data, and generation of test scripts [7], and are used to conduct different types of testing, such as functional, performance, and usability testing [8].

Nevertheless, the functions of MBT tools vary largely from one to another. Users without prior knowledge struggle to choose appropriate tools corresponding to their testing needs among the wide list. The MBT tools require input in terms of different models (e.g., UML, PetriNet, and BPMN) and focus on different MBT aspects with different generation strategies for data, test cases, and test scripts. Moreover, most of the existing MBT tools support mainly automatic test case generation rather than test data generation and test script generation due to two reasons: first, the test case generation requires complicated strategies involving various test selection criteria from MBT models, and the generation results highly reply on the selected criteria and strategies; second, the test case generation brings many more testing benefits, as the main efforts spent on traditional testing lie in manual preparation of test cases. To provide users with a common understanding and systematic comparison, this chapter reviews the identified set of MBT tools, focusing on the test case generation aspect. The MBT tools have been previously reported and compared in several surveys: the first report is presented in [9] in 2002 to illustrate the basic principles of test case generation and relevant tools; a later comparison is made in [10] from perspectives of modeling, test case generation, and tool extensibility; since more and more MBT tools rely on state-based models, the review in [11] illustrates the state-based MBT tools following criteria of test coverage, automation level, and test construction. The most recent work [12] presents an overview of MBT tools focusing on requirement-based designs and also illustrates an example by use of the representative tools. Due to the increasing popularity of MBT, the existing MBT tools have rapidly evolved and new available tools have emerged every year and MBTaaS is one of these tools.

3.1 Model-Based Testing

MBT is an application of model-based design for generating test cases and executing them against the SUT for testing purpose. The MBT process can be generally divided into five steps [13]

- *Step 1: Creation of the MBT models.* In the first step, users create the MBT models from requirement/system specifications. The specifications define the testing requirements or the aspects to test of the SUT (e.g., functions, behaviors, and performances). The created MBT models usually represent high-level abstractions of the SUT and are described by formal languages or notations (e.g., UML, PetriNet, and BMPN). The formats of the MBT models depend on the characteristics of the SUT (e.g., function-driven or data-driven system, deterministic or stochastic system) and the required input formats of the MBT tools.

- *Step 2: Generation of test cases.* In the second step, abstract test cases from the MBT models are generated automatically, based on test selection criteria. The test selection criteria guide the generation process by indicating the interesting focus to test, such as certain functions of the SUT or coverage of the MBT model (e.g., state coverage, transition coverage, and data flow coverage). Applying different criteria to the same MBT model will generate different sets of test cases that could be complementary. Abstract test cases without implementation details of the SUT are generated from the MBT models.

- *Step 3: Concretization of the test cases.* In the third step, the abstract test cases (produced in step 2) are concretized into executable test cases with the help of mappings between the abstraction in the MBT models and the system implementation details. Executable test cases contain low-level implementation details and can be directly executed on the SUT.

- *Step 4: Execution of the test cases.* The executable test cases are executed on the SUT either manually or within an automated test execution environment. To automate the test execution, system adapters are required to provide channels connecting the SUT with the test execution environment. During the execution, the SUT is, respectively, stimulated by inputs from each test case, and the reactions of the SUT (e.g., output and performance information) are collected to assign a test verdict. For each test case, a test verdict is assigned indicating if a test passes or fails (or is inconclusive).

- *Step 5: Results analysis.* At the end of the testing, the testing results are reported to the users. For nonsatisfactory test verdicts, the MBT process records traces to associate elements from specifications to the MBT models and then to the test cases, which are used to retrieve the possible defects.

3.2 MBT Approaches for the IoT Domain

The related work we discuss in this chapter addresses the panel of existing MBT approaches with a view on the IoT systems. The MBT can be offline, whereby the test cases are generated in a repository for future execution or online, whereby the test cases are generated and executed automatically [14]. In this chapter, we discuss both: offline MBT techniques with respect to functional and security testing and online MBT techniques. Nevertheless, recent studies show that current MBT approaches, although they can be suitable for testing the IoT systems need to be adapted [15].

3.2.1 Functional Testing

Standards are a well-known method used to increase the user's confidence and trust in a product. MBT approaches to support conformance testing to standards have been applied to different domains, such as electronic transactions and notably for CC certifications, for instance, on the global platform (GP). Authors in [15] propose testing the framework for the IoT services, which has been evaluated on the IoT testbed, for test design using behavioral modeling, based on state machines and test execution in TTCN-3. In addition, the IoT behavioral models (IoTBM) represent a chosen explicit model, describing the service logic, used protocols, and interfaces as well as the interactions with the connected services. IBM reports as well on the application of MBT for compliance to standards based on the IBM tool suite (Rhapsody and Jazz) [16]. Nevertheless, we target generic and reusable solutions by the IoT community.

3.2.2 Security Testing

Mode-based security testing (MBST) is a dynamic application security testing (DAST) approach consisting of dynamically checking the security of the systems's security requirements [17]. Furthermore, MBST has recently gained relative importance in academia and in industry, as it allows for systematic and efficient specification and documentation of security test objectives, security test cases, and test suites, as well as automated or semiautomated test generation [18]. Existing work in MBST focuses mainly on functional security and vulnerability testing. Functional security testing aims are ensuring the correct implementation of the security requirements. While, vulnerability testing is based on risk assessment analysis, aiming to discover potential vulnerabilities. Authors in [14] discuss extensively the existing approaches on both these aspects of MBST. Existing approaches address various domains, such as electronic transactions, cryptographic components, but not directly the IoT domain.

3.2.3 Online Testing

Online techniques can indeed address the uncertain behavior of the IoT systems, but lack application and maturity for the IoT domain, as current techniques mostly address a testbed setup for execution in a real-life environment, without dealing with the test cases conception.

Several online tools exist in the MBT literature. For instance, UPPAAL for Testing Real-time systems ONline (UPPAAL TRON) [19], supports the generation of test cases from models, and their online execution on the SUT.

Authors in [20] expressed the system and its properties as Boolean equations, and then used an equation library to check the equations on the fly. The model checker used on a model with faults produces counter examples, seen as negative abstract test cases.

Shieferdecker et al. designed a mutation-based fuzzing approach that uses fuzzing operators on scenario models specified by sequence diagrams. The fuzzing operators perform a mutation of the diagrams resulting in an invalid sequence. Shieferdecker et al. use online behavioral fuzzing which generates tests at run time [21].

3.3 MBT Tools

In this section, we present a representative set of academic and industrial MBT tools for test case generation that we investigated for usage in the IoT domain and discuss their characteristics. A digest of the tools is presented in Table 1 and provides a comparison of the different MBT tools based on the testing type, test selection, and test generation technology. It also gives an indication of the test generation process (online, offline, or both).

3.3.1 CompleteTest

CompleteTest [22] is an academic tool for safety intensive critical systems. This tool takes a function block diagram (FBD) as an input model and integrates the UPPAAL [23] model checker to perform symbolic reachability analysis on FBD models for test case generation. A set of coverage criteria, including decision coverage and condition coverage, are used to guide the generation process. This tool presents a simulation environment to simulate the abstract test cases against the FBD models, and also a search-based algorithm to generate executable test cases in C.

3.3.2 DIVERSITY

DIVERSITY [24] is an open-source Eclipse-based tool for formal analysis. It takes models defined in xLIA (eXecutable Language for Interaction & Assemblage) [24] as input and generates test cases in TTCN-3 which is a standardized testing language for multiple testing purposes, developed by

Table 1 MBT Tools Characterization

Tools	Test Type	Test Generation		
		Test Selection	Technology	Process
CertifyIt	Functional and security	Structural coverage, test case specification	Search based, model checking	Both
CompleteTest	Functional and security	Structural coverage	Model checking, search based	Offline
DIVERISTY	Functional and security	Structural coverage, random and stochastic, test case specification	Symbolic execution	Offline
FMBT	Functional	Structural coverage, test case specification	Search based	Both
HTG	Functional	Data coverage	Random generation, search based	Offline
Lurette	Functional	Random and stochastic	Random generation	Online
MISTA	Functional	Structural coverage, random and stochastic	Search based, random generation	Both
Modbat	Functional	Random and stochastic	Random generation, search based	Both
MoMuT	Functional and security	Fault based	Search based	Offline
Pragamadev	Functional and security	Structural coverage, random and stochastic	Symbolic execution	Offline
Tcases	Functional	Data coverage	Constraint solving	Offline
TCG	Functional and security	Structural coverage, random and stochastic	Search based, random generation	Offline
VERA	Security	Structural coverage	Search based	Both

the European Telecommunication Standards Institute (ETSI). The symbolic execution algorithm [25] is used by DIVERSITY to use symbolic values for inputs rather than actual inputs to generate multiple test cases consecutively. Moreover, DIVERSITY provides functionality for validation of the MBT models to detect unexpected behaviors, such as deadlocks or overdesign of the SUT.

3.3.3 FMBT

FMBT [26] is an open-source tool developed by Intel that generates test cases from models written in the AAL/Python pre/postcondition language. FMBT is capable of both online and offline testing on Linux platforms. It provides the necessary interfaces to test a wide range of objects from individual C++ classes to GUI applications and distributed systems containing different devices. For now, FMBT supports all the MBT steps in commands without graphic interfaces.

3.3.4 HTG

HTG [27] is an academic test case generation tool for hybrid systems. HTG uses a hybrid automaton model or SPICE netlists [28] as input and generates test cases in C++. A data coverage measure based on star discrepancy [29] is used to guide the test generation and ensure the test cases are relatively equally distributed over the possible data space. The generated test cases can be applied to numeric simulation and circuit simulation domains.

3.3.5 Lurette

Lurette [30] is an automatic test generator for reactive systems. It focuses on environmental modeling using Lutin [31] to perform guided random exploration of the SUT environment while considering the feedback. For reactive systems, the test cases are realistic input sequences generated from the deterministic or nondeterministic environmental model. The generation process is online, as their elaboration must be intertwined with the execution of the SUT: the SUT output is used as the environment input. The test verdict is automated using Lustre oracles [32].

3.3.6 MISTA

MISTA [33] is an open-source tool that generates test cases from models of finite-state machines or function nets. Both control-oriented and data-oriented models can be built by MISTA. The formats of the test cases cover several languages (Java, C, C++, C#, PHP, Python, HTML, and VB) and test frameworks (xUnit, Selenium IDE, and Robot framework). In addition, MISTA supports both online and offline testing.

3.3.7 Modbat

Modbat [34] is an open-source tool based on extended finite-state machines specialized for testing the APIs of software. A Scala-based domain-specific language is used to create the models with features for probabilistic and non-deterministic transitions, component models with inheritance, and exceptions. Test cases are generated as sequences of method calls to the API that can be directly executed against the SUT.

3.3.8 MoMuT

MoMuT is a set of model-based test case generation tools that work with the UML state machine, timed automata, requirement interfaces, and action systems [35]. This tool features a fault-based test case generation strategy [36] that allows mutations to be made on models and generates richer test cases from both original and mutated models to detect if models contain certain user-selectable or seeded faults. A fault localization mechanism is included in MoMuT for debugging purposes when a test case fails.

Table 1 is an abstract of the MBT tools that offer the IoT testing capabilities. It represents the different testing characterizations for each presented MBT tool: their test type, test selection criteria, the technology used for test generation, and the generation process.

3.3.9 PragmaDev

PragmaDev Studio [37] is a commercial tool with complete support for all the MBT steps. This toolset allows users to create the MBT models in SDL and correspondingly generates the test cases in TTCN-3. PragmaDev Studio integrates with the core of DIVERSITY and uses the symbolic execution algorithm for test case generation and the MBT model validation. Graphical interfaces are provided for all supported functionalities, and especially, a tracer is designed for the testing results analysis to trace elements from requirements, models, and test cases via a standard graphical representation. PragmaDev Studio has published a free version for users with small MBT projects.

3.3.10 Tcases

Tcases [38] is a combinatorial testing tool to test system functions and generate input testing data. An XML document defining the SUT as a set of functions is required as input as well as the data space for variables of the functions. The test generation is guided by a predefined data coverage level, through which the number of generated test cases can be controlled. For a set of input variables, Tcases can generate n–wise test cases [39]. The test cases are in stored XML and can be transformed to JUnit test cases via an integrated convector.

3.3.11 TCG

TCG [40] is an open-source plugin of the LoTuS modeling tool [41] to generate test cases from both probabilistic and nonprobabilistic models. After test generation, TCG can proceed to a second test selection from the first generation result to provide a refined set of test cases. The test generation supports the structural model coverage criteria and statistical methods, while the test selection uses five different techniques: test purpose, similarity of paths, weight similarity of paths, postprobable path, and minimum probability of the path.

3.3.12 VERA

VERA [42] is an academic tool for vulnerability testing, which allows users to define attacker models by means of extended finite-state machines, and correspondingly generates test cases targeting generic vulnerabilities of Web applications. To efficiently perform the tests, VERA also provides a library containing common vulnerability test patterns for modeling.

3.3.13 CertifyIt

Smartesting CertifyIt [43] is a commercial tool for test case generation from models of IBM RSAD [44]. The input models include the UML state machine and class diagram, while the generated test cases can be exported in a test environment, such as HP Quality Center and IBM Quality Manager, or as HTML, XML, Perl/Python Script, and Java classes for Junit. In addition, CertifyIt can publish the test cases in script format to facilitate test execution, and the traceability is also well maintained for results analysis.

Finally, we increment the classical usage of the CertifyIt tool for functional and security testing and integrate it into a MBT as a Service (MBTAAS) environment, which delivers immense value for the IoT community. The adaptation of CertifyIt for the IoT domain has already shown its value, as discussed by the authors in [3] and [45]. In this context, we enrich it by integrating it with the MBTAAS tool.

4. MBT FOR IoT CONFORMANCE TESTING

Interoperability is one of the key challenges in the IoT testing because of the substantial number of devices deployed and expected to be deployed. Our experience showed that unit testing is not sufficient to check a standardized IoT platform implementation and ensure against interoperability issues. The reason is that the system's functionalities are not seen globally, but from an elementary coding level, a function or even a statement, is that the full spectrum of functionalities, as required by the standard, may not be validated through this type of testing.

On the contrary to unit testing, MBT has its entry point at the standard level, which defines the full spectrum of functionalities. Thus, it ensures the test coverage of these functionalities and provides the oracle to validate the implementation's conformance to the standard.

This section aims to illustrate how MBT can respond to the challenge of interoperability and present it through a real use case based on a real interoperability standard, FIWARE [46]. FIWARE has a community of over 1000 start-ups and is being adopted in over 80 cities mainly not only in Europe but also around the world. Ensuring the interoperability to such a large number of companies using the FIWARE standard is primordial to its successful adoption.

We apply our approach to the IoT Services Enablement Generic Enablers (GEs) implementing the FIWARE specifications, and more specifically to the data handling GE. Thus, we verify the implementations for compliance to the standard. The FIWARE platform [46] provides a rather simple yet powerful set of APIs that ease the development of smart applications in multiple vertical sectors. The specifications of these APIs are public and royalty free. Besides an open-source reference implementation (Generic Enabler implementation), each of the FIWARE components are publicly available. Generic Enablers offer many general-purpose functions and provide high-level APIs in:

- Cloud hosting: enables entrepreneurs and individuals to test the technology as well as their applications on FIWARE.
- Data/context management: provides solutions to innovative applications that require management, processing and exploitation of context information, as well as data streams in real-time and on massive scales. The context information is any relevant information regarding the environment and its users.

- Internet of Things services enablement: allows things to become available, searchable, accessible, and usable contextual resources.
- Applications, services, and data delivery: support the creation of an ecosystem of applications, services, and data.
- Security of data: provides security to the IoT application, for example, OAuth 2.0 [47].
- Interface to networks and devices (I2ND): defines an enabler space for providing GEs to run an open and standardized network infrastructure.
- Advanced web-based user interface: brings components that will provide a simple, uniform way to create rich networked 2D and 3D applications that run in a browser.

4.1 MBT Specificities for IoT Conformance Testing

FIWARE uses the open mobile alliance (OMA) NGSI-9/10 standardized RESTful interfaces [48]. This standardized interface is evolving in parallel with the IoT technology, to be able to take into consideration its novelties. For instance, the OMA released the NGSI interfaces—version 1 (NGSI v1) and FIWARE is aiming to contribute to the standard to release NGSI v2. Thus, it needs to be easy to apply regression testing when the standard evolves. The MBT offers a relatively safe, effective, and systematic way to create regression test suites and new test suites to cover the new functionalities, based on the models [49].

In addition, RESTful systems communicate over the hypertext transfer protocol (HTTP) with the same HTTP verbs (GET, POST, PUT, DELETE, etc.) that web browsers use to retrieve web pages and to send data to remote servers. Their purpose is to exchange information about the availability of contextual information.

In our context of testing the implementation of the data handling GE, we are interested in applying conformance testing on NGSI-9/10. Moreover, although it is often assumed that the GEs are correctly implemented and are used or are part of the services provided by FIWARE, noncompliance to the FIWARE specifications may lead to several dysfunctional interoperability issues. To ensure the interoperability, the GEs should be thoroughly tested to assess their compliance to these specifications. MBT is considered to be a lightweight formal method to validate software systems. It is formal because it works off formal (that is, machine-readable) specifications (or models) of the software SUT (usually called the implementation or just SUT). It is lightweight because, contrary to other formal methods, such as B and

Z notation, MBT does not aim at mathematically proving that the implementation matches the specifications under all possible circumstances. In MBT, the model is an abstraction of the SUT and represents the tested perimeter based on the specifications, which is not necessarily the entire system. As such, it allows to systematically generate from the model a collection of tests (a "test suite") that, when run against the SUT, will provide sufficient confidence that the system behaves accordingly to the specification. Testing follows the paradigm of not proving the absence of errors, but showing their presence. MBT, on the other hand, scales much better and has been used to test life-size systems in very large projects [7].

In FIWARE, the MBT conformance model is a representation of the standard. The model includes all the behaviors of the NGSI-v1 and generates the complete suite of test cases that are needed to test for implementation compliance to that standard. Changing an MBT model designed for conformance testing implies that the standard has changed.

In the following sections, we provide an overview of the MBT approach applied for conformance testing of one FIWARE GE implementation.

4.1.1 MBT Model

This approach considers a subset of the UML to develop the MBT models, composed of two types of diagrams for the structural modeling: class diagrams and object diagrams, and OCL to represent the system's behavior. From one MBT model, various test selection criteria can be applied, as shown in Fig. 2. For this study, we used the Smartesting MBT approach

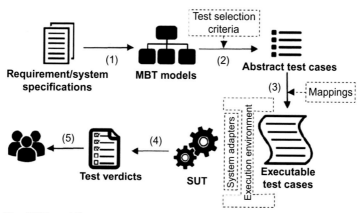

Fig. 2 The MBT workflow.

and its test generation tool—CertifyIt, which uses the coverage-based test selection criteria.

Tested requirements are manually extracted from the specification. Each of the requirements is than tagged into the model, which serves as a basis to maintain traceability to the specification. Based on the specification information and tested requirements perimeter, the structure of the system is represented in a class diagram, which is a static view of the system limited to elements used for test generation.

The FIWARE class diagram (Fig. 3) is an abstraction of the IoT system:

- Its entities, with their attributes. The class *Sut* represents the SUT where things can register to it. The CEP statement *CepStatement* can also be registered to the *Sut* and the *Sut* supports *Subscription* to its registered things.
- Operations that model the API of the SUT. The operation names give a self-explanation of their action, for example, the class *registerContext* enables a sensor to register itself in the IoT platform.
- Points of observations that may serve as oracles are explained in the following paragraphs (for instance, an observation returns the current state of the user's connection to a website).

The class diagram, providing a static view of the SUT, is instantiated by an object diagram. The object diagram provides the initial state of the system and also all objects that will be used in the test input data as parameters for the operations in the generated tests.

It contains the input data based on the partition equivalence principle. The input data are instances of the classes and they are represented in the form of objects, as introduced previously. Each input data is identified by an enumeration, representing an equivalent partition of data from which the test could be derived. For instance, from Fig. 4, to test the Espr4FastData API, we may consider an initial state of the system with a SUT having nothing connected to it and some entities to be used for registering, creating the CEP statements, subscriptions, etc.

Finally, the dynamic view of the system or its behaviors are described by the object constraint language (OCL) constraints written as pre/postconditions in the operations in a class within a class diagram, as depicted in Fig. 5. The test generation engine sees these behavioral objects as test targets. The operations can have several behaviors, identified by the presence of the conditional operator, if-then-else. The precondition is the union of the operation's precondition and the conditions of a path that is necessary to traverse for reaching the behaviors postcondition. The postcondition

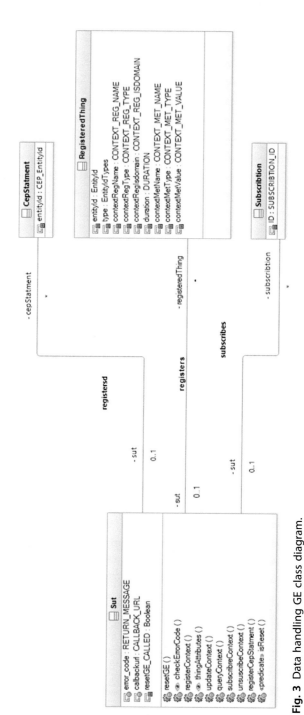

Fig. 3 Data handling GE class diagram.

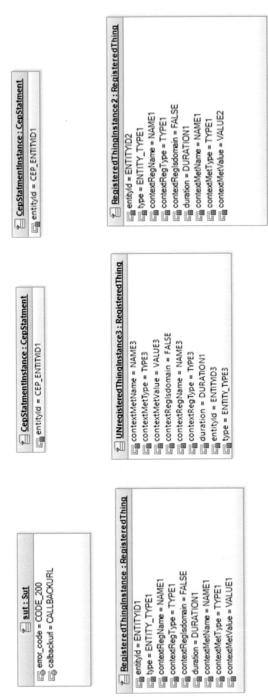

Fig. 4 Excerpt of the data handling object diagram.

```
EspR4FastData_GETestSuite    resetGE

1 ---@REQ: resetEspR4FastData
2 if IN_URL = CALLBACK_URL::INVALIDURL
3 then
4       ---@AIM: Invalid url
5       self.error_code = RETURN_MESSAGE::URL_ERROR
6 else
7       if not self.registeredThing->isEmpty()
8       then
9             ---@AIM: success
10            self.registeredThing->isEmpty() and
11            self.error_code = RETURN_MESSAGE::CODE_200
12      else
13            ---@AIM: success and isEmpty
14            self.error_code = RETURN_MESSAGE::CODE_200
15      endif
16 endif and
17 result = self.error_code and
18 self.resetGE_CALLED = true and
19 checkErrorCode()
```

Fig. 5 Object constraint language (OCL) example.

corresponds to the behavior described by the action in the "then" or "else" clause of the conditional operator. Finally, each behavior is identified by a set of tags which refers to a requirement covered by the behavior. For each requirement, two types of tags exist

- @REQ—a high-level requirement
- @AIM—its refinement

Both are followed by an identifier. One advantage of the used MBT tool is the possibility to automatically deduct the test oracle. A specific type of operation, called observation, defines the test oracle. The tester with these special operations can define the system points or variables to observe, for instance, a function returns a code. Thus, based on these observations, the test oracle is automatically generated for each test step. The next session provides us an inside view on the test case generation process.

4.1.2 Test Case Generation

Based on the chosen test selection criteria, CertifyIt extracts the test objectives into a test suite, called with the CertifyIt terminology a *smartsuite*.

In our case study, two smartsuites named *EspR4FastData_GETestSuite* and *UserTestScenarios* were created. The *EspR4FastData_GETestSuite* test suite indicates that the generation should be done for all defined test objectives whereas for the *UserTestScenarios_SmartSuite*, they are user-defined scenarios of tests to define some very specific cases with the help of the MBT model. Based on the object diagram and the OCL postcondition, the

CertifyIt tool automatically extracts a set of test targets, which is the tool comprehensible form of a test. The test targets are used to drive the test generation. As discussed previously, each test has a set of tags associated with it, which ensures the coverage and traceability to the specification. Fig. 6 gives a snapshot of a test in the CertifyIt tool. On the left side, the tool lists all generated tests clustered per covered requirement. On the right side, the test case can be visualized and for each step a test oracle is generated. As discussed, the tester with the observation manually defines the system points to observe when calling any function. Figure, *checkResult*, observes the return code of each function with respect to the activated requirement. In addition, on the bottom-right of the figure for each test case and test step, it is possible to observe the test targets (set of tags).

In addition, Fig. 7 illustrates the generated tests with CertifyIt, in exported HTML. More specifically, the shown test covers the behavior that resets the GE implementation using the *resetGE* function. As discussed previously, each test has a set of associated tags for which it ensures the coverage.

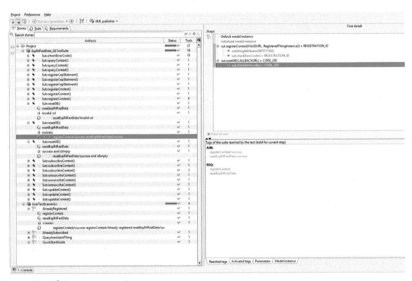

Fig. 6 CertifyIt test case view.

Steps	Actions		Requirements, aims and custom tags	
Step 1 (sut)	*resetGE*		REQ	resetEspR4FastData
	This operation completely destroy the CEP threads, the CEP data, and all the application level data. EXAMPLE : http://localhost/EspR4FastData-3.3.3/admin/espr4fastdata (queried with HTTP Delete method)		AIM	resetEspR4FastData/Invalid url
⊕ 1.1	@synthesis@ Check that the error code is URL_ERROR @/synthesis@ Check if an error code URL_ERROR is returned			

Fig. 7 Implementation abstract test case (HTML export).

The export of this test, as illustrated in Fig. 7, in the rightmost column, maps to each step the covered requirements or more precisely the test objective tagged by @REQ and @AIM in the model. The export, as shown in the middle column of the figure, contains guidelines for test reification, giving directions on how the abstract test case should be adapted into a concrete script, which is described in the next section.

Within the considered limits of the implementation, the tool extracted 26 "test targets" and generated 22 tests in around 10 s to cover the test targets. Each story in the CertifyIt tool corresponds to one test target. However, one test covers one or more test targets. Moreover, the tool's engine generates fewer tests then test targets, because it uses the "light merge" of test methods, which considers that one test covers one or more test objectives (all test objectives that have been triggered by the test steps). For instance, the test shown in Fig. 6 covers two test objectives. The "light merge" of the tests shortly and on a high-level means that the generator will not produce separate tests for the previously reached test targets.

The generated tests are abstract and to execute them on the SUT, they should be further adapted. For our study, a SoapUI exporter was created, which publishes tests into SoapUI XML projects using the SoapUI library.

4.1.3 Test Case Reification and Execution

In the used MBT approach, to execute tests on the SUT, the activity of test cases reification, also named test case adaptation, is performed first; it consists of creating an adaptation layer to fulfill the gap between the abstract test case generated from the MBT model, the concrete interface, and the initial set of test data of the SUT. As classically required for the test adaptation activities, exporting the abstract test cases (ATS) into executable scripts is the first step to perform, in this case, the ATS are exported into XML files which can be imported and executed on the SoapUI. To illustrate the test adaptation activities, let us consider the following simple test case given in Fig. 7:

```
sut.resetGE(INVALIDURL)
    -> sut.checkErrorCode () = URL_ERROR
```

This test has one step that calls the function to reset the data with a URL set to "INVALIDURL." CertifyIt automatically generates the test oracle, by using an observation operation defined in the model, in our example, it is called "checkErrorCode." For instance, the expected result of the test is the error *URL_ERROR*. Further, we created a SoapUI exporter, which from the CertifyIt tool publishes the abstract test cases into concrete SoapUI projects. These projects can be imported and executed on SoapUI.

Fig. 8 shows an example of four executed test cases after importing the SoapUI project. On the left side, the test cases are illustrated. On the right side, it is possible to observe the test case execution status for each step. The green color on a test step represents its successful execution status while the red color represents a failed test step. On the top, a global test pass/fail result is summarized.

4.2 Experimentation and Results

The use case showed the feasibility of our MBT solution for the IoT conformance testing and we could show the benefits of applying the MBT approach in terms of improving the APIs interoperability, thus ensuring the conformance to the specifications. In terms of project planning, it takes 16 person-hours to create the MBT model. More specifically, it takes 4 person-hours to model the static view of the system (the class diagram) suitable for testing and 12 person-hours to model the dynamic view of the system (to write the OCL postconditions). These metrics abstract the domain

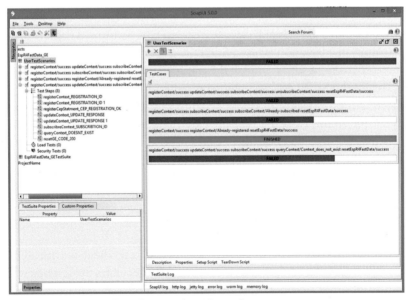

Fig. 8 SoapUI test case launch example with result status.

knowledge from the FIWARE standards and the implementation of the specification itself. If the MBT approach is integrated within the project, the testing teams already possess this knowledge. In addition, the time spent to create the adaptation layer is linked to the developers/testers experience and it is considered as negligible. One major advantage in applying the MBT approach on the FIWARE data enabler implementation is that the test repository remains stable, while the project requirements and specification may evolve over time. Another advantage concerns the testing exhaustiveness. It is indeed impossible to test any system exhaustively. Nevertheless, it is possible, as we did, to generate a test suite that covers the test objectives in an automated and systematic way. The CertifyIt tool further allows generating reports to justify the test objective coverage, which can be easily used for auditing. This couple of examples shows the usefulness of an automated and systematic use of an MBT approach on the IoT applications that should comply to specific standards.

4.3 Lessons Learnt

MBT is a suitable approach for emerging technologies and especially for the IoT platforms where the maturity level and the technology development are still ongoing. First, automated and systematic test suites are conceived covering the defined test objectives. Second, the MBT further allows generating reports to justify the test objective coverage, which can be used for auditing purposes. These examples show the usefulness of such an MBT approach on applications that should comply to a standard.

5. MBT FOR IoT SECURITY TESTING

End-to-end security is used successfully today, for example, in online banking applications. Correct and complete end-to-end security in the growing IoT is required to ensure data privacy, integrity, etc. Introducing MBT to this problem may raise certain questions on the feasibility of the models being generic enough to withstand the wide range of IoT applications. It is impossible to provide an exhaustive list of all application domains of the IoT. The IoT components involved in a solution have in most common cases, a wide range of different components, such as hardware (devices, sensors, actuators, etc.), software, and network protocols. Security on their endpoints (client–server or client–client for peer-to-peer) is an absolute requirement for secure communications. Such a solution contains the following components:

- Identity: This component encompasses the known and verifiable entity identities on both ends.
- Protocols (for example, TLS): Protocols are used to dynamically negotiate the session keys, and to provide the required security functions (for example, encryption and integrity verification) for a connection. Protocols use cryptographic algorithms used to implement these functions.
- Cryptographic algorithms (for example, advanced encryption standard [AES] and secure hash algorithm [SHA-1]): These algorithms use the previously mentioned session keys to protect the data in transit, for example, through encryption or integrity checks.
- Secure implementation: The endpoint (client or server) that runs one of these protocols mentioned previously must be free of bugs that could compromise security.
- Secure operations: Users and operators must understand the security mechanisms, and how to manage exceptions. For example, web browsers warn about invalid server certificates, but users can override the warning and still make the connection. This concern is a non-technical one, but is of critical concern today.

In addition, for full end-to-end security, all these components should be seen globally to ensure their security. For instance, in networks with end-to-end security, both ends can typically (depending on the protocols and algorithms used) rely on the fact that their communication is not visible to anyone else, and that no one else can modify the data in transit.

5.1 MBT Specificities for IoT Security Testing

In this section, to illustrate the approach, we are focusing on MBT applied to two security components: algorithms and secure operation. For a real case study, we use the steps taken by ARMOUR [50], the H2020 European project. The project introduces different experiences to test the different components. More specifically, each experience proposes individual security tests covering different vulnerability patterns for one or more involved components. In ARMOUR, experience 1 (Exp1: Bootstrapping and group sharing procedures) covers the security algorithms for encryption/decryption and protocols to secure communication. Experience 7 (Exp7: Secure IoT platforms) covers the IoT data storage and retrieval in an IoT platform. Both experiences can wisely be integrated together. Indeed, it makes sense to pair the secure transfer and storage of data within an IoT platform, thus ensuring that the confidentiality and integrity of the data is maintained from the moment it is produced by a device, until it is read by a data consumer.

Fig. 9 depicts the E2ES scenario which combines Exp1 and Exp7. The IoT platform used in this particular scenario by Exp7 is the oneM2M standard implementation. Both data consumers/producers retrieve keys from a credential manager (CM) and the IoT platform role is to support the data storage/retrieval. However, a single security problem in any of the components involved in this scenario can compromise the overall security.

One specificity of testing such a security scenario, MBT-wise, is that each experience has its own MBT model and the E2ES MBT model is a combination of the Exp1 and Exp7 models.

A wise integration of the experiments is made from each experience model. We call it an integration model which is used by a Business Process Model Notation (BPMN) model to generate the abstract test cases as shown in Fig. 10. A standard BPMN provides businesses with the capability of understanding their internal business procedures in a graphical notation and gives organizations the ability to communicate these procedures in a standard manner. The test cases then follow the normal MBT procedure, that is, test generation and then publication. In this use case, the published

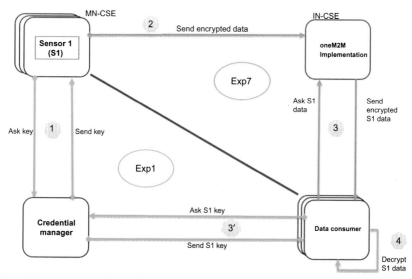

Fig. 9 Large-scale end-to-end security scenario.

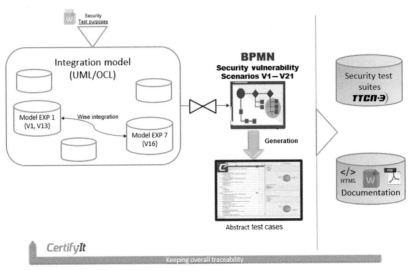

Fig. 10 End-to-end security MBT description.

tests are written as TTCN-3 [51] executables for test automation and can be exported as documentation for manual testing.

5.1.1 MBT Model

Each of the IoT systems are composed of producers, consumers, an IoT platform, and eventually a security service, gateways, and/or sniffer. The entities communicate through message requests (of different types) and respond to messages through message responses. This generic model, which we call the IoT MBT generic model (Fig. 11), can be created automatically from the MBT tool (*CertifyIt*). This model can then be customized with respect to the IoT SUT, for instance, choosing suitable names, add different types of message exchanges, configure response status codes, test patterns, delete unnecessary model elements, etc.

Specifically, oneM2M is a global organization that creates requirements, architecture, API specifications, security solutions, and interoperability for machine-to-machine (M2M) and the IoT technologies. It was established through an alliance of standards organizations to develop a single horizontal platform for the exchange and sharing of data among all applications and provides functions that M2M applications across different industry segments commonly need. The service layer is developing into a critical component for future IoT architectures. The consortium consists of eight of the worlds

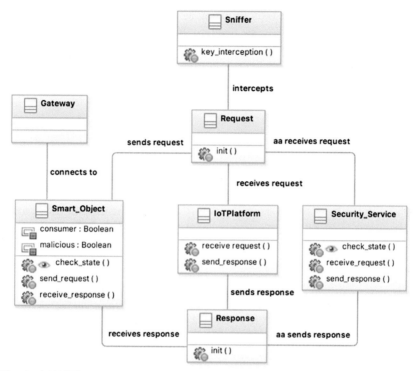

Fig. 11 IoT MBT generic model.

preeminent standards development organizations, notably: ARIB (Japan), ATIS (USA), CCSA (China), ETSI (Europe), TIA (USA), TSDSI (India), TTA (Korea), and TTC (Japan). These partners collaborate with industry fora or consortia and over 200 member organizations to produce and globally maintain applicable technical specifications for a common M2M/IoT service layer. The oneM2M standard employs a simple horizontal, platform architecture that fits within a three-layer model comprising applications, services, and networks.

In the same way that consumers want to ensure that their personal and usage data are not misused, any number of stakeholders will also want to ensure that their data is protected and the services are securely delivered. Unlike the traditional internet, a typical IoT system:

- Inputs information about our behaviors, thus it directly exposes our privacy.
- Outputs adaptations to our environment, thus it potentially affects our safety.

Security is a chain that is only as strong as its weakest link; hence, all stages of a devices/services lifecycle need to be properly secured throughout its lifetime. In oneM2M, there are three layers of protection for the communications:

- Security Association Establishment Framework (SAEF): Adjacent entities.
- EndtoEnd Security of Primitive (ESPrim): Originator Hosting platform.
- EndtoEnd Security of Data (ESData): Data producer to data consumer.

We used in our use case, a oneM2M compliant platform (OM2M [52]) to test the end-to-end security with the MBT approach introduced in Fig. 10, and as a security service, we will be using a "Credential Manager" (CM). The CM oversees establishing a secure channel with the smart object to deliver the group key in a secure way. Fig. 12 shows the modifications made to the generic MBT model to represent the wise integration between Exp1 and Exp7.

Here is an enumeration of the behaviors emerging from the Exp1 and Exp7 integration and that the model implements:

1. The smart object requests its group key to the CM.
2. The CM extracts the set of attributes of the entity from the request payload and it generates the group key associated with such a set.
3. The CM sends to the smart object its group key.
4. The sniffer intercepts the messages exchanged between the smart object and the CM and tries to read the group key.

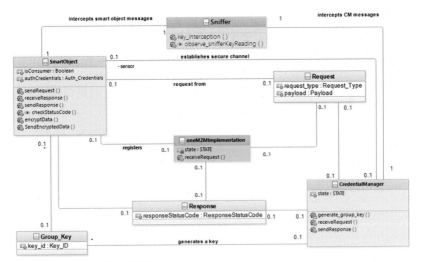

Fig. 12 Integration of the MBT model.

The model also includes test oracles. For example: If the sniffer cannot read the group key, the test will be satisfactory. Otherwise, the test will be unsatisfactory.

To drive the test generation, a BPMN model (Fig. 13) is created.

The BPMN model drives the test generation by giving some guidelines to the MBT tool for test coverage. The BPMN will ensure to test the business scenario required: test the end-to-end security of the process from key retrieval to secure encrypted data in the IoT platform.

Once all the MBT material described is in place, we proceed to the test case generation step.

5.1.2 Test Case Generation

Section 4.1.2 gave us some initial insight of the test case generation from the model perspective. The generation process for MBT the IoT security uses the same principles, and we will extend the study approach to the generation within the CertifyIt tool. The BPMN model (Fig. 13) allows one to generate one test case related to the end-to-end security scenario. The generated test case is shown in Fig. 14. Fig. 14 shows the test case details on its right-hand side. We recognize the steps introduced in the BPMN model as:

1. The sensor sends a request to the CM asking for a KEY.
2. The CM receives the request.
3. The CM generates a KEY and sends it to the sensor.
4. The sensor receives the KEY.
5. The sensor encrypts its data with the received KEY.
6. The sensor sends the key to the oneM2M implementation.
7. The oneM2M implementation receives the encrypted value.

The generated test case is still an abstract test as the enumerated values taken from the model are present. For example, the test oracle for the fourth test step is the *"KEY_GENERATION_SUCCESS"* code. All enumerated values need to be adapted to obtain a concrete executable test. The next session shows how this process is conducted.

5.1.3 Test Case Reification

After successfully generating the tests, a conversion (or, in the CertifyIt terminology publish) into a target language is needed. The default publishers in CertifyIt are XML, Junit, and HTML. The XML publisher is the base publisher that will convert all test cases and other model data to an XML file that then can be used by other publishers to generate tests in other formats. The

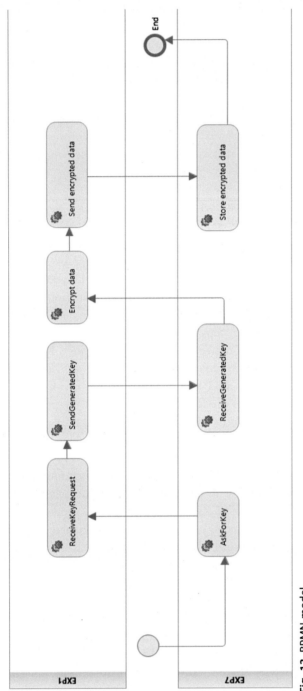

Fig. 13 BPMN model.

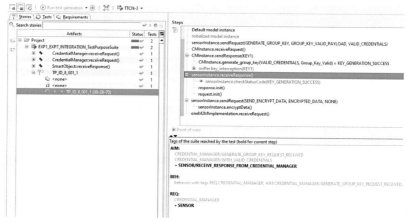

Fig. 14 Test case generation with the MBT tool.

publisher in JUnit generates tests that can be executed in Java with a glue code, i.e., a code that will be an adapter between the SUT and the JUnit tests converting the abstract values into concrete ones. The HTML publisher serves mainly for documentation purposes because its output is simply a webpage where all steps for every test case are shown. The TTCN-3 publisher works on the same principle: the TTCN-3 generated tests are joined with an adapter file where all types, templates, functions, and components are defined. Then, we proceed to the next phase of test compilation. The TTCN-3 example generated from the model is shown in Fig. 15. It follows the same structure as that generated from the model. The TTCN-3 tests start with the variables declaration. Then, the static part of the preamble is used to configure the test component. The second part of the preamble and the test body is derived from the abstract test case. The send/receive instructions are part of the test body and are detected by the naming convention that they are in fact templates, not functions. The messages used in the receive part are taken from the "observer" functions in the model.

Once the abstract test suite (ATS) in the TTCN-3 format is obtained, executing it is the next step. To achieve this, a compiler transforming the TTCN-3 code into an intermediate language, like C++ or Java, is needed. The TTCN-3 code is abstract in a sense that it does not define the way to communicate with the SUT, or how the TTCN-3 structures will be transformed into a real format. This is a task for the system adapter (SA) and the Codec, which are described in detail in the next section. There are few commercial and noncommercial compilers that are available for download from

```
testcase tc_CSE_SEC_BI_002_02() runs on M2M system M2MSystem {

    // Local variables
    var M2MResponsePrimitive v_response;
    var RequestPrimitive v_request;

    // Test component configuration
    f_cf01Up();

    f_preamble_registerAe();//c_CRUDNDi);

    f_createRequestPrimitive(int2, -1, v_request);

    mcaPort.send(m_request(v_request));
    tc_ac.start;
    alt {
        [] mcaPort.receive(mw_responseKO) -> value v_response {
            tc_ac.stop;
            setverdict(pass, testcasename() &
            ": Test passed with response code " &
            int2char(enum2int(v_response.responsePrimitive_.responseStatusCode)));
        }
        [] mcaPort.receive(mw_responseOK) {
            tc_ac.stop;
            setverdict(fail, testcasename() & ": Error!");
        }
        [] tc_ac.timeout {
            setverdict(inconc, testcasename() & ": No answer!");
        }
    }

    // Postamble
    f_postamble_deleteResources();

}//end tc_CSE_SEC_BI_002_02
```

Fig. 15 TTCN-3 abstract test suite example.

the official TTCN-3 webpage [51]. Usually, all the compilers follow this
procedure:
- Compile TTCN-3 to a target language (Titan uses C++,
  TTWorkbench uses Java).
- Add the Codec and SA written in the target language.

In the scope of this study, we show the Titan test case execution tool as it is
an open–source project.

5.1.4 Test Case Execution

The goal of the testing process, as described in Fig. 10, is to execute the
MBT–generated tests on the SUT.

As described in Fig. 16, after the publication of our tests in TTCN-3, we
use the Titan compiler to compile them to C/C++ code. To this code, we
add the C/C++ code of the SA and the Codec. Using a makefile generated
by Titan, we can compile the complete suite into a test executable. Its

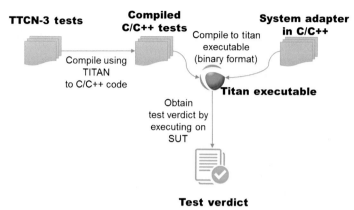

Fig. 16 Test execution.

execution can be controlled by Titan via a configuration file explained in the following paragraphs. At the end of the execution, we obtain a log of the execution, with a status for every test case executed. The test cases results can be traced back to their test purposes and test patterns.

Module parameters, abbreviated as modulepars, are a feature of TTCN-3 described in its specification. They are similar to constants and are used to define the parameters of a test suite. These parameters can be, for example, the address of the SUT, the timeout period, or some details of the SUT that can be chosen depending on the SUT implementation. These parameters also are called Protocol Implementation eXtra Information for Testing (PIXIT). Titan can use these module parameters inside its configuration file to provide configuration of the ATS without the need to recompile it. In the same file the user can configure the SA, for example, if it uses a UDP test port, through which system port of the test system it should send or receive the data.

TTCN-3 provides a whole structure inside a module to control the execution order of the test cases. Inside the control part, the tester has full control over the execution. He can create and start one or more components that can run the tests in parallel, and can use conditions and loops to create an execution of the test suite that is closely adapted to the needs of the SUT or the tester itself. Also, Titan widens this control by implementing a higher level of control inside the configuration file. Here, the user can specify, if there is more than one module with control parts, which module will be first to execute, which will be last, and the exact order of the test cases, etc.

Every TTCN-3 compiler has the right to implement its own method to store the test execution results in the form of a logging file. The Titan open-source compiler uses a proprietary log format from Ericsson by default, but its logging interface is very rich and thus it is not difficult to create new logging formats adapted to a user's needs. The logging facility in Titan can produce logs with various levels of details to manage different requirements for the console log and log file. The logging parameters can be configured in Titan's configuration file by setting the ConsoleMask and FileMask options, respectively, for the level of information that should be shown on the console and saved in the log file. TTCN-3 has strict control of a test case status. The five different values for test statuses are "none," "pass," "inconc" (short of inconclusive), "fail," and "error." Every time a test case is started, its status is automatically assigned to "none." The tester, through the test case, can change it to "pass," "inconc," or "fail," depending on the conditions that are defined in the test case. To prevent a failed test becoming valid, when the test case is executing and the status is set to "fail," it cannot be reverted to "pass" or "inconc." The same is true for "inconc," it cannot revert to "pass." A special status is the "error" status and it is reserved for any runtime errors to distinguish them from SUT errors.

5.2 Experimentation and Results

To test the end-to-end security scenario efficiently, the need for a large-scale test framework is a primordial factor as it emulates a real implementation of an IoT service. The MBT Testing Framework for Large-Scale IoT Security Testing (Fig. 17) consists of seven modules from three tool suites: CertifyIt tool suite, Titan, and FIT IoT-LAB[53]. The FIT IoT-LAB is part of the future Internet of things (FIT) that has received 5.8 million Euros in funding from the French Equipex research grant program. FIT IoT offers the large-scale testbed on which the test cases are executed. The test configuration is considered in the modeling and the test execution environment in the FIT IoT-LAB by the test engineers.

After the MBT-generated test cases and their publication in the TTCN-3 test scripts, they are deployed in the FIT IoT-LAB ready to be executed by Titan. Titan will first compile, then adapt, and finally execute the test cases on a SUT within the FIT IoT-LAB. Results of the feasibility of the demonstrated approach have been achieved and the large-scale scenario could be demonstrated through the ARMOUR experiments. We could test the complete chain of events described in the BPMN scenario and record

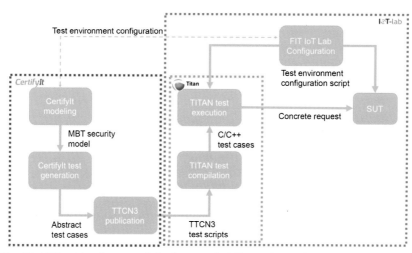

Fig. 17 MBT testing framework for large-scale IoT security testing.

the messages exchanged in the testbed through the use of its integrated sniffer. The sniffer revealed that substantial portions of the messages exchanged (92%) where not ciphered. This result is further used to benchmark the security of an IoT deployment. Our experiments can safely assess that the data transiting between the sensor and the IoT platform is ciphered. This is far from optimal to conclude that the deployment is secure. We discuss more details on how our results can be used to benchmark an IoT deployment in the next session.

5.3 Lessons Learnt

The experiments concluded within ARMOUR have as one of the main objectives to define an approach for benchmarking security and trust technologies for experimentation on large-scale IoT deployments. In this respect, it is a major necessity to provide tools for the IoT stakeholders to evaluate the level of preparedness of their system to the IoT security threats. Benchmarking is the typical approach to this and ARMOUR will establish a security benchmark for end-to-end security in the large-scale IoT, by building up the ARMOUR large-scale testing framework and process. Additionally, ARMOUR experiments will be benchmarked using the ARMOUR benchmarking methodology. Several dimensions will be considered including

- Security attacks detection.
- Defense against attacks and misbehavior.

- Ability to use trusted sources and channels.
- Levels of security and trust conformity, etc.

Benchmark results of the large-scale IoT security and trust solutions experiments will be performed and the result datasets will be made available via the FIESTA [54] testbed. Additional benchmarks of reference secure and trusted IoT solutions will be performed not only to establish a baseline ground-proof for the ARMOUR experiments but also to create a proper benchmarking database of secure and trusted solutions appropriate for the large-scale IoT. To define this methodology and the different dimensions to be considered, ARMOUR will

- Consider the methodology defined in ARMOUR for generic test patterns and test models based on the threats and vulnerability identified for the IoT and the procedure to provide test generations based on the previous one.
- Identify metrics per functional block (authentication, data security, etc.) to perform various micro- and macrobenchmarking scenarios on the target deployments.
- Collect metrics from the evaluation and then use the metrics to categorize them (taxonomy) into different functional aspects, and based on this, provide a label approach to the security assessment.

Microbenchmarks provide useful information to understand the performance of subsystems associated with a smart object. Microbenchmarks are useful to identify possible performance bottlenecks at the architectural level and allow embedded hardware and software engineers to compare and assess the various design trade-offs associated with component level design. Macrobenchmarks provide statistics relating to a specific function at the application level and may consist of multiple component level elements working together to perform an application layer task. The final objective is to provide the security benchmarking and assurance that should include a measure of our level of confidence in the conclusion. As an example of such a measure of confidence used to evaluate the security of an IoT product or system, we mention the evaluation assurance level (EAL) from the CC. The EAL is a discrete numerical grade (from EAL1 through EAL7) assigned to the IoT product or system following the completion of a CC security evaluation, which is an international standard in effect since 1999. These increasing levels of assurance reflect the fact that incremental assurance requirements must be met to achieve CC certification. The intent of the higher assurance levels is a higher level of confidence that the systems security features have been reliably implemented. The EAL level does not measure the security of the system; it rather simply states at what level the system was tested. The EALs include the following:

- EAL 1 Functionally tested.
- EAL 2 Structurally tested.
- EAL 3 Methodically tested and checked.
- EAL 4 Methodically designed, tested, and reviewed.
- EAL 5 Semiformally designed and tested.
- EAL 6 Semiformally verified design and tested.
- EAL 7 Formally verified design and tested.

6. ONLINE MBT FOR IoT ROBUSTNESS TESTING

Robustness testing, or fuzzing, provides many diverse types of valid and invalid input to software to make it enter an unpredictable state or disclose confidential information. It works by automatically generating input values and feeding them to the software package. Fuzzing can use different input sources. The developer running the test can supply a long- or short-list of input values and can write a script that generates the input values. Also, fuzz testing software can generate input values randomly or from a specification; this is known as generation fuzzing. Traditionally, fuzzing is used and the results show that it is very effective at finding security vulnerabilities, but because of its inherently stochastic nature, the results can be highly dependent on the initial configuration of the fuzzing system.

"Practical model-based testing: A tools approach" [7] defines MBT as follows: *We define MBT as the automation of the design of black-box tests. The difference from the usual black-box testing is that rather than manually writing tests based on the requirements documentation, we instead create a model of the expected SUT behavior, which captures some of the requirements. Then the MBT tools are used to automatically generate tests from that model.*

For online MBT IoT robustness testing, MBeeTle is a specific tool which is designed to combine MBT and fuzzing. MBeeTle is an independent tool based on the CertifyIt technology, whose purpose is to randomly generate test cases for test fuzzing.

Our experience in using a Fuzzing approach based on models with a tool such as MBeeTle, showed several advantages, such as:

- Minimal initial configuration is required to start a fuzzing campaign.
- Minimal supervision of the fuzzing campaign is required.

- Uniqueness determination is handled through an intelligent backtrace analysis.
- Automated test case minimization reduces the effort required to analyze the results.

The approach is complementary to other MBT testing approaches and can reveal unexpected behavior due to its random nature.

6.1 MBT Specificities for the IoT Robustness Testing

The MBT for the IoT robustness testing model is based on the same MBT models used for security functional and vulnerability testing; it generates test cases using behavioral fuzzing. Contrary to the previous two approaches where the test case generation aims to cover the test objectives produced from the test purposes, the MBeeTle fuzzing tool generates weighted random test cases to cover the test objectives based on the expected behavior of the system. The tool relies on the principle to rapidly generate as high as possible many fuzzed tests with a substantial number of steps in each period using a weighted random algorithm. The generated tests are valid with respect to the constraints in the MBT model. Thus, contrary to most fuzzers, the produced test cases on the one hand are syntactically correct with respect to the systems inputs. On the other hand, since it uses a weighted random algorithm and measures the coverage of the behaviors, it avoids duplication of the generated tests, which makes the test evaluation and assessment easier. Another advantage of this MBT fuzzing is its rapidity and contrary to classical functional testing approach, it explores the system states in various uncommon contexts and potentially placing the system into a failed state. Fig. 18 depicts at a high level the test generation algorithm. Each step covers a test objective, which corresponds to one behavior of the system identified in the model in the OCL expression in an operation with specific preexisting tags @REQ/@AIM (illustrated in the figure). A step is generated by

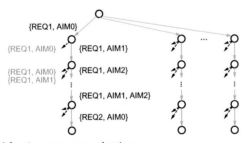

Fig. 18 Behavioral fuzzing test step selection.

randomly choosing the available states possibly leading to an error from a given context of the system (for instance, start exploring the systems behavior when administrator operations are called from normal user connection). More specifically, a test step is selected only if it activates a new behavior (tag), otherwise the algorithm continues the exploration to conceive a test step that activates a new tag. The test generation stops either when the test generation time allocated to the generation engine has elapsed or by fulfilling the user conditions (for instance, all tags in the model are covered, or a stop is signaled).

Moreover, MBT fuzzing is a complementary approach to functional and security (functional and vulnerability) testing, offering multifold benefits:

- it activates behaviors in uncommon contexts that potentially lead to system failures, unknown by domain experts.
- it rapidly generates tests, and since the algorithms are random, they are extremely quick and feedback to the user is sent rapidly (hundreds of tests in a few minutes).
- it reuses the same MBT model used for functional and security testing.
- it reuses the same adaptation layer, thus no extra effort is needed for executing the test cases.

Nevertheless, the random algorithms lack power in constructing complex contexts of the system, which is bypassed using existing valid test sequences produced using functional testing, as a preamble of the test case. Thus, it lowers its impact on the quality of the test cases.

6.1.1 MBT Model

The MBT fuzzing tool (MBeeTle) uses the same model used for functional and security testing as described in step 1 of Section 3.1. On a technical level, the model does not change and this is important in terms of model reusability and time-saving. On the other hand, important changes are made into the test case generation and execution.

6.1.2 Test Case Generation, Reification, and Execution

The MBT robustness testing or MBT fuzzing uses an *online* test generation methodology. The online generation consists of executing directly produced test steps by the tool on the SUT. For each generated test step, the tool waits for its result before generating another step. If the result is not the one expected by the model, the generation process is interrupted. Fig. 19 describes the online generation methodology.

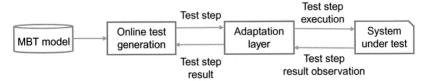

Fig. 19 Online generation methodology.

To communicate with the SUT in its own language and protocol, the online test generator sends its generated test steps to an adaptation layer that will translate the tests into executable steps. The result of each action taken on the SUT is traced back to the generator to decide if it should continue the test generation by exploring the model or to interrupt the process depending on the status of the last step result.

6.2 Future Works

The methodology introduced is fairly new and its validation is an ongoing work as of April 2017 in the ARMOUR H2020 project described previously. The main goal is to experiment with the MBT robustness testing on the large-scale testbed FIT IoT-LAB. Thus, with respect to the obtained results, we can assert the effectiveness of this methodology in real-time and on large-scale deployments.

7. MODEL-BASED TESTING AS A SERVICE

The IoT platforms tend nowadays to be significantly more service-oriented to user applications. The question of testing and validation of the IoT platforms can be tackled with the same *"as a service"* approach. This section presents the general architecture of the *model-based testing as a service* (MBTAAS) approach. We then present in more details, how each service works individually to publish, execute, and present the tests/results.

7.1 MBTAAS Architecture

An overview of the general architecture can be found in Fig. 20, where we find the four main steps of the MBT approach (MBT modeling, test generation, test implementation, and execution).

However, to the difference of a classical MBT process, MBTAAS implements several web services, which communicate with each other to realize the testing steps. A web service, uses web technology, such as HTTP, for

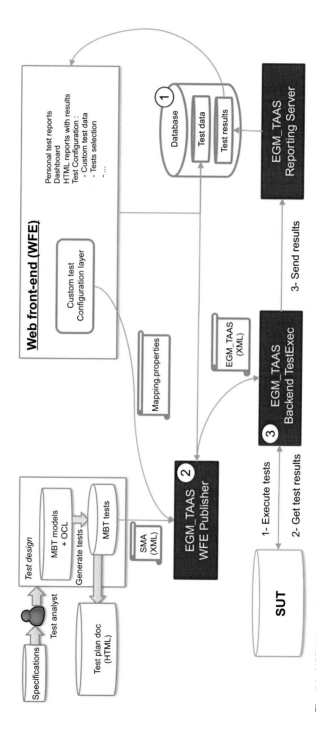

Fig. 20 MBTAAS architecture.

machine-to-machine communication, and more specifically for transferring machine-readable file formats such as XML[a] and JSON.[b]

In addition to the classical MBT process, the central piece of architecture is the database service ① that is used by all the other services. Its involvement is explained as each service is described individually. Nevertheless, the database service can be separated from all the other services, and it can be running in the cloud where it is easily accessible. The database stores essential information, such as test data (input data for test execution) and test results. The entry point of the system is the web front-end service (customization service). This service takes a user input to customize a testing session and it communicates it to a publication service ②. The purpose of the publisher service is to gather the MBT results file and user custom data to produce a customized test description file (EGM_TAAS file). This file is then sent to an execution service ③ which controls the execution of the customized tests. It stimulates the SUT with input data to obtain a response as the SUT output data. The execution service then finally builds a result response and sends it to a last service, the reporting service. The reporting service is configured to feed the database service with the test results. These test results are used by the web front-end service to submit them to the end user. This testing architecture is executed at a modular level to respond to the heterogeneity of an IoT platform. In the following sections, a detailed description of each service is provided.

7.2 Customization Service

To provide user-friendly testing as a service environment, MBTAAS implements a graphical web front-end service to configure and launch the test campaigns. The customization service is a website where identified users have a private dashboard. The service offers a preconfigured test environment. The user input can be reduced to a minimum, that is: defining a SUT endpoint (URL). User specific test customization offers a wide range of adjustments to the test campaign. The web service enables

- *Test selection*: from the proposed test cases, a user can choose to execute only a part of them.
- *Test data*: preconfigured data are used for the tests. The user can add his own specific test data to the database and choose it for a test. It is a test per test configurable feature.

---

[a] https://www.w3.org/XML/.
[b] http://www.json.org.

- *Reporting*: by default, the reporting service will store the result reported on the web front-end service database (more details on this in Section 7.5). The default behavior can be changed to fit the user needs, for example, having the results in another database, tool.

After completion of the test configuration and depressing the launch test button, a configuration file is constructed. The configuration files, as seen in Fig. 21: *Configuration File excerpt*, defines a set of {key = value}. This file is constructed with default values that can be overloaded with user-defined values.

The configuration file is one of the three components that the publisher service needs to generate the test campaign. The next section describes the publication process in more detail.

7.3 Publication Service

The publisher service, as its name states, publishes the abstract tests generated from the model into a concrete test description file. It requires three different inputs (Fig. 20) for completion of its task: the model, the configuration file, and the test data. The model provides test suites containing abstract tests. The concretization of the abstract tests is made with the help of a database and a configuration file. For example, the abstract value ENDPOINT_URL taken from the model is collected from the configuration file.

The concrete test description file is for this approach, an XML file that instantiates abstract tests. The test description file has two main parts, general information and at least one test suite (Fig. 22). A test suite is composed of one or more test cases, and a test case itself is composed of one or more test

```
14   ###################################################################
15   ################   REQUIRED PARAMETERS  ############################
16   ###################################################################
17
18   #NAME OF THE OWNER OF THE REPORT
19   OWNER=EGM_TE_XML_PUBLISHER
20   #REPORT LOCATION AFTER TESTS (FOR EGM_TAAS_BACKEND)
21   REPORT_LOCATION=http://193.48.247.210:8081/report
22   #HOW TO REPORT (FOR EGM_TAAS_BACKEND)
23   REPORT_TYPE=POST_URL
24   #URL OF SUT TO TEST WITH THE PORT (FULL PATH)
25   ENDPOINT_URL=http://193.48.247.246:1026
26   #URL of EGM_TAAS backend that will execute the tests
27   EGM_TAAS_BACKEND = localhost:8080/executeTests
28   #Name of the Model file to be Used by EGM_TAAS_BACKEND
29   EGM_TAAS_MODEL = OrionCB_GE.xml
30   #Where to Output the results in the EGM_TAAS_BACKEND
31   EGM_TAAS_OUTPUT = tmp
```

Fig. 21 Configuration file excerpt.

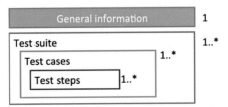

Fig. 22 Published file parts.

steps. This hierarchy respects the IEEE 829-2008, Standard for Software and System Test Documentation.

The general informational part of the file is useful to execute the tests (Section 7.4) and report the results. Here are some of the most important parameters that can be found in that part of the file:

- *Owner*: used for traceability purposes. It allows one to know who is the detainer of the test to present it in his personal cloud dashboard.
- *Sut_endpoint*: the hostname/IP address of the SUT. This address should be reachable from the execution service point of view.
- *Location*: the hostname/IP address of the reporting service (Section 7.5).

The test suites contain all useful test information. In this case, the IoT platform proposes HTTP-based RESTful applications. The mandatory information required to succeed a RESTful query are the URL (SUT endpoint) and the HTTP method (GET, POST, PUT, DELETE). The test suite and test cases purpose is the ordering and naming of the test, but the test information and test data are stored in the test steps. Each test step has its own configuration. Once the file is published, it is sent to the execution service to execute the tests.

7.4 Execution Service

The execution service is the functional core of the MBTAAS architecture. The execution service will run the test and collect the results depending on the configuration of the received test description file. The FIWARE RESTful interface tests are executed with the REST execution module. Each test is run against the SUT and a result report (Listing 1) is constructed on the test step completion. The result report contains information on the date-time of execution, time spent executing the step and some other test specific values. The "endpoint" represents the URL value and validity of where the step should be run. An invalid endpoint would allow one to skip the type check on the endpoint value and thus allow ne to acquire an execution error. The response of each step is validated within the "assertion_list" tags (test oracle).

It validates each assertion depending on the assertion type and values against the received response.

```
Listing 1: Test step result report
< teststep  name = " UpdateEntity1 " >
    < executionResults >
         < timestamp > { TIMESTAMP } < / timestamp >
         < executionTimeMs > 22 < / executionTimeMs >
    < / executionResults >
    < endpoint >
         < value > { IP } : { PORT }/ upContext < / value >
         < isinvalid > false < / isinvalid >
    < / endpoint >
    < method > POST < / method >
    < headers > { HEADERS } < / headers >
    < payload > { PAYLOAD } < / payload >
    < assertion_list >
         < assertion >
             < type > JSON < / type >
             < key > code < / key >
             < value > 404 < / value >
             < result > false < / result >
         < / assertion >
    < / assertion_list >
    < result > false < / result >
    < response > {
         " errorCode "  :  {
             " code "  :  " 400 " ,
             " reasonPhrase "  :  " Bad  Request " ,
             " details "  :  " JSON  Parse  Error "
         }
    }
    < / response >
< / teststep >
```

Fig. 23 shows an excerpt of the execution service log. To execute one test step, the endpoint (URL) must be validated. Then, a RESTful request is created and executed. A response is expected for the assertions to be evaluated. Finally, an overall test result is computed. The overall assertion evaluation algorithm is as simple as: "All test assertions must be true," which implies if one assertion is false, the test is marked as failed.

```
[egm.modelTools.HttpRequestExecuter] Validating url : http://███████████:1026/v1/updateContext
[egm.modelTools.HttpRequestExecuter] URL is VALID
[egm.modelTools.HttpRequestExecuter] Starting Jetty HTTP Client
[egm.modelTools.HttpRequestExecuter] Jetty HTTP Client Started with Success
[egm.modelTools.HttpRequestExecuter] Creating request : URL = http://███████████:1026/v1/updateContext, HTTPMETHOD = POST
[egm.modelTools.HttpRequestExecuter] Request created
[egm.modelTools.HttpRequestExecuter] Request status code: 200
[egm.modelTools.HttpRequestExecuter] Response content: {
    "errorCode" : {
       "code" : "400",
       "reasonPhrase" : "Bad Request",
       "details" : "JSON Parse Error"
    }
}

[egm.modelTools.HttpRequestExecuter] Stopping Jetty HTTP Client
[egm.modelTools.HttpRequestExecuter] Jetty HTTP Client Stopped
[egm.modelTools.HttpRequestExecuter]

[egm.model.Assertion] Asserting...expression to be assert: is key "code" contains value: "404"
[egm.modelTools.TestStepResponseParser] is JSON data key "code" contains value: "404"
[egm.modelTools.TestStepResponseParser] no value of : "404" has been found for id: "errorCode"
[egm.modelTools.TestStepResponseParser] no value of : "404" has been found for id: "code"
[egm.model.TestStep] Execution Result of step UpdateEntity148 : false
```

Fig. 23 Execution snapshot.

Once the execution service has finished all test steps, their results are gathered within one file, this file is called test results, and it is sent to the reporting service.

7.5 Reporting Service

After executing the test, a file containing the test description alongside the results is sent to and received by the reporting service. The reporting service configuration is made on the web front-end service. The configuration is passed with the test configuration file where the publisher service retranscribes that information to the file sent to the execution service. The execution service then includes the reporting configuration in the report file where it is used in the reporting service once it receives it. By default, the reporting service will save the results in the database service ①. For this use case, the database is implemented as a MySQL database. This database is then used by the web front-end service to present the test results to the user.

MBTAAS has proven to be generic enough to be integrated with little effort for the IoT conformance, security, and robustness testing.

8. LESSONS LEARNT AND DISCUSSION

The work and results presented in this chapter is a combination of multiple work done in the domain of MBT for the IoT systems testing. We provided a demonstration of the capabilities of an MBT approach adapted to the IoT platform conformance and security testing. Results of their application are encouraging to move forward and bring the service layer to it.

The initial cost in applying MBT in general is usually high, due to: (i) the time spent on the models conception and (ii) the first implementation of an adaptation layer for test execution that can be complex depending on the SUT. This high initial cost of implementing MBT can be rewarded if we consider that the project requirements and specification may evolve in its development cycle. In this later case, the MBT for the IoT platforms can keep the test repository stable. This leads us to conclude that MBT is a suitable approach for emerging technologies and standards, especially on the IoT platforms where the maturity level is still increasing while technology development is still ongoing. Being able to generate tests automatically and in a systematic way, as we did, makes it possible to constitute a test suite covering the defined test objectives. The MBT further offers the means to format the results into automated reports to justify the test objective coverage, which can be easily used for certification and labeling purposes. These several examples show the usefulness of an automated and systematic use of an MBT approach on applications that should comply to specific standards. Regarding the security point of view, where new vulnerabilities are emerging faster than security experts can predict them, MBT allows the integration of new vulnerability patterns with little cost to an ongoing test campaign. Combined with a user-friendly and easily accessible interface through a service-oriented solution, the first experiments with MBTAAS show that it is a promising and powerful tool. The reusability of the service approach is a strong and solid factor for its adoption. The only changes to be made concern the test environment and the adaptation layer, to make it compatible with any IoT component under test. One major concern is to provide the tests to the community in the easiest way possible. It provides to all (not only testers) the capacity to test their components remotely using an online service. The service provides to users a test-friendly interface that simplifies the test execution tools installation and the various technical challenges with respect to system incompatibilities, such as networking access problems.

The newly introduced MBT fuzzing method does not follow the standard fuzzing testing methods. Its goal in the IoT testing is to extensively use the MBT model to generate random test steps, thus ensuring a deeper exploration of requirements coverage. There are no malformed test steps in this process. The objective is to ensure the correct behavior of the tested SUT over its lifetime span as it enables stress testing over an extended period of time. The new method has yet to prove its effectiveness in a large-scale testbed.

9. CONCLUSION

This chapter presented the state of the art on MBT of the IoT systems and described specific challenges and MBT solutions for the IoT testing.

More specifically, it provided answers that address the key challenges of scalability, interoperability, and security of the IoT with real use cases. The solutions that we proposed have a common execution approach: MBTAAS. The MBT used for the IoT conformance testing shows that the specificities in generating test cases to validate the correct behavior of an SUT with respect to its standard lies in the model creation. In this case, the model is a representation of the standard. Any test executed on the SUT that fails could represent a gap between the specification and implementation, or a compliance from the SUT toward the standard it is implementing. The MBT used for the IoT security testing diverted from the traditional MBT setup with its specific test case reification environment. Finally, we had a first view on the specificities of an MBT for the IoT robustness testing approach. It illustrates how MBT fuzzing could be applied to the IoT systems. The tooling we presented gives the possibility to reuse the already created model during the functional/security testing phase; thus, no additional time is spent for modeling. This approach is considered as complimentary to security or conformance testing, as we expect that the detected inconstancies will differ from the others.

The standardization institutes and certification entities that make available tomorrows standards have the process of testing as the highest importance. They provide standard implementers a set of test cases to assess the conformance, scalability, and security of their products. Test cases traceability after execution is also a major factor as it is used to benchmark the test results. With security threats constantly evolving, and to maintain high confidence and trust, the proposed test cases must be continuously maintained. This is where MBT finds its interest and added value. The traceability starts with the specification, then the requirements, and finally to the test cases which are an embedded feature to that of the classical MBT approach. The test case traceability provided by MBT gives the possibility to implement test generation strategies which aim to cover a maximum of the specification. The automated test generation is low time-consuming comparing to hand written tests, and thus makes the maintenance of generated test cases cheaper. MBT has major advantages in terms of test coverage, test maintenance, etc. Those advantages are accelerating the adoption of MBT over time.

The MBT approach for test case generation and execution is relatively new. Nowadays, it is becoming more and more visible and accepted compared to 10/15 years ago. The approach was previously reserved for a smaller number of use cases due to the high difficulty of taking in hand the modeling. Modeling requires highly skilled testers and at the time the profession was not common in the software industry. Only highly critical companies had the privilege of modeling their concepts to test them. The IoT sector is reaching this critical level nowadays. Security intrusions are more sophisticated than ever and occur more frequently then we realize. The need for quality assurance in the field of the IoT testing is essential to accompany its growth. With MBT, the evolution of the security threads is added into the model and they are by this process considered in the test generation process. The process saves time when it is not started from scratch. The models for conformance testing are reusable and adapted to security testing. MBT shows its benefits in the medium to long term as it has a high initial cost and it has been one of the strongest roadblocks for MBT adoption. The MBT approach requires highly skilled testers and it is complicated to change the daily habits of testers in the field. It requires a significant time investment that most are not willing to partake. One other roadblock for the MBT approach adoption is the complexity of the existing tools. Nevertheless, MBT is still relatively new and we see more and more tools emerging. However, these tools require a high degree of formation and skill for practical usage. The test generation tools are complicated and the complexity of test generation can take a significant amount of time (NP-complete problems). In the future, the MBT approach requires significant evolution to survive and show a rise in its adoption. New techniques have already been developed by tool makers, as seen in Section 6, MBeeTle is one of the emerging tools for online testing that responds to the need for a real-time analysis of the test cases on the IoT deployments. This new tool allows for a loss of non-determinism in a real-time system.

To conclude, the technology for automated model-based test case generation has matured to the point where the large-scale deployments of this technology are becoming commonplace. The prerequisites for success, such as qualification of the test team, integrated tools chain availability and methods are now identified, and a wide range of commercial and open-source tools are available, as introduced in this chapter. Although MBT will not solve all testing problems in the IoT domain, it is extremely useful and brings considerable progress within the state of the practice for functional software testing effectiveness, increasing productivity, and improving

functional coverage. Moreover, the MBT approach has shown its importance in the IoT by introducing an ease of access to conformance, security,
and robustness testing. The technology is constantly evolving, making it tester friendly and considerably reducing the high initial cost that is typically the
case. Because of its easier accessibility for testers, as well as more fine-grained
integration with the large-scale IoT testbeds, MBT should expect a larger
approval rate.

ACKNOWLEDGMENTS

This research was supported by the ARMOUR project Grant ID 688237 of the Horizon
2020 Call ICT-12-2015 Integrating experiments and facilities in FIRE+. We would like
to thank Dr. Cedric ADJIH, Dr. José Luis Hernández RAMOS, and Sara Nieves Matheu
GARCIA for sharing their valuable insights with us.

REFERENCES

[1] DYN DDoS Attack Report, 2017, http://www.computerworld.com/article/
    3135434/security/ddos-attack-on-dyn-came-from-100000-infected-devices.html (last
    visited April 2017).
[2] R.V. Binder, B. Legeard, A. Kramer, Model-based testing: where does it stand?
    Commun. ACM 58 (2) (2015) 52–56, ISSN 0001-0782, https://doi.org/
    10.1145/2697399.
[3] G. Baldini, A. Skarmeta, E. Fourneret, R. Neisse, B. Legeard, F.L. Gall, Security cer-
    tification and labelling in Internet of Things, in: 2016 IEEE 3rd World Forum on Inter-
    net of Things (WF-IoT)2016, pp. 627–632.
[4] L.B. Kramer A, Model-Based Testing Essentials—Guide to the ISTQB Certified
    Model-Based Tester: Foundation Level, Wiley, 2016.
[5] Capgemini, World Quality Report 2016-17, 2016.
[6] Model-Based Engineering Forum, 2017, http://modelbasedengineering.com/ (last vis-
    ited April 2017).
[7] M. Utting, B. Legeard, Practical Model-Based Testing—A Tools Approach, Morgan
    Kaufmann, San Francisco, CA, 2006, ISBN 0123725011.
[8] A. Spillner, T. Linz, H. Schaefer, Software Testing Foundations: A Study Guide
    for the Certified Tester Exam: Foundation Level, ISTQB Compliant, 2014, Santa
    Barbara, CA.
[9] A. Hartman, Model Based Test Generation Tools, Agedis Consortium, 2002, http://
    www.agedis.de/documents/ModelBasedTestGenerationTools_cs.pdf.
[10] C. Budnik, R. Subramanyan, M. Vieira, Peer-to-Peer Comparison of Model-Based
    Test, in: GI Jahrestagung, 2008.
[11] M. Shafique, Y. Labiche, A Systematic Review of Model Based Testing Tool Support,
    Technical Report SCE-10-04, 2010, Carleton University, Canada.
[12] R. Marinescu, C. Seceleanu, H.L. Guzen, P. Pettersson, A Research Overview of
    Tool-Supported Model-based Testing of Requirements-Based Designs, Elsevier,
    2015, pp. 89–140.
[13] M. Utting, A. Pretschner, B. Legeard, A taxonomy of model-based testing approaches.
    Soft. Test. Verif. Reliab. 22 (5) (2012) 297–312, ISSN 1099-1689, https://doi.org/
    10.1002/stvr.456.

[14] M. Utting, B. Legeard, F. Bouquet, E. Fourneret, F. Peureux, A. Vernotte, Chapter Two—Recent Advances in Model-Based Testing, in: Advances in Computers, vol. 101, Elsevier, 2016, pp. 53–120.

[15] E.S. Reetz, Service testing for the Internet of things (Ph.D. thesis), University of Surrey, 2016.

[16] IBM, Model-Based Testing (MBT) at InterConnect 2016? Yes! https://www.ibm.com/developerworks/community/blogs/35dfcb99-111b-423a-aaa4-50f3fddae141/entry/Model_Based_Testing_MBT_at_InterConnect_2016_Yes?lang=en, (accessed 06.03.17).

[17] I. Schieferdecker, J.A. Gromann, A. Rennoch, Model Based Security Testing *Selected Considerations* (keynote at SECTEST at ICST 2011), http://www.avantssar.eu/sectest2011/pdfs/Schieferdecker-invited-talk.pdf, (accessed 25.09.12).

[18] I. Schieferdecker, J. Grossmann, M. Schneider, Model-Based Security Testing. in: MBT, 2012, pp. 1–12, https://doi.org/10.4204/EPTCS.80.1.

[19] K.G. Larsen, M. Mikucionis, B. Nielsen, Online testing of real-time systems using UPPAAL, in: J. Grabowski, B. Nielsen (Eds.), Formal Approaches to Software Testing. FATES 2004, Lecture Notes in Computer Science, vol. 3395, Springer, Berlin, Heidelberg, 2005.

[20] A. Kriouile, W. Serwe, Using a formal model to improve verification of a cache-coherent system-on-chip, in: C. Baier, C. Tinelli (Eds.), Tools and Algorithms for the Construction and Analysis of Systems, Lecture Notes in Computer Science, vol. 9035, Springer, Berlin, Heidelberg, 2015, pp. 708–722.

[21] M. Schneider, J. Gromann, I. Schieferdecker, A. Pietschker, Online model-based behavioral fuzzing, in: 2013 IEEE Sixth International Conference on Software Testing, Verification and Validation Workshops, 2013, pp. 469–475.

[22] CompleteTest, 2017, http://www.completetest.org/about/ (last visited April 2017).

[23] K.G. Larsen, P. Pettersson, W. Yi, UPPAAL in a nutshell, Int. J. Softw. Tools Technol. Transfer 1 (1997) 134–152.

[24] Eclipse Formal Modeling Project, 2017, https://projects.eclipse.org/proposals/eclipse-formal-modeling-project (last visited April 2017).

[25] A. Faivre, C. Gaston, P. Gall, Symbolic Model Based Testing for Component Oriented Systems, Springer, 2007.

[26] FMBT, 2017, https://01.org/fmbt/ (last visited April 2017).

[27] HTG, 2017, https://sites.google.com/site/htgtestgenerationtool/home (last visited April 2017).

[28] Netlist Syntax, 2017, http://fides.fe.uni-lj.si/spice/netlist.html (last visited April 2017).

[29] S.M. LaValle, M.S. Branicky, S.R. Lindemann, On the relationship between classical grid search and probabilistic roadmaps. Int. J. Robot. Res. (2004), https://doi.org/10.1177/0278364904045481.

[30] Lurrette, 2017, http://www-verimag.imag.fr/Lurette,107.html (last visited April 2017).

[31] P. Raymond, Y. Roux, E. Jahier, Lutin: a language for specifying and executing reactive scenarios, EURASIP J. Embed. Syst. 2008 (1) (2008) 753821.

[32] Lustre V6, 2017, http://www-verimag.imag.fr/Lustre-V6.html (last visited April 2017).

[33] MISTA—Model-Based Testing, 2017, (last visited April 2017) http://cs.boisestate.edu/dxu/research/MBT.html.

[34] Modbat, 2017, https://people.kth.se/artho/modbat/ (last visited April 2017).

[35] Momut, 2017, https://momut.org/ (last visited April 2017).

[36] W. Herzner, R. Schlick, A. Austrian, H. Br, J. Wiessalla, F.F. Aachen, Towards fault-based generation of test cases for dependable embedded software.

[37] PragmaDev—Modeling and Testing Tools, 2017, http://pragmadev.com/ (last visited April 2017).

[38] Tcases, 2017, https://github.com/Cornutum/tcases (last visited April 2017).

[39] J. Sanchez, A Review of Pair-Wise Testing, Springer, 2016.

[40] L.L. Muniz, U.S.C. Netto, P.H.M. Maia, TCG—A Model-based Testing Tool for Functional and Statistical Testing, ICEIS (2015, .

[41] LoTuS, 2017, http://jeri.larces.uece.br/lotus/ (last visited April 2017).

[42] Vera, 2017, http://www.spacios.eu/index.php/spacios-tool/ (last visited April 2017).

[43] Smartesting CertifyIt, 2017, http://www.smartesting.com/en/certifyit/ (last visited April 2017).

[44] IBM—Rational Software Architect Designer, 2017, http://www-03.ibm.com/software/products/en/ratsadesigner (last visited April 2017).

[45] R. Neisse, G. Baldini, G. Steri, A. Ahmad, E. Fourneret, B. Legeard, Improving Internet of Things device certification with policy-based management, in: 2017 Global Internet of Things Summit (GIoTS), 2017.

[46] FIWARE, 2017, https://www.fiware.org/ (last visited April 2017).

[47] OAuth 2.0, 2017, https://oauth.net/2/ (last visited April 2017).

[48] O.M. Alliance, NGSI Context Management, 2012, http://technical.openmobilealliance.org/Technical/release_program/docs/NGSI/V1_0-20120529-A/OMA-TS-NGSI_Context_Management-V1_0-20120529-A.pdf.

[49] E. Fourneret, J. Cantenot, F. Bouquet, B. Legeard, J. Botella, SeTGaM: generalized technique for regression testing based on UML/OCL models, in: 8th International Conference on Software Security and Reliability, SERE, 20142014, pp. 147–156.

[50] ARMOUR H2020 Project, 2017, http://www.armour-project.eu/ (last visited April 2017).

[51] The Testing and Test Control Notation Version 3 (TTCN-3), http://www.ttcn-3.org/ (last visited April 2017).

[52] OM2M, oneM2M Standard Implementation, 2017, http://www.eclipse.org/om2m/ (last visited April 2017).

[53] FIT IoT Lab, Large Scale Testbed, 2017, https://www.iot-Iab.info/(last visited April 2017).

[54] FIESTA Testbed, 2017, http://fiesta-iot.eu/ (last visited April 2017).

ABOUT THE AUTHORS

Abbas Ahmad, a PhD student at Easy Global Market, has a master's degree in computer sciences from the University of Franche-Comté. He specialized his master's degree in "Design and development of secure software." He has accomplished a 6-month internship at Airbus Operations SAS, in the R&D department where he developed IATT (Isami Analyst Testing Tool), a nonregression test tool. He also has an ISTQB® (International Software Testing Qualifications Board) foundation level certification, an ISTQB® Model-Based Tester foundation, a FIWARE developer's week Brussels diploma, and a TTCN-3 certificate in test automation. With a strong knowledge of software and application testing, he also has been a FIWARE friendly tester during 4 months, where he tested the FIWARE cloud portal and some Generic Enablers. One of his achievements is wining first prize of the Internet of Things Hackathon during the IoT Week (Lisbon 2015). He is now running a PhD thesis in the field of model-based testing for IoT systems.

Fabrice Bouquet studied computer sciences and received his PhD degree in University of Provence, France in 1999. He is a lecturer during high years, and since 2008, he has been a full professor at the University of Franche-Comté, France. His research interests in validation of complex systems from requirements to model include operational semantic, testing, model transformation, and functional and nonfunctional properties with applications in vehicle, aircraft, smart objects, and energy.

Elizabeta Fourneret is a Project Manager at Smartesting. She has strong experience in the field of security testing based on models. She leads several R&D projects at Smartesting oriented on functional and security testing for critical systems, such as HSM (Hardware Security Modules), and IoT systems. Within the scope of the ongoing H2020 ARMOUR project on large-scale experiments of IoT Security & Trust she is leading the Large-Scale Testing Framework development.

She has received her PhD degree in Computer Science in the domain on model-based regression testing of security-critical systems at the National Research Institute in Computer

Science and Automation—INRIA—Nancy, France and her MSc degree in Software Verification and Validation at the University of Franche-Comté in Besançon, France.

Bruno Legeard is a Professor at UBFC, Institut FEMTO-ST, Scientific Advisor, and cofounder of Smartesting. Bruno has more than 20 years' expertise in Model–Based Testing/Model-Based Security Testing (MBT/MBST) and its introduction in the industry. His research activities mainly concern features about automation of Model-Based Test case generation using AI techniques. His research results in more than 100 scientific and industrial publications based on MBT and MBST. He is the author of three books disseminating Model-Based Testing in the industry. "Practical Model-Based Testing" has more than 1200 citations.

CHAPTER TWO

Advances in Software Model Checking

Junaid H. Siddiqui, Affan Rauf, Maryam A. Ghafoor
Lahore University of Management Sciences, Lahore, Pakistan

Contents

Abstract

Society is becoming increasingly dependent on software which results in an increasing cost of software malfunction. At the same time, software is getting increasingly complex and testing and verification are becoming harder and harder. Software model checking is a set of techniques to automatically check properties in a model of the software. The properties can be written in specialized languages or be embedded in software in the form of exceptions or assertions. The model can be explicitly provided in a specification language, can be derived from the software system, or the software system itself can be used as a model. Software model checkers check the given properties in a large number of states of the model. If a model checker is able to verify a property on *all* model states, it is proven that the property holds and the model checker works like a theorem prover. If a model checker is unable to verify a property on *all* model states, the model checker is still an efficient automated testing technique.

This chapter discusses advances in software model checking and focuses on techniques that use the software as its model and embedded exceptions or assertions as the properties to be verified. Such techniques are most likely to be widespread in the software industry in the coming decade due to their minimal overhead on the programmer and due to recent advances in research making these techniques scale.

1. INTRODUCTION

As software pervades our societies and lives, the need for software reliability grows more and more. With computers controlling airplanes and running financial markets, the cost of even a single software failure can be quite high. Failures already cost the U.S. economy $59.5 billion every year [1]. It is estimated that software flaws cost the global economy $175 billion [2] every year. It is further estimated that a substantial portion of this cost could be eliminated by an improved testing infrastructure. To meet the ever-increasing demand for reliability, we must create methodologies that deliver more reliable software at a lower cost.

With an exponential increase in the complexity of software, manual testing is become increasing less effective. Commercial software is often released with a large number of known bugs. Manual testing exercises a very small number of states the software can be in.

Formal verification of software strives to formulate the verification problem mathematically and solve it. There are two essential ways to achieve it: using theorem proving and using model checking.

Theorem proving attempts to build a formal proof of the correctness of the system with the help of the programmer. For example, loop invariants are deduced or provided by the programmer that remain true regardless of the number of iterations of the loop. The theorem prover uses these invariants to prove properties of the system. In essence, the properties are proved to be true for every state of the system. However, theorem proving has been limited to small programs and often requires extensive and laborious input from the programmer.

On the other hand, model checking checks properties of the system on a large number of states, thus verifying the properties for only these states. The work of a model checker on a single state is easy, but generalizing the result to *all* states is difficult. For example, in a particular state, a program loop would have run a fixed number of times, and it is apparent if a particular error state is reached or not.

There are two fundamental variations of model checking: bounded model checking and abstraction-based model checking. In bounded model checking, the number of states is bounded, e.g., by limiting the maximum number of conditional statements on a path. The finite number of states can be searched using either stateful search or stateless search. Section 2 discusses bounded model checking. Bounded model checking is an effective technique to find software bugs but it cannot prove the absence of bugs. Advances in bounded model checking enable identifying equivalent states, or treating multiple states as one, resulting in checking more states in less time. An extreme of treating multiple states as one is to divide the problem in a fixed number of abstract states with each abstract state representing possibly infinite states. The divisions are done based on a certain property, e.g., if a Boolean variable is true or false. Any properties checked on these finite states are true for all states, but any errors found may not be real errors in a concrete system.

A lot of model checking work is concentrated on checking properties of concurrent systems and the state space explosion resulting from thread interleavings. External inputs are often provided by an alternate test input generation system and the model checker exercises all given input choices. Symbolic execution is a technique to systematically check properties of systems over all input choices. It converts the program path in an SMT (satisfiability modulo theory) formula to solve and generate inputs that are guaranteed to exercise different paths. Due to loops, most programs have an infinite number of paths. Bounded symbolic execution exercises a bounded number of states. Recent advances have brought abstraction-based techniques to symbolic execution in some specific cases.

Model checking is often used for increasing software quality. While testing suffers from the problem of generating a good test suite and verification or proving program correctness is limited to small programs, model checking enables systematic exploration of a large number of program states, thus providing better guarantees of software quality than testing but still applies to much larger programs and verifies more general properties than verification.

1.1 Properties

The properties to be checked in a software system can be broadly classified into *safety* properties and *liveness* properties [3]. Safety properties imply that something bad does not happen while liveness properties specify that something good will eventually happen. For verification, a program is considered "safe" if no computation can reach a bad state and for liveness properties, every computation reaches the good state.

In this survey, we will focus on safety properties. Safety properties can be formulated as a reachability problem. Particular error locations in the program (e.g., assert statements inserted by the programmer) signify a bad state. If we guarantee that an assert statement is unreachable, we have shown that the program satisfies a safety property.

In bounded model checking, we verify that the safety properties are held in the bounded number of states but we cannot prove that the properties are held in general. In symbolic model checking, we verify that the safety property is held in general but if a violation is found, we cannot always prove that the violation is true for the specific system under test.

1.2 Soundness and Completeness

A sound algorithm for safety property verification declares programs safe with respect to some safety property that are truly safe. A complete algorithm for safety property verification declares all truly safe programs with respect to some safety property as safe. No sound and complete algorithm can solve the verification problem in general due to the undecidability of the halting problem [4]. Most model checking algorithms are sound but not complete. Bounded model checking explores a subset of the state space hoping to find a counterexample to the safety property (e.g., reaching the assert statement). If a counterexample is found, the safety property is violated and the algorithm reports that. However, if no counter example is found, the program may be safe or may be unsafe. Another approach is to check a superset of the state space by making abstractions. If the superset is safe, the program is truly safe. However, if the superset is unsafe, the actual program may be safe or may be unsafe.

1.3 Schedule Nondeterminism and Input Nondeterminism

Programs change their state as they execute, usually in a deterministic manner. However, when executing a conditional statement, the program state can possibly change in two different ways. The inputs to the program indirectly determine if one or both branches are feasible to execute. This is called input nondeterminism. Also, for concurrent programs, the state can change in n different ways if there are n active threads. This is called schedule nondeterminism.

Fundamentally, state search for both schedule nondeterminism and input nondeterminism is identical. However, the techniques to prune equivalent states or use symbolic states (that represent a number of concrete states) are different for both. Traditionally, techniques focusing on schedule nondeterminism have been called model checkers while those focusing on input nondeterminism in a symbolic way are called symbolic execution tools. Traditional model checkers handle input nondeterminism by exploring all given input choices without symbolic abstraction resulting in state space explosion.

2. EXPLICIT-STATE BOUNDED MODEL CHECKING

Bounded model checking checks all states of the system under test within some bounds. The states are concrete or explicit, i.e., actual states the program can be in. Explicit-state model checking for any program with loops has to be bounded because the number of explicit states are infinite in a program with loops. Model checking is a search of the graph formed by program states and transitions. For concurrent systems, it deals with exploration of all possible interleaving of processes leading to state space explosion for larger set of processes. In general, concurrent events are modeled by exploring all possible interleaving of events relative to each process which results in a large set of paths with many states on each path. Each path represents one unique interleaving of processes. There are two fundamental techniques to search the state space: stateless and stateful.

2.1 Stateful Search

In stateful search, the graph of program states and transitions is explicitly made and explored. The graph is finite due to the bounds on the bounded search. If no state reaches the error location, the given safety property is verified.

Stateful search keeps a worklist of states, and in each iteration, takes an element from the worklist and applies the next transitions to it. The updated state or states are added back to the worklist. The choice of next element from the worklist determines whether a depth-first or a breadth-first technique is used. Many model checkers use a heuristic-based choice that favors exploring new lines of code in anticipation of finding a fault.

Even within small bounds, the number of states grows exponentially and the resulting problem is called state space explosion. Dealing with the number of states is one of the key research challenges in model checking. To reduce the number of states, equivalence groups are found such that a bug in any state of the group will be reflected in all states of the group. The two primary techniques for this are symmetry reduction and partial order reduction (POR).

2.2 Symmetry Reduction

Symmetry reduction techniques identify symmetries in process order, communication order, and the structure of specifications to identify equivalence classes of states where a single representative state is enough to find bugs in any of the states in the equivalence class [5, 6]. Identifying all symmetries in model checking is not feasible. Hence, many techniques are based on heuristics and attempt to identify many symmetries in limited time.

2.3 Partial Order Reduction

POR is an optimization technique that exploits the fact that many paths are redundant, as they are formed due to the execution of independent transitions in different orders. Another way to understand it is considering that many concurrent operations are commutative, i.e., they result in the same state. POR is the identification and pruning of transitions that will result in identical states. Model checkers that include state comparison to avoid duplicate states still benefit from POR because of the early exclusion of transition rather than generating and comparing states.

Model checkers employ POR to avoid unnecessary exploration of schedules. POR addresses the problem of state explosion for concurrent systems by reducing the size of state space to be searched by model checking software. POR [7–9] on shared variables and objects is a well-researched field. It exploits the commutativity of concurrently executed transitions, which results in the same state when executed in different orders. It considers only a subset of paths representing a restricted set of behaviors of the application while guarantees that ignored behaviors do not add any new information.

2.4 Execution-Based Search

Verisoft [10] introduced execution-based model checking which is a special case of bounded model checking. Instead of modeling the semantics of a programming language or a specification language and implementing the effects of each transition in the model checker, an execution-based model checker just uses the programming language run time to execute the system on a specific set of user inputs and a specific thread schedule.

The advantage of execution-based search is that any error state found is accompanied by a complete execution trace with inputs and schedule to reach the error state. Furthermore, no translation of the software into a modeling language or translation of programming language semantics into state transitions is required.

Execution-based model checkers mostly use inputs generated from an external test suite (generated automatically using another technique or manually) and exercise different schedules on those inputs. Input choices are not explored using the model checker as it results in an enormous number of states. Covering input domains efficiently requires using symbolic states using, e.g., symbolic execution.

The key challenge of execution-based model checkers is storing the state. Being execution-based, their state is the entire program state. Java Path-Finder (JPF) [11], another execution-based model checker, stores state-deltas to optimally store state. On the other hand, Verisoft performs stateless execution-based search.

2.5 Stateless Search

Verisoft also introduced the idea of stateless search. Instead of explicitly storing a worklist of states and applying one transition to the state, stateless search works by forcing the program on a particular schedule. At every non-deterministic choice, e.g., thread schedule or user input, the execution queries the scheduler, and the scheduler provides the next number in the schedule, an integer in a finite domain of active threads or finite domain of valid input values.

After an execution is finished, the next schedule is formed by effectively incrementing the last schedule. The very last nondeterministic choice is changed to the next value in the domain and if the choices are exhausted, the second last nondeterministic choice is changed. Eventually, all choices are explored.

Like other tools in the execution-based category, Verisoft cannot practically explore input domains as it results in a very large number of states. It focuses on exploring input schedules.

The key advantage of stateless search is the minimal requirements on state storage. However, the main disadvantage is the inability to identify duplicate states or adapt complex heuristic-based search orders that are possible when states are explicitly stored.

2.6 Tools and Recent Advancements

The first famous explicit-state model checkers with stateful search were the SPIN model checker [12] and the Murphi model checker [13]. Verisoft introduced stateless execution-based model checking while JPF provides stateful execution-based model checking.

Verisoft introduced stateless execution-based model checking. It observes processes that communicates using interprocess communication (IPC) primitives. By trapping calls to IPC primitives, the Verisoft scheduler is able to influence the schedule in which different processes proceed.

JPF is an explicit-state model checker which explores different interleavings of threads by selecting a thread nondeterministically from a set of live threads. In order to explore all possible schedules, JPF uses the choice generator on thread operations. JPF generates schedules on the basis of calls to initialization and termination of these threads by executing the statements of the thread until the thread finishes the execution.

JPF executes Java byte code instructions from the application on its own virtual machine (VM). JPF has listeners which get notified by the VM whenever a specific operation is performed. These listeners can be used to perform a specific operation and/or to further control the execution of the program. JPF explores states of the program along all possible paths and whenever end of path is reached, it backtracks to explore other unexplored paths.

Other popular execution-based model checking tools include CMC and CHESS. CMC (C model checker) uses the OS scheduler to explore different interleavings. It hashes the state to identify similar states in an attempt to prioritize new states. CMC has been used to successfully find bugs in real systems [14, 15]. CHESS is another execution-based model checker but it bounds on the number of context switches instead of states [16, 17]. This enables exercising very long traces and increase the bug finding ability of the model checker.

JPF-SE [18] exploits JPF to explore and gather program constraints. It combines model checking with symbolic execution to build path conditions that are later used for test input generation. JFuzz [19] is whitebox, fuzzy testing tool that is also built on the top of the JPF and used for test input generation of Java programs. Model checking is also employed to verify android application. JPF-Android [20] is an extension of JPF for android applications. JPF has been used to verify distributed application using MPI [21].

2.7 Parallel Model Checking

Stern and Dill's parallel Murphi [22] is an example of a parallel model checker. It keeps the set of visited states shared between parallel workers so that the same parts of the state space are not searched by multiple workers. Keeping this set synchronized between the workers results in expensive communication and thus the algorithm does not scale well.

A similar technique was used by Lerda and Visser [23] to parallelize the JPF model checker [24]. Parallel version of the SPIN model checker [25] was produced by Lerda and Sisto [26]. SPIN is extended to take full advantage of multicore machines [27]. It is parallelized [28] in order to avail all cores of CPU to efficiently verify system properties. More work has been done in load balancing and reducing worker communication in these algorithms [29–31].

Parallel Randomized State Space Search for JPF by Dwyer et al. [32] takes a different approach with workers exploring randomly different parts of the state space. This often speeds up time to find first error with no worker communication. However, when no errors are present, every worker has to explore every state.

Ranged model checking [33] enables more effective checking of Java programs using the JPF model checker. The key insight is that the order in which JPF makes nondeterministic choices defines a total ordering of execution paths it explores in the program it checks. Thus, two in-order paths define a range for restricting the model checking run by defining a start point and an end point for JPF's exploration. Moreover, a given set of paths can be linearly ordered to define consecutive (essentially), nonoverlapping ranges that partition the exploration space and can be explored separately. While restricting the run of a model checker is a well-known technique in model checking, this work conceptually restricts the run using vertical boundaries rather than the traditional approach of using a horizontal boundary, i.e., the search depth bound.

2.8 Model Checking in Other Domains

Model checking has also been used for web applications [34–36] to model check interleaving of web requests such that all valid schedules of web requests are exercised to verify that error states are not reachable.

For model checking of database operations, Paleari et al. [37] found dependent operations through a log from a single execution, ignoring program semantics, but they did not perform model checking of other possible orders. This made it label many operations as dependent when they really were not. Zheng and Zhang [38] used static analysis to find race conditions when accessing external resources. Static analysis inherently has more false positives (labeling more operations dependent).

Emmi and Majumdar [39] presented concolic execution to generate both input data for the program as well as suitable database records to systematically explore all paths of the program, including those paths whose execution depend on data returned by database queries. Marcozzi et al. [40] describe symbolic execution of SQL statements along with other constraints in program for generating test inputs in order to test database application. Other researchers have also used symbolic techniques to test databases [41–45]. Effective POR for model checking database applications [46] improves POR precision for database operations.

3. SYMBOLIC EXECUTION

When dealing with external inputs, explicit-state model checking produces a lot of different but unnecessary states. For example, the symbolic expression $x < 5$ represents all states in which x is less than 5. An explicit-state model checker has to exercise all these states individually even if the states exercise the same path in code.

Symbolic execution is a technique to use symbolic states for program input. At every conditional statement, two symbolic states are formed with different clauses. When the program ends, there is a clause in the state for every conditional statement on the path executed by the program.

Symbolic execution relies on SAT (satisfiability theory) or SMT (satisfiability modulo theory) solvers to solve the set of clauses in each state, also called *path conditions*. Solving the path condition is NP-complete but heuristic-based solvers are able to solve most path conditions in a reasonable time. If paths reaching error states are solvable, a safety property violation is found.

Symbolic execution was developed over three decades ago [47, 48]. A key idea used in symbolic execution is to build *path conditions*—given a path, a path condition represents a constraint on the input variables, which is a conjunction of the branching conditions on the path. Thus, a solution to a (feasible) path condition is an input that executes the corresponding path. A common application of symbolic execution is indeed to generate test inputs, say to increase code coverage. Automation of symbolic execution requires constraint solvers or decision procedures [49, 50] that can handle the classes of constraints in the ensuing path conditions.

A lot of progress has been made during the last decade in constraint solving technology, in particular SAT [51] and SMT [49, 50] solving. Moreover, raw computation power is now able to support the complexity of solving formulas that arise in a number of real applications. These technological advances have fueled the research interest in symbolic execution, which today not only handles constructs of modern programming languages and enables traditional analyses, such as test input generation [52–55], but also has nonconventional applications, for example, in checking program equivalence [56], in repairing data structures for error recovery [57], and in estimating power consumption [58].

Forward symbolic execution is a technique for executing a program on symbolic values [48, 59]. There are two fundamental aspects of symbolic execution: (1) defining semantics of operations that are originally defined for concrete values and (2) maintaining a *path condition* for the current program path being executed—a path condition specifies necessary constraints on input variables that must be satisfied to execute the corresponding path.

Program 1: To find middle of three integers

```
1  int mid(int x, int y, int z) {
2    if (x<y) {
3      if (y<z) return y;
4      else if (x<z) return z;
5      else return x;
6    } else if (x<z) return x;
7    else if (y<z) return z;
8    else return y; }
```

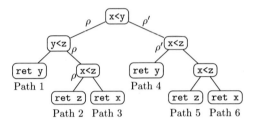

Fig. 1 Execution tree of Program 1.

Program 1 contains a function that finds the middle of three integer values. Fig. 1 shows the execution tree of this program and shows the six path conditions explored. The path conditions for each of these paths can be solved using off-the-shelf SAT solvers for concrete tests that exercise the particular path. For example, path 2 can be solved to $X = 1$, $Y = 3$, and $Z = 2$. Following are all the six paths explored:

```
Path 1: [X < Y < Z] L2 -> L3
Path 2: [X < Z < Y] L2 -> L3 -> L4
Path 3: [Z < X < Y] L2 -> L3 -> L4 -> L5
Path 4: [Y < X < Z] L2 -> L6
Path 5: [Y < Z < X] L2 -> L6 -> L7
Path 6: [Z < Y < X] L2 -> L6 -> L7 -> L8
```

3.1 Stateful Symbolic Execution

PREfix [60] is among the first systems to show the bug finding ability of symbolic execution on real code. Generalized symbolic execution [52] defines symbolic execution for object-oriented code and uses *lazy initialization* to handle pointer aliasing. Generalized symbolic execution used the JPF model checker and hence was an explicit-state technique.

KLEE [55] is the most recent tool from the EGT/EXE family. KLEE is open-sourced and has been used by a variety of users in academia and industry. KLEE works on LLVM byte code [61]. It works on unmodified programs written in C/C++ and has been shown to work for many off-the-shelf programs. KLEE, like JPF, is an explicit-state tool. It uses LLVM libraries to interpret code line by line and update its effect on the symbolic variables stored in the state. Like JPF, it stores differences in state to efficiently store state.

3.2 Stateless Symbolic Execution

Just like bounded model checking techniques, symbolic execution can be implemented in an stateless tool. In a stateless execution-based approach, the state comprises of concrete variable values (like explicit-state model checkers) in addition to the symbolic formulas for each variable. The concrete values satisfy the corresponding symbolic formula. The effect of each program statement is updated on both the concrete values and the symbolic values. When a conditional statement is encountered, the program proceeds in whatever direction the concrete values take it and the conditional clause is added to the path condition. When an execution finishes, the last clause in the path condition is negated (or the path before if the last one has already been negated) to arrive at a path condition for a yet unexplored path. The SAT solver is used to solve this path condition and the concrete values thus found guide the next execution.

Traditional or forward symbolic execution uses SAT solver to find if error paths are truly feasible or not while stateless or back-tracking-based symbolic execution uses the SAT solver to find concrete values that will drive the next execution on a different path.

Symbolic execution guided by concrete inputs has been a topic of extensive investigation during the last 7 years. DART [53] combines concrete and symbolic execution to collect the branch conditions along the execution path. DART negates the last branch condition to construct a new path condition that can drive the function to execute on another path. DART focuses only on path conditions involving integers. To overcome the path explosion in large programs, SMART [62] introduced interprocedural static analysis techniques to compute procedure summaries and reduce the paths to be explored by DART. CUTE [54] extends DART to handle constraints on references. EGT [63] and EXE [64] also use the negation of branch predicates and symbolic execution to generate test cases. They increase the precision of symbolic pointer analysis to handle pointer arithmetic and bit-level memory locations.

3.3 Symbolic Bounded Model Checking

Symbolic bounded model checking uses a search algorithm similar to explicit-state bounded model checking but uses symbolic states representing a number of explicit states. Unlike symbolic execution, the symbolic states do not have a mapping over program paths. The symbolic states are formed over program variables and a single formula is formed for all paths in a symbolic state.

Symbolic execution enumerates paths and uses the solver to solve those path conditions, whereas symbolic bounded model checking unroll the control flow graph up till a particular depth and form a single formula for a given symbolic state about reachability of an error location. In this way the program computations are all formulated and backtracked inside the solver.

CBMC [65, 66] is a bounded model checker for C. It is based on checking assertion violations. Errors are converted into assertions though automated instrumentation. For loops, it considers an unwinding bound. It now includes support for concurrent programs and weak memory models [67].

Other early bounded model checkers in this category include F-Soft [68], Saturn [69], and Calysto [70]. They all perform symbolic bounded model checking generating constraints in prepositional logic and use SAT solvers to solve them.

ESBMC [71] builds on the front end of CBMC. It supports different theories and SMT solvers to exploit high-level information to reduce formula size.

Scaling bounded model checking for large programs is a challenging problem. Blitz [72] uses information from control flow and data flow of the program as well as information from unsatisfiability proofs to prune the set of variables and procedures to generate smaller instances.

Due to complexities of a high-level language, it becomes difficult to apply model checking directly on the source code. LLBMC [73] thus uses LLVM compiler framework to convert C/C++ source code to LLVM intermediate code. Then it is converted into logical formulas and after simplification fed to solvers.

Concurrency errors manifest themselves with a few context switches. Therefore, context-bounded verification tools limit the number of context switches. Sequentialization is a technique that translates a concurrent program into an equivalent nondeterministic sequential program by replacing control nondeterminism with data nondeterminism. Lazy sequentialization [74] performs better by using lesser data nondeterminism.

3.4 Ranged and Incremental Symbolic Execution

Due to the nature of various model checking algorithms, every code change invalidates prior results and a new analysis has to be run. Due to the pace at which software changes, analysis results often become obsolete very soon. However, a lot of work from prior runs can be efficiently utilized since most of the code remains unchanged. Incremental techniques try to leverage this pattern and make incremental analysis faster.

Ranged symbolic execution [75] introduces an approach to scale symbolic execution for test input generation. The key idea of ranged symbolic execution is that the *state* of the analysis can be represented highly compactly: a *test input* is all that is needed to effectively encode the state of a symbolic execution run. Ranged symbolic execution uses two test inputs to define a *range*, i.e., the beginning and end, for a symbolic execution run. Ranged symbolic execution enables distributing the path exploration in a parallel setting with multiple workers.

Ranged symbolic execution enables symbolic exploration between two given paths. For example, if path 2 and path 4 of the earlier example are given, it can explore paths 2, 3, and 4. It only needs the concrete solution that satisfies the corresponding path condition. Therefore it is efficient to store and pass paths. This is shown in Fig. 2 and the general idea of dividing search in ranges is shown in Fig. 3.

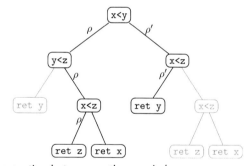

Fig. 2 Symbolic execution between paths ρ and ρ'.

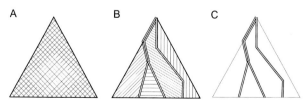

Fig. 3 High-level overview of dividing symbolic execution into nonoverlapping ranges for independent symbolic executions. (A) Standard symbolic execution. (B) Ranged symbolic execution (four ranges). (C) Only boundaries redundantly analyzed.

Incremental symbolic execution applies symbolic execution in increments based on versions of code [21]. This technique is based entirely on dynamic analysis and patches completely automated test suites based on the code changes. Constraint solving for unchanged code is eliminated by checking constraints using the test suite of a previous version. Checking constraints is orders of magnitude faster than solving them. It identifies ranges of paths, each bounded by two concrete tests from the previous test suite. Exploring these path ranges covers all paths affected by code changes up to a given depth bound.

Directed incremental symbolic execution [76] leverages differences among program versions to optimize symbolic execution of *affected* paths that may exhibit modified behavior. The basic motivation is to avoid symbolically executing paths that have already been explored in a previous program version that was symbolically executed. A reachability analysis is used to identify affected locations, which guide the symbolic exploration.

Directed Incremental Symbolic Execution (DiSE) [76] uses static analysis to find the code blocks that are effected by a change and then uses this information in dynamic analysis to prune the execution tree for symbolic execution. DiSE only generate affected path conditions because it preforms symbolic execution after statically analyzing the both programs CFGs. If a user wants a complete test suite using DiSE, he/she needs to check what path conditions get obsolete, which is not clear how to do.

Katch [77] combines static and dynamic analysis for increased coverage of software patches. Katch uses heuristics basic on static analysis to select tests from a manual test suite that are most likely to hit modified code and then it uses heuristics based on dynamic analysis to change the test inputs to increase chances of hitting modified code. The dynamic analysis is built upon symbolic execution. Since the problem of code reachability is undecidable in general, Katch hopes that the heuristics will lead to modified code in most cases.

Green [78] is a cache for storing path conditions and their solutions. In Green caching technique, they slice the path condition for the purpose of reducing it to be checked for satisfiability, then they perform heuristic-based canonization to maximize the chance of finding matches with other path conditions and lastly they store this form in the store. On finding matches they server the results from their store instead of calling the solver. While Green is not an incremental symbolic execution tool in itself, it enables more efficient checking in incremental situations. The reason of that benefit is that in an incremental setting, many path conditions will match from the last execution and thus the solving can be avoided. Path conditions are complex structures and even though Green canonicalizes them, it requires some work to compare them and map the variables. As the cache size increases, comparing becomes more difficult. Validating a path condition, on the other hand, is a quick operation and requires no lookup.

GreenTrie [79] extends Green framework, which supports constraint reuse based on the logical implication relations among constraints.

Memoized symbolic execution [80] reuses results of a previous run of symbolic execution by storing them in a trie-based data structure and reusing them by maintaining and updating the trie in the next run of symbolic execution on the modified program. The savings achieved by Memoise for regression analysis relies on the position of the change, and may vary quite a lot between various kinds of changes. We believe our technique is more effective due to low cost of storing concrete tests vs storing and comparing symbolic execution results in tries.

Conc-iSE [81] proposes incremental symbolic execution for concurrent software. It generates new tests by exploring only the executions affected by code changes between two program versions. They perform interthread and interprocedural change-impact analyses to check if a statement is affected by the changes and then use the information to choose executions that need to be explored again.

3.5 Parallel Symbolic Execution

A couple of recent research projects have proposed techniques for parallel symbolic execution [82, 83]. ParSym [82] parallelized symbolic execution by treating every path exploration as a unit of work and using a central server to distribute work between parallel workers. While this technique implements a direct approach for parallelization [64, 84], it requires communicating symbolic constraints for every branch explored among workers, which incurs a higher overhead. In contrast, static partitioning [83] uses an initial shallow run of symbolic execution to minimize the communication overhead during parallel symbolic execution. The key idea is to create preconditions using conjunctions of clauses on path conditions encountered during the shallow run and to restrict symbolic execution by each worker to program paths that satisfy the precondition for that worker's path exploration. However, the creation of preconditions results in different workers exploring overlapping ranges, which results in wasted effort.

KleeNet [85] uses KLEE to find interaction bugs in distributed applications by running the distributed components under separate KLEE instances and coordinating them using a network model. KleeNet performs separate symbolic execution tasks of each component of the distributed application in parallel.

3.6 Other Advancements in Symbolic Execution

Hybrid concolic testing [86] uses random search to periodically guide symbolic execution to increase code coverage. However, it explores overlapping ranges when hopping from symbolic execution in one area of code to another, since no exploration boundaries are defined (other than time out).

Staged symbolic execution [87, 88] is a technique to apply symbolic execution in stages, rather than the traditional approach of applying it all at once, to compute abstract symbolic inputs that can later be shared across different methods to test them systematically. Staged symbolic execution conceptually divides symbolic execution in "horizontal" slices called "stages" that can be executed sequentially. On the other hand, ranged symbolic execution conceptually divides symbolic execution in "vertical" slices called "ranges" that can be explored in parallel.

Symbolic execution for Alloy models [89] applies the idea of symbolic execution on analyzing and solving models written in the Alloy modeling language [90].

Symbolic Execution of stored procedures in database management systems [91] automatically generates test cases and corresponding database states for stored procedures. They treat values in database tables as symbolic, model the constraints on data imposed by the schema and by the SQL statements executed by the stored procedure. An SMT solver finds values that will drive the stored procedure on a particular execution path.

Postconditioned symbolic execution tries to mitigate the path explosion problem by a new redundancy removal method. At each branching location, in addition to determine the feasibility of a particular branch, it also checks whether the branch is subsumed by previous explorations [92].

Zou et al. [93] transform the problem of detecting redundant constraints in a path condition to the redundancy identification problem in linear programming (LP). They used an efficient algorithm to eliminate the redundant constraints in LP and transforms the result back to a path condition.

Wang et al. [94] mitigate the path explosion problem by predicting and eliminating the redundant paths based on symbolic value.

Feist et al. [95] enable the capability of enumerating paths within a given program slice. They propose a new path selection criterion which is based on the probability of a path to exit the program slice. This minimizes the number of queries to the SMT solvers.

Symbolic PathFinder [96] uses lazy initialization which "allows symbolic execution to effectively deal with heap-allocated data structures." BLISS [97] improves it "by taking advantage of precomputed relational bounds on the interpretation of class fields in order to reduce the number of spurious structures even further".

Braione et al. [98] propose "a language HEX to specify invariants of partially initialized data structures, and a decision procedure that supports the incremental evaluation of structural properties in HEX. Used in combination with the symbolic execution of heap manipulating programs, HEX prevents the exploration of invalid states, thus improving the efficiency of program testing and analysis, and avoiding false alarms that negatively impact on verification activities."

Ibing [99] "uses path merging based on interpolation with unsatisfiable cores to mitigate the exponential path explosion problem."

4. ABSTRACTION-BASED MODEL CHECKING

Abstraction-based model checking for checking safety properties performs abstract reachability analysis. The analysis is performed on abstract states that capture some but not all of the state of the program. Each concrete program operation becomes an abstract operation performing a subset of the effects in the real program. Abstract states can be formed by removing variables from the state or dividing their values into domains (e.g., positive, zero, negative). The goal of abstraction is to make the number of states finite while retaining the variables relevant to proving a particular property.

Abstract model checking performs the same model checking algorithm as previously discussed on the simplified graph of abstract states and abstract transitions. Unlike explicit-state model checking a counterexample found (e.g., error state reachable) may not be a real counterexample. On the other hand, if there is no way to reach the error state in the abstract graph, then we are guaranteed that there is no way to reach the error state in the real program. Thus abstraction-based model checking is sound in general but not complete.

In the most common form, predicate abstraction [100, 101] is used to make an abstract version of a program with respect to a set of given predicates. If n Boolean predicates are used, the program state can be represented using an n bit vector. Thus the program state is finite in this case. However, there are still an exponential number of program states. These can be further abstracted using techniques that require polynomial number of states. Thus a program transition picks a more general representative n bit vector instead of the most precise n bit vector to represent that state. The resulting performance gain is at the cost of precision.

4.1 Abstraction Refinement

Abstraction-based model checking can return spurious counterexamples. Refinement focuses to identify if the counterexample is true in the real program.

Abstracting too much increases the chances of finding spurious counterexamples and abstracting too little increases the number of states exponentially. Counterexample guided refinement techniques attempt to get the best of both worlds by starting with too much abstraction and as soon as a counterexample is found, find a way to add more details to the state such that the counterexample no longer remains valid. Thus precision is added on an as-needed basis. CEGAR (counter-example guided abstraction refinement) [102–104] proposed the idea of iteratively refining abstract domains.

4.2 Tools and Recent Advancements

SLAM and BLAST are well-known model checkers that use predicate-based abstraction and the CEGAR loop. SLAM [105, 106] was first to use the idea of counter examples to refine the abstraction to uncover bugs in device drivers. SLAM uses Boolean program to represent the abstract intermediate program. C programs can be converted into Boolean programs given a set of Boolean predicates. Each Boolean predicate became a variable in the Boolean program. Each program statement is translated into its effects on the Boolean variables. Symbolic model checking of the Boolean program is then performed, followed by CEGAR refinement of counterexamples. SLAM has been used for device driver verification [107].

BLAST [108] is also based on principle of CEGAR and it constructs coarse abstraction of the program and then explores the state space. Failure of verification of the program indicates presence of either a bug in the system or inaccurate abstraction. It revises abstraction by lazily adding predicates present on path exercised by failed execution to include program properties. Lazy abstraction increases the performance of abstraction-based model checking without compromising precision.

CPAChecker [109] is a tool for software verification based on CPA (configurable program analysis) framework. It provides an interface to configure different parameters related to abstraction of the program. It enables comparison of different techniques in same setting that eventually helps in arguing validity of the property. Further CPAChecker is extended to support recursive programs and floating point arithmetic [110].

Other model checkers based on abstraction include the MAGIC model checker [111], the F-SOFT model checker [68], and the IMPACT model checker [112]. The MAGIC model checker extends abstraction-based model checking to concurrent programs. The additional schedule nondeterminism requires a two-stage abstraction algorithm. The F-SOFT model checker introduces various optimizations and better abstract domains to abstraction-based model checking [113, 114]. Finally, the IMPACT model checker directly uses interpolation in the lazy abstraction framework and showed significant performance improvements.

5. CONCLUSIONS

Software model checking has advanced in the last few decades from research ideas working on small specialized examples to working on real large programs finding actual bugs. Execution-based model checkers enable using the software system as its own model without the need to write specifications or perform an error-prone translation of the program into its specifications. Furthermore, model checkers can use programmer inserted assertions as a specification of safety properties, thus eliminating the need to provide a separate set of specifications.

Explicit-state model checkers keep an explicit-program state in memory while symbolic model checkers keep a symbolic state that represents a number of explicit states. Stateless model checkers backtrack over a schedule to exercise all program states and stateful model checkers explicitly keep all to be explored program states in memory. Symmetry reduction and POR enable identifying equivalent states, thus covering a reduced number of states while maintaining the bug finding ability. Finally, abstract model checkers divide the state space into a finite number of abstract states to be able to prove properties instead of focusing on finding bugs.

Counterexample refinement techniques attempt to identify if the bugs found in an abstract state are real program bugs or not.

Software model checking will likely gain much wider significance in the coming decades. It hits the sweet spot between theorem proving targeting both soundness and completeness and random testing targeting neither. By aiming at one of soundness or completeness and more importantly following a systematic process of exploring software states, a lot of arbitrariness associated with random testing is avoided and a lot of scalability is gained.

Software model checking still suffers from state space explosion and the cost of systematically exploring a number of states. However, in the past decade, it has progressively been applied to larger and more complex software and the trend is likely to continue in the coming years. Advances in software model checkers introduce heuristics to make the search more effective, address scalability, e.g., by making the process parallel or incremental, or increase soundness or completeness in specific cases. Software model checkers are now able to verify properties in very large programs and expected to become integrated in the average developer's tool chain in the coming decades.

Specific reasons why model checking is not widely used in software industry include time taken for analysis, false positives and false negatives, and limitations of model checking for software requiring external interaction. All of these are active areas of research with scalability receiving the most attention as it is the biggest hurdle. New data sources need to be tapped to add supplemental information that can direct the systematic search and make it more scalable. Some ideas recently explored include taking hints from software histories, bug repositories, comments, and design patterns.

Apart from scalability, another key challenge for software model checking is checking of user interfaces and interacting with external non-deterministic sources. Progress has been made in this direction in recent years with techniques attempting to verify user interfaces, to verify the impact of language translations on user interfaces, and to interact with external systems in a more deterministic way, e.g., using proxies or by remembering past responses. Despite the progress, model checking software with external interfaces remains a challenging area and the coming decade is likely to bring new ideas and better solutions.

Model checking will also need to catchup with new and resurgent software paradigms including agent-driven, event-driven, and functional paradigms. Event-driven has received the most attention in recent years, especially in the context of web systems, but there are still challenges in handling the large number of event sequences possible and separating the most important sequences in a systematic way.

While model checking is becoming increasingly useful for verification, it will likely play a much broader role in software development in the coming decades. Techniques likes automated fault localization, automated repair, programming by sketching, software construction from specifications, and other emerging ideas often depend on systematic search techniques as an enabling technology and these techniques are slowing gaining prominence. One example is recent approaches in automated repair that apply repair while software is being written and give repair suggestions as hints just like semantic compiler hints. This requires a highly scalable and heuristic systematic search. Model checking needs to cope with these challenges, and once it does, it will likely become a fundamental part of the software development process.

Model checking has grown in scalability and new applications but attempts to combine the various model checking approaches have been few and far between. This is likely one of the most promising directions of radical improvements in scalability of software model checking and its applications to significantly larger systems. A fundamental limitation is that researchers are often specialized in a particular model checking technique, whereas model checking techniques have been heavily borrowing ideas from programming languages and compiler design community and from hardware model checking and from pure mathematics that underlies many of the programming language calculus. A more interdisciplinary approach taking ideas from a variety of domains with a unified focus is needed to take software model checking to the next level.

REFERENCES

[1] National Institute of Standards and Technology, The Economic Impacts of Inadequate Infrastructure for Software Testing, Planning report 02-3, 2002.

[2] J. Woodcock, First steps in the verified software grand challenge, Computer 39 (2006) 57–64.

[3] B. Alpern, F.B. Schneider, Recognizing Safety and Liveness, Technical report, Ithaca, NY, 1986.

[4] A. Turing, On computable numbers with an application to the Entscheidungsproblem, Proc. Lond. Math. Soc. s2-42 (1) (1937) 230–265.

[5] E.M. Clarke, T. Filkorn, S. Jha, Exploiting symmetry in temporal logic model checking, in: Proceedings of the 5th International Conference on Computer Aided Verification, CAV '93, Springer-Verlag, London, UK, ISBN 3-540-56922-7, 1993, pp. 450–462, http://dl.acm.org/citation.cfm?id=647762.735503.

[6] E.A. Emerson, A.P. Sistla, Symmetry and model checking, Formal Methods Syst. Des. 9 (1) (1996) 105–131, ISSN 1572-8102, https://doi.org/10.1007/BF00625970.

[7] C. Flanagan, P. Godefroid, Dynamic partial-order reduction for model checking software, in: ACM Sigplan Notices, vol.40, ACM, New York, NY, 2005, pp. 110–121.

[8] P. Godefroid, J. van Leeuwen, J. Hartmanis, G. Goos, P. Wolper, Partial-Order Methods for the Verification of Concurrent Systems: An Approach to the State-Explosion Problem, vol. 1032, Springer-Verlag, Berlin, Heidelberg, 1996.

[9] E.M. Clarke, O. Grumberg, M. Minea, D. Peled, State space reduction using partial order techniques, Int. J. Softw. Tools Technol. Transfer 2 (3) (1999) 279–287.

[10] P. Godefroid, Model checking for programming languages using VeriSoft, in: Proc. 24th Symposium on Principles of Programming Languages (POPL), 1997, pp. 174–186, http://portal.acm.org/ft_gateway.cfm?id=263717&type=pdf&coll= GUIDE&dl=GUIDE&CFID=54069747&CFTOKEN=82180120.

[11] W. Visser, K. Havelund, G. Brat, S. Park, Model checking programs, in: Proc. 15th International Conference on Automated Software Engineering (ASE), 2000, p. 3, http://portal.acm.org/ft_gateway.cfm?id=786967&type=external&coll=GUIDE& dl=GUIDE&CFID=52387804&CFTOKEN=40229725.

[12] G.J. Holzmann, The model checker SPIN, IEEE Trans. Softw. Eng. 23 (5) (1997) 279–295, ISSN 0098-5589, https://doi.org/10.1109/32.588521.

[13] D.L. Dill, The Murphi verification system, in: Proceedings of the 8th International Conference on Computer Aided Verification, CAV '96, Springer-Verlag, London, UK, ISBN 3-540-61474-5, 1996, pp. 390–393, http://dl.acm.org/citation.cfm? id=647765.735832.

[14] M. Musuvathi, D.R. Engler, Model checking large network protocol implementations, in: Proceedings of the 1st Conference on Symposium on Networked Systems Design and Implementation—Volume 1, NSDI'04, USENIX Association, Berkeley, CA, 2004, pp. 12–12. http://dl.acm.org/citation.cfm?id=1251175.1251187.

[15] J. Yang, P. Twohey, D. Engler, M. Musuvathi, Using model checking to find serious file system errors, ACM Trans. Comput. Syst. 24 (4) (2006) 393–423, ISSN 0734-2071, https://doi.org/10.1145/1189256.1189259.

[16] M. Musuvathi, S. Qadeer, Iterative context bounding for systematic testing of multithreaded programs, in: Proceedings of the 28th ACM SIGPLAN Conference on Programming Language Design and Implementation, PLDI '07, ACM, New York, NY, ISBN 978-1-59593-633-2, 2007, pp. 446–455, https://doi.org/ 10.1145/1250734.1250785.

[17] S. Qadeer, J. Rehof, Context-bounded model checking of concurrent software, Springer, Berlin, Heidelberg, ISBN 978-3-540-31980-1, 2005, pp. 93–107, https:// doi.org/10.1007/978-3-540-31980-1_7.

[18] S. Anand, C.S. Păsăreanu, W. Visser, JPF-SE: a symbolic execution extension to Java PathFinder, in: International Conference on Tools and Algorithms for the Construction and Analysis of Systems, Springer, 2007, pp. 134–138.

[19] K. Jayaraman, D. Harvison, V. Ganesh, A. Kiezun, jFuzz: a concolic whitebox fuzzer for java, in: Proceedings of the First NASA Formal Methods Symposium, April, 2009, pp. 121–125.

[20] H. van der Merwe, B. van der Merwe, W. Visser, Verifying android applications using Java PathFinder, ACM SIGSOFT Softw. Eng. Notes 37 (6) (2012) 1–5.

[21] W.U. Rehman, M.S. Ayub, J.H. Siddiqui, Verification of MPI Java programs using software model checking, in: Proceedings of the 21st ACM SIGPLAN Symposium on Principles and Practice of Parallel Programming, PPoPP '16, ACM, New York, NY, ISBN 978-1-4503-4092-2, 2016, pp. 55:1–55:32, https://doi.org/10.1145/ 2851141.2851192.

[22] U. Stern, D.L. Dill, Parallelizing the Murphi verifier, in: CAV '97: Proceedings of the 9th International Conference on Computer Aided Verification, Springer-Verlag, London, UK, ISBN 3-540-63166-6, 1997, pp. 256–278.

[23] W. Visser, K. Havelund, G. Brat, S. Park, Model checking programs, in: ASE '00: Proceedings of the 15th IEEE International Conference on Automated Software Engineering, IEEE Computer Society, Washington, DC, ISBN 0-7695-0710-7, 2000, p. 3.

[24] F. Lerda, W. Visser, Addressing dynamic issues of program model checking, in: SPIN '01: Proceedings of the 8th International SPIN Workshop on Model Checking of Software, Springer-Verlag, Inc., New York, NY, ISBN 3-540-42124-6, 2001, pp. 80–102.

[25] G.J. Holzmann, The model checker SPIN, IEEE Trans. Softw. Eng. 23 (5) (1997) 279–295, ISSN 0098-5589, https://doi.org/10.1109/32.588521.

[26] F. Lerda, R. Sisto, Distributed-memory model checking with SPIN, in: Proceedings of the 5th and 6th International SPIN Workshops on Theoretical and Practical Aspects of SPIN Model Checking, Springer-Verlag, London, UK, ISBN 3-540-66499-8, 1999, pp. 22–39.

[27] G.J. Holzmann, D. Bosnacki, The design of a multicore extension of the SPIN model checker, IEEE Trans. Softw. Eng. 33 (10) (2007) 659–674.

[28] G.J. Holzmann, Parallelizing the spin model checker, in: International SPIN Workshop on Model Checking of Software, Springer, 2012, pp. 155–171.

[29] M.D. Jones, J. Sorber, Parallel search for LTL violations, Int. J. Softw. Tools Technol. Transf. 7 (1) (2005) 31–42, ISSN 1433-2779, https://doi.org/10.1007/s10009-003-0115-8.

[30] R. Kumar, D. Eric, G. Mercer, D. Quinn, O. Snell, D. Bryan, S. Morse, B.Y. University, B.Y. University, W. Embley, G.R. Bryce, A. Dean, Load balancing parallel explicit state model checking, Electr. Notes Theor. Comput. Sci. 128 (2005) 19–34.

[31] R. Palmer, G. Gopalakrishnan, A distributed partial order reduction algorithm, in: FORTE '02: Proceedings of the 22nd IFIP WG 6.1 International Conference Houston on Formal Techniques for Networked and Distributed Systems, Springer-Verlag, London, UK, ISBN 3-540-00141-7, 2002, p. 370.

[32] M.B. Dwyer, S. Elbaum, S. Person, R. Purandare, Parallel randomized state-space search, in: ICSE '07: Proceedings of the 29th International Conference on Software Engineering, IEEE Computer Society, Washington, DC, ISBN 0-7695-2828-7, 2007, pp. 3–12, https://doi.org/10.1109/ICSE.2007.62.

[33] D. Funes, J.H. Siddiqui, S. Khurshid, Ranged model checking, SIGSOFT Softw. Eng. Notes 37 (6) (2012) 1–5, ISSN 0163-5948, https://doi.org/10.1145/2382756.2382799.

[34] B. Petrov, M. Vechev, M. Sridharan, J. Dolby, Race detection for web applications, in: Proceedings of the 33rd ACM SIGPLAN Conference on Programming Language Design and Implementation, PLDI '12, ISBN 978-1-4503-1205-9, 2012, pp. 251–262.

[35] M. Martin, M.S. Lam, Automatic generation of XSS and SQL injection attacks with goal-directed model checking, in: Proceedings of the 17th Conference on Security Symposium, SS'08, 2008, pp. 31–43.

[36] S. Artzi, A. Kiezun, J. Dolby, F. Tip, D. Dig, A. Paradkar, M.D. Ernst, Finding bugs in web applications using dynamic test generation and explicit-state model checking, IEEE Trans. Softw. Eng. 36 (4) (2010) 474–494, ISSN 0098-5589.

[37] R. Paleari, D. Marrone, D. Bruschi, M. Monga, On race vulnerabilities in web applications, in: Proceedings of the 5th International Conference on Detection of Intrusions and Malware, and Vulnerability Assessment, DIMVA '08.

[38] Y. Zheng, X. Zhang, Static detection of resource contention problems in server-side scripts, in: Proceedings of the 34th International Conference on Software Engineering, ICSE '12, ISBN 978-1-4673-1067-3, 2012, pp. 584–594.

[39] M. Emmi, R. Majumdar, K. Sen, Dynamic test input generation for database applications, in: Proceedings of the 2007 International Symposium on Software Testing and Analysis, 2007, pp. 151–162.

[40] M. Marcozzi, W. Vanhoof, J.-L. Hainaut, Testing Database Programs Using Relational Symbolic Execution, Technical report, University of Namur, 2014.

[41] K. Pan, X. Wu, T. Xie, Database state generation via dynamic symbolic execution for coverage criteria, in: Proceedings of the Fourth International Workshop on Testing Database Systems, ACM, 2011, p. 4.

[42] S.A. Khalek, S. Khurshid, Systematic testing of database engines using a relational constraint solver, in: 2011 IEEE Fourth International Conference on Software Testing, Verification and Validation (ICST), 2011, pp. 50–59.

[43] K. Pan, X. Wu, T. Xie, Generating program inputs for database application testing, in: 2011 26th IEEE/ACM International Conference on Automated Software Engineering (ASE), IEEE, 2011, pp. 73–82.

[44] G. Wassermann, D. Yu, A. Chander, D. Dhurjati, H. Inamura, Z. Su, Dynamic test input generation for web applications, in: Proceedings of the 2008 International Symposium on Software Testing and Analysis, ISSTA '08, ISBN 978-1-60558-050-0, 2008, pp. 249–260.

[45] V. Felmetsger, L. Cavedon, C. Kruegel, G. Vigna, Toward automated detection of logic vulnerabilities in web applications, in: Proceedings of the 19th USENIX Conference on Security, USENIX Security'10, ISBN 888-7-6666-5555-4, 2010, p. 10.

[46] M.A. Ghafoor, M.S. Mahmood, J.H. Siddiqui, Effective partial order reduction in model checking database applications, in: 2016 IEEE International Conference on Software Testing, Verification and Validation (ICST), 2016, pp. 146–156, https://doi.org/10.1109/ICST.2016.25.

[47] L.A. Clarke, A system to generate test data and symbolically execute programs, IEEE Trans. Softw. Eng. 2 (3) (1976) 215–222, http://portal.acm.org/ft_gateway.cfm?id=1313532&type=external&coll=Portal&dl=GUIDE&CFID=50149523&CFTOKEN=17299335.

[48] J.C. King, Symbolic execution and program testing, Commun. ACM 19 (7) (1976) 385–394, http://portal.acm.org/ft_gateway.cfm?id=360252&type=pdf&coll=Portal&dl=GUIDE&CFID=50149523&CFTOKEN=17299335.

[49] C. Barrett, C. Tinelli, CVC3, in: Proc. 19th International Conference on Computer Aided Verification (CAV), 2007, pp. 298–302.

[50] L. de Moura, N. Bjørner, Z3: an efficient SMT solver, in: International Conference on Tools and Algorithms for the Construction and Analysis of Systems (TACAS), 2008, pp. 337–340, https://doi.org/10.1007/978-3-540-78800-3_24.

[51] N. Sörensson, N. Een, An extensible SAT-solver, in: Proc. 6th International Conference on Theory and Applications of Satisfiability Testing (SAT), 2003, pp. 502–518.

[52] S. Khurshid, C.S. Pasareanu, W. Visser, Generalized symbolic execution for model checking and testing, in: Proc. 9th International Conference on Tools and Algorithms for the Construction and Analysis of Systems (TACAS), 2003, pp. 553–568, http://citeseerx.ist.psu.edu/viewdoc/summary?doi=10.1.1.8.8862.

[53] P. Godefroid, N. Klarlund, K. Sen, DART: directed automated random testing, in: Proc. 2005 Conference on Programming Languages Design and Implementation (PLDI), 2005, pp. 213–223, http://portal.acm.org/ft_gateway.cfm?id=1065036&type=pdf&coll=Portal&dl=GUIDE&CFID=50151101&CFTOKEN=70334910.

[54] K. Sen, D. Marinov, G. Agha, CUTE: a concolic unit testing engine for C, in: Proc. 5th Joint Meeting of the European Software Engineering Conference and Symposium on Foundations of Software Engineering (ESEC/FSE), 2005, pp. 263–272, http://portal.acm.org/ft_gateway.cfm?id=1081750&type=pdf&coll=Portal&dl=GUIDE&CFID=50151101&CFTOKEN=70334910.

[55] C. Cadar, D. Dunbar, D.R. Engler, KLEE: unassisted and automatic generation of high-coverage tests for complex systems programs, in: Proc. 8th Symposium on Operating Systems Design and Implementation (OSDI), 2008, pp. 209–224.

[56] D.A. Ramos, D.R. Engler, Practical, low-effort equivalence verification of real code, in: Proc. 23rd International Conference on Computer Aided Verification (CAV), 2011, pp. 669–685.

[57] B. Elkarablieh, I. Garcia, Y.L. Suen, S. Khurshid, Assertion-based repair of complex data structures, in: Proc. 22nd International Conference on Automated Software Engineering (ASE), 2007, pp. 64–73.

[58] C. Seo, S. Malek, N. Medvidovic, Component-level energy consumption estimation for distributed java-based software systems, in: Proc. 11th International Symposium on Component-Based Software Engineering, 2008, pp. 97–113.

[59] L.A. Clarke, Test data generation and symbolic execution of programs as an aid to program validation (Ph.D. thesis), University of Colorado at Boulder, 1976.

[60] W.R. Bush, J.D. Pincus, D.J. Sielaff, A static analyzer for finding dynamic programming errors, Softw. Pract. Experience 30 (7) (2000) 775–802, https://doi.org/10.1002/(SICI)1097-024X(200006)30:73C775::AID-SPE3093E3.0.CO;2-H.

[61] V. Adve, C. Lattner, M. Brukman, A. Shukla, B. Gaeke, LLVA: a low-level virtual instruction set architecture, in: Proc. 36th International Symposium on Microarchitecture (MICRO), 2003, pp. 205–216.

[62] P. Godefroid, Compositional dynamic test generation, in: Proc. 34th Symposium on Principles of Programming Languages (POPL), 2007, pp. 47–54.

[63] C. Cadar, D. Engler, Execution generated test cases: how to make systems code crash itself, in: Proc. International SPIN Workshop on Model Checking of Software, 2005, pp. 2–23.

[64] C. Cadar, V. Ganesh, P.M. Pawlowski, D.L. Dill, D.R. Engler, EXE: automatically generating inputs of death, in: Proc. 13th Conference on Computer and Communications Security (CCS), 2006, pp. 322–335, https://doi.org/10.1145/1180405.1180445.

[65] E. Clarke, D. Kroening, K. Yorav, Behavioral consistency of C and Verilog programs using bounded model checking, in: Proceedings of the 40th Annual Design Automation Conference, DAC '03, ACM, New York, NY, ISBN 1-58113-688-9, 2003, pp. 368–371, https://doi.org/10.1145/775832.775928.

[66] E. Clarke, D. Kroening, F. Lerda, A tool for checking ANSI-C programs, in: International Conference on Tools and Algorithms for the Construction and Analysis of Systems, Springer, 2004, pp. 168–176.

[67] D. Kroening, M. Tautschnig, CBMC-C bounded model checker, in: International Conference on Tools and Algorithms for the Construction and Analysis of Systems, Springer, 2014, pp. 389–391.

[68] F. Ivančić, Z. Yang, M.K. Ganai, A. Gupta, P. Ashar, Efficient SAT-based bounded model checking for software verification, Theor. Comput. Sci. 404 (3) (2008) 256–274, ISSN 0304-3975, https://doi.org/10.1016/j.tcs.2008.03.013.

[69] Y. Xie, A. Aiken, Scalable error detection using Boolean satisfiability, in: Proceedings of the 32nd ACM SIGPLAN-SIGACT Symposium on Principles of Programming Languages, POPL '05, ACM, New York, NY, ISBN 1-58113-830-X, 2005, pp. 351–363, https://doi.org/10.1145/1040305.1040334.

[70] D. Babic, A.J. Hu, Calysto: scalable and precise extended static checking, in: Proceedings of the 30th International Conference on Software Engineering, ICSE '08, ACM, New York, NY, ISBN 978-1-60558-079-1, 2008, pp. 211–220, https://doi.org/10.1145/1368088.1368118.

[71] L. Cordeiro, B. Fischer, J. Marques-Silva, SMT-based bounded model checking for embedded ANSI-C software, IEEE Trans. Softw. Eng. 38 (4) (2012) 957–974.

[72] C.Y. Cho, V. D'Silva, D. Song, Blitz: compositional bounded model checking for real-world programs, in: 2013 IEEE/ACM 28th International Conference on Automated Software Engineering (ASE), IEEE, 2013, pp. 136–146.

[73] S. Falke, F. Merz, C. Sinz, The bounded model checker LLBMC, in: 2013 IEEE/ACM 28th International Conference on Automated Software Engineering (ASE), IEEE, 2013, pp. 706–709.

[74] O. Inverso, E. Tomasco, B. Fischer, S. La Torre, G. Parlato, Lazy-CSeq: a lazy sequentialization tool for C, in: International Conference on Tools and Algorithms for the Construction and Analysis of Systems, Springer, 2014, pp. 398–401.

[75] J.H. Siddiqui, S. Khurshid, Scaling symbolic execution using ranged analysis, in: Proceedings of the ACM International Conference on Object Oriented Programming Systems Languages and Applications, OOPSLA '12, ACM, New York, NY, ISBN 978-1-4503-1561-6, 2012, pp. 523–536, https://doi.org/10.1145/2384616.2384654.

[76] S. Person, G. Yang, N. Rungta, S. Khurshid, Directed incremental symbolic execution, in: Proc. 2011 Conference on Programming Languages Design and Implementation (PLDI), 2011, pp. 504–515.

[77] P.D. Marinescu, C. Cadar, KATCH: high-coverage testing of software patches, in: Proc. 12th Joint Meeting of the European Software Engineering Conference and Symposium on Foundations of Software Engineering (ESEC/FSE), 2013.

[78] W. Visser, J. Geldenhuys, M.B. Dwyer, Green: reducing, reusing and recycling constraints in program analysis, in: Proc. 2012 Joint Meeting of the European Software Engineering Conference and Symposium on Foundations of Software Engineering (ESEC/FSE), 2012.

[79] X. Jia, C. Ghezzi, S. Ying, Enhancing reuse of constraint solutions to improve symbolic execution, in: Proceedings of the 2015 International Symposium on Software Testing and Analysis, ACM, 2015, pp. 177–187.

[80] G. Yang, C.S. Păsăreanu, S. Khurshid, Memoized symbolic execution, in: Proc. 2012 International Symposium on Software Testing and Analysis (ISSTA), ISSTA 2012, ISBN 978-1-4503-1454-1, 2012, pp. 144–154, https://doi.org/10.1145/2338965.2336771.

[81] S. Guo, M. Kusano, C. Wang, Conc-iSE: incremental symbolic execution of concurrent software, in: 2016 31st IEEE/ACM International Conference on Automated Software Engineering (ASE), IEEE, 2016, pp. 531–542.

[82] J.H. Siddiqui, S. Khurshid, ParSym: parallel symbolic execution, in: Proc. 2nd International Conference on Software Technology and Engineering (ICSTE), 2010, pp. V1405–V1409.

[83] M. Staats, C. Păsăreanu, Parallel symbolic execution for structural test generation, in: Proc. 19th International Symposium on Software Testing and Analysis (ISSTA), 2010, pp. 183–194.

[84] P. Godefroid, M.Y. Levin, D.A. Molnar, Automated whitebox fuzz testing, in: Proc. Network and Distributed System Security Symposium (NDSS), 2008.

[85] R. Sasnauskas, O. Landsiedel, M.H. Alizai, C. Weise, S. Kowalewski, K. Wehrle, KleeNet: discovering insidious interaction bugs in wireless sensor networks before deployment, in: Proc. 9th International Conference on Information Processing in Sensor Networks (ISPN 2010), 2010, pp. 186–196.

[86] R. Majumdar, K. Sen, Hybrid concolic testing, in: Proc. 29th International Conference on Software Engineering (ICSE), 2007, pp. 416–426.

[87] J.H. Siddiqui, S. Khurshid, Staged symbolic execution, in: Proceedings of the 27th Annual ACM Symposium on Applied Computing, SAC '12, ACM, New York, NY, ISBN 978-1-4503-0857-1, 2012, pp. 1339–1346, https://doi.org/10.1145/2245276.2231988.

[88] J.H. Siddiqui, S. Khurshid, Scaling symbolic execution using staged analysis, Innov. Syst. Softw. Eng. 9 (2) (2013) 119–131, ISSN 1614-5046, https://doi.org/10.1007/s11334-013-0196-9.

[89] J.H. Siddiqui, S. Khurshid, Symbolic execution of alloy models, in: Proceedings of the 13th International Conference on Formal Methods and Software Engineering, ICFEM '11, Springer-Verlag, Berlin, Heidelberg, ISBN 978-3-642-24558-9, 2011, pp. 340–355, http://dl.acm.org/citation.cfm?id=2075089.2075119.

[90] D. Jackson, Software Abstractions: Logic, Language, and Analysis, The MIT Press, 2006.

[91] M.S. Mahmood, M. Abdul Ghafoor, J.H. Siddiqui, Symbolic execution of stored procedures in database management systems, in: Proceedings of the 31st IEEE/ACM International Conference on Automated Software Engineering, ASE 2016, ACM, New York, NY, ISBN 978-1-4503-3845-5, 2016, pp. 519–530, https://doi.org/10.1145/2970276.2970318.

[92] Q. Yi, Z. Yang, S. Guo, C. Wang, J. Liu, C. Zhao, Eliminating path redundancy via postconditioned symbolic execution, IEEE Trans. Softw. Eng. (2017), https://doi.org/10.1109/TSE.2017.2659751, http://ieeexplore.ieee.org/document/7835264/.

[93] Q. Zou, W. Huang, J. An, W. Fan, Redundant constraints elimination for symbolic execution, in: Information Technology, Networking, Electronic and Automation Control Conference, IEEE, 2016, pp. 235–240.

[94] H. Wang, T. Liu, X. Guan, C. Shen, Q. Zheng, Z. Yang, Dependence guided symbolic execution, IEEE Trans. Softw. Eng. 43 (3) (2017) 252–271.

[95] J. Feist, L. Mounier, M.-L. Potet, Guided dynamic symbolic execution using subgraph control-flow information, in: International Conference on Software Engineering and Formal Methods, Springer, 2016, pp. 76–81.

[96] C.S. Păsăreanu, N. Rungta, Symbolic PathFinder: symbolic execution of Java bytecode, in: Proceedings of the IEEE/ACM international Conference on Automated Software Engineering, ACM, 2010, pp. 179–180.

[97] N. Rosner, J. Geldenhuys, N.M. Aguirre, W. Visser, M.F. Frias, BLISS: improved symbolic execution by bounded lazy initialization with sat support, IEEE Trans. Softw. Eng. 41 (7) (2015) 639–660.

[98] P. Braione, G. Denaro, M. Pezzè, Symbolic execution of programs with heap inputs, in: Proceedings of the 2015 10th Joint Meeting on Foundations of Software Engineering, ACM, 2015, pp. 602–613.

[99] A. Ibing, Dynamic symbolic execution with interpolation based path merging, in: Int. Conf. Advances and Trends in Software Engineering, 2016.

[100] T. Agerwala, J. Misra, Assertion Graphs for Verifying and Synthesizing Programs, Technical report, Austin, TX, 1978.

[101] S. Graf, H. Saïdi, Construction of abstract state graphs with PVS, in: Proceedings of the 9th International Conference on Computer Aided Verification, CAV '97, Springer-Verlag, London, UK, ISBN 3-540-63166-6, 1997, pp. 72–83, http://dl.acm.org/citation.cfm?id=647766.733618.

[102] T. Ball, S. Rajamani, Boolean Programs: A Model and Process for Software Analysis, Technical report, 2000) https://www.microsoft.com/en-us/research/publication/boolean-programs-a-model-and-process-for-software-analysis/.

[103] E. Clarke, O. Grumberg, S. Jha, Y. Lu, H. Veith, Counterexample-Guided Abstraction Refinement, in: Springer, Berlin, Heidelberg, ISBN 978-3-540-45047-4, 2000, pp. 154–169, https://doi.org/10.1007/10722167_15.

[104] H. Saidi, Model checking guided abstraction and analysis, in: Proceedings of the 7th International Symposium on Static Analysis, SAS '00, Springer-Verlag, London, UK, ISBN 3-540-67668-6, 2000, pp. 377–396, http://dl.acm.org/citation.cfm?id=647169.760067.

[105] T. Ball, S.K. Rajamani, The SLAM project: debugging system software via static analysis, in: ACM SIGPLAN Notices, vol. 37, ACM, 2002, pp. 1–3.

[106] T. Ball, V. Levin, S.K. Rajamani, A decade of software model checking with SLAM, Commun. ACM 54 (7) (2011) 68–76.

[107] T. Ball, E. Bounimova, B. Cook, V. Levin, J. Lichtenberg, C. McGarvey, B. Ondrusek, S.K. Rajamani, A. Ustuner, Thorough static analysis of device drivers, in: Proceedings of the 1st ACM SIGOPS/EuroSys European Conference on Computer Systems 2006, EuroSys '06, ACM, New York, NY, ISBN 1-59593-322-0, 2006, pp. 73–85, https://doi.org/10.1145/1217935.1217943.

[108] D. Beyer, T.A. Henzinger, R. Jhala, R. Majumdar, The software model checker Blast, Int. J. Softw. Tools Technol. Trans. 9 (5–6) (2007) 505–525.

[109] D. Beyer, M.E. Keremoglu, CPAchecker: a tool for configurable software verification, in: International Conference on Computer Aided Verification, Springer, 2011, pp. 184–190.

[110] M. Dangl, S. Löwe, P. Wendler, CPAchecker with support for recursive programs and floating-point arithmetic, in: International Conference on Tools and Algorithms for the Construction and Analysis of Systems, Springer, 2015, pp. 423–425.

[111] S. Chaki, E. Clarke, A. Groce, O. Strichman, Predicate Abstraction with Minimum Predicates, in: Springer, Berlin, Heidelberg, 2003, pp. 19–34.

[112] K.L. McMillan, Lazy abstraction with interpolants, in: Proceedings of the 18th International Conference on Computer Aided Verification, CAV'06, Springer-Verlag, Berlin, Heidelberg, ISBN 3-540-37406-X, 978-3-540-37406-0, 2006, pp. 123–136, https://doi.org/10.1007/11817963_14.

[113] C. Wang, Z. Yang, A. Gupta, F. Ivančic, Using counterexamples for improving the precision of reachability computation with polyhedra, in: Proceedings of the 19th International Conference on Computer Aided Verification, CAV'07, Springer-Verlag, Berlin, Heidelberg, ISBN 978-3-540-73367-6, 2007, pp. 352–365, http://dl.acm.org/citation.cfm?id=1770351.1770405.

[114] Mixed Symbolic Representations for Model Checking Software Programs, Proceedings of the Fourth ACM/IEEE International Conference on Formal Methods and Models for Co-Design, MEMOCODE '06, IEEE Computer Society, Washington, DC, ISBN 1-4244-0421-5, 2006, pp. 17–26, https://doi.org/10.1109/MEMCOD.2006.1695896.

ABOUT THE AUTHORS

Junaid Haroon Siddiqui is an assistant professor of Computer Science at Lahore University of Management Sciences in Pakistan. Previously, he received his Ph.D. from The University of Texas at Austin. His research interests are static and dynamic techniques in program verification.

Affan Rauf is a Ph.D. candidate in the School of Science and Engineering, Lahore University of Management Sciences, Pakistan. His work interests include automated test case generation, symbolic execution, and software model checking.

Maryam Abdul Ghafoor is a Ph.D. candidate at Lahore University of Management Sciences, Pakistan. She received MS degree in Computer Science from the same school in 2015. Her research interests include program analysis, symbolic execution, automated software testing, and verification.

CHAPTER THREE

Emerging Software Testing Technologies

Francesca Lonetti, Eda Marchetti
ISTI–CNR, Pisa, Italy

Contents

Advances in Computers, Volume 108
ISSN 0065-2458
https://doi.org/10.1016/bs.adcom.2017.11.003

Abstract

Software testing encompasses a variety of activities along the software development process and may consume a large part of the effort required for producing software. It represents a key aspect to assess the adequate functional and nonfunctional software behavior aiming to prevent and remedy malfunctions. The increasing complexity and heterogeneity of software poses many challenges to the development of testing strategies and tools. In this chapter, we provide a comprehensive overview of emerging software testing technologies. Beyond the basic concepts of software testing, we address prominent test case generation approaches and focus on more relevant challenges of testing activity as well as its role in recent development processes. An emphasis is also given to testing solutions tailored to the specific needs of emerging application domains.

1. INTRODUCTION

The testing phase is an important and critical part of software development, consuming even more than half of the effort required for producing deliverable software [1]. Unfortunately, often due to time or cost constraints, the testing is not developed in the proper manner or is even skipped. The testing activity in fact is not limited to the detection of "bugs" in the software, but it encompasses the entire development process.

The testing planning starts during the early stages of requirement analysis and proceeds systematically, with continuous refinements during the course of software development until the completion of the coding phase, with the beginning of the test cases execution. This last step represents the biggest part of software cost that can be evaluated in terms of: the cost of designing a suitable set of test cases which can reveal the presence of bugs; the cost of running those tests, which also requires a considerable amount of time; the cost of detecting them, i.e., the development of a proper "oracle," which can identify the manifestation of bugs as soon as possible; and the cost of correcting them. All these activities have in common the same testing purpose: evaluating the product quality for increasing the software engineering confidence in the proper functioning of the software. However, it must be made clear that testing cannot show the absence of defects; it can only reveal that software defects are present [2].

In this chapter, we refer to the definition of the "Software Testing" introduced in [3]: "*Software Testing consists of the dynamic verification of the behaviour of a program on a finite set of test cases, suitably selected from the usually infinite executions domain, against the specified expected behaviour.*" As shown by this definition, testing deals with dynamic verification of system quality, which also involves the code execution, as will be better described in this chapter. Generally, the techniques applicable for quality evaluation can be divided into two sets: static techniques, which do not involve code execution, and dynamic techniques, to which testing belongs to, which instead require running code. The static techniques are applicable to all during the process development for different purposes such as to check the adherence of the code to the specification or to detect defects in code by its inspection or review. Instead, the latter approach more properly observes failures as they show up. In particular, dynamic analysis techniques involve the execution of the code and the analysis of its responses in order to determine its validity and detect errors. The behavioral properties of the program are also observed. Other examples of dynamic analysis include simulation, sizing and timing analysis, and prototyping, which may be applied throughout the life cycle [1]. Here, we briefly present the static techniques (Section 2.1.1), preferring to concentrate on testing, which is the main topic of this chapter.

Before continuing the presentation it is important to clarify the terminology relative to the terms "fault," "defect," and "failure" that we will use. Although their meanings are strictly related, there are some distinctions between them. As discussed in [1], a failure is the manifested inability of the program to perform the function required, i.e., a system malfunction evidenced by incorrect output, abnormal termination or unmet time and space constraints. The cause of a failure, i.e., the missing or incorrect code, is a fault. In particular, a fault may remain undetected until some stirring up event activates it. In this case it brings the program into an intermediate unstable state, called error, which if propagated to the output causes a failure. The process of failure manifestation is therefore *Fault-Error-Failure*, which can be iterated recursively: a fault can be caused by a failure in some other interacting system.

Testing reveals failures and a consequent analysis stage is needed to identify the faults that caused them. In particular, it is possible that many different failures can result from a single fault, and the same failure can be caused by different faults.

The chapter is organized as follows: in Section 2 we present basic concepts of testing including types of test (static and dynamic), test levels, and objectives characterizing the testing activity; in Section 3 we address most prominent test generation techniques and tools; Section 4 targets challenging aspects of software testing, whereas Section 5 identifies the role of software testing in relevant software processes. Finally, Section 6 outlines needs and trends of software testing for emerging application domains, and Section 7 draws discussion and conclusions.

2. BASIC CONCEPTS OF TESTING

The one term testing involves different concepts, refers to a full range of test techniques, even quite different from one other, and embraces a variety of aims. In this section we provide some basic concepts of software testing useful in the remaining of this chapter.

2.1 Type of Tests

Since the 1980, the widespread use of modern technologies has led a large part of the software engineering to focus its attention on quality, usability, safety, and other characteristic attributes of software applications. In particular, interest was captured both by the process for software development and by its results. Using their experience software engineering researchers have gradually arrived at the conviction that only the joint between a mature and well-established development process with specific techniques for the quantitative evaluation of the attributes of interest of the artifacts produced can guarantee high quality and reliable applications. Therefore, research has been split into two sets, with of course some natural intersections and points of contact: the former interested to the process (software process improvement [SPI]) [4] and the latter focused on the product.

Frameworks such as CMM [5], SPICE [6], and RUP [7] (detailed in Section 5) are the products of the SPI research work belonging to the former set. They capture the good practices for the process assessment and are de facto references used by thousands of organizations. Considering the latter set, generally the techniques applicable to the product can be divided into two groups: static techniques, which do not involve code execution, and the dynamic techniques, which instead require code running, to which testing belongs. In the remaining of this section more details about these two techniques are provided.

2.1.1 Static Techniques

Static techniques are based on the (manual or automated) examination of project documentation, of software models and code, and of other related information about requirements and design. Thus, static techniques can be employed all along software development, and their earlier usage is of course highly desirable.

Considering a generic development process (see Section 5 for more details), they can be applied [1]

- at the requirements stage for checking language syntax, consistency and completeness, as well as the adherence to established conventions;
- at the design phase for evaluating the implementation of requirements and detecting inconsistencies (for instance between the inputs and outputs used by high-level modules and those adopted by submodules);
- during the implementation phase for checking that the form adopted for the implemented products (e.g., code and related documentation) adheres to the established standards or conventions, and that interfaces and data types are correct.

Traditional static techniques include [8] *software inspection*, that is, the step-by-step analysis of the documents (deliverables) produced, against a compiled checklist of common and historical defects; *software reviews*, that is, the process by which different aspects of the work product are presented to project personnel (managers, users, customer, etc.) and other interested stakeholders for comment or approval; *code reading*, that is, the desktop analysis of the produced code for discovering typing errors that do not violate style or syntax; *algorithm analysis and tracing*, that is, the process in which the complexity of algorithms employed and the worst-case, average-case, and probabilistic analysis evaluations can be derived.

The processes implied by the above techniques are heavily manual, error-prone, and time-consuming. To overcome these problems, researchers have proposed static analysis techniques relying on the use of formal methods [9]. The goal is to automate as much as possible the verification of the properties of the requirements and the design. Toward this goal, it is necessary to enforce a rigorous and unambiguous formal language for specifying the requirements and the software architecture.

In the middle between static and dynamic analysis techniques, there is symbolic execution [10]. It involves the execution of a program by replacing variables with symbolic values (see Section 3.4).

2.1.2 Dynamic Techniques

Dynamic techniques [3] obtain information of interest about a program by observing some executions. Standard dynamic analysis include testing (on which we focus in the rest of the chapter) and profiling. Essentially, a program profile records the number of times some entities of interest occur during a set of controlled executions. These data can be used to derive measures of coverage or frequency. Other specific dynamic techniques also include simulation, sizing and timing analysis, and prototyping [1].

Testing properly said is based on the execution of the code on valued inputs. Of course, although the set of input values can be considered infinite, those that can be run effectively during testing are finite. It is in practice impossible, due to the limitations of the available budget and time, to exhaustively exercise every input of a specific set even when not infinite. In other words, by testing we observe some samples of the program behavior. A test strategy therefore must be adopted to find a trade-off between the number of chosen inputs and overall time and effort dedicated to testing purposes. Different techniques can be applied depending on the target and the effect that should be reached. We will describe test selection strategies in Section 3.

2.2 Objectives of Testing

Software testing can be applied for different purposes, such as verifying, that the functional specifications are implemented correctly, or that the system shows specific nonfunctional properties such as performance, reliability, and usability. A (certainly noncomplete) list of relevant testing objectives includes [1,3]

- Acceptance/qualification testing: it is the final test action prior to deploying a software product. Its main goal is to verify that the software respects the customer requirement. Generally, it is run by or with the end users to perform those functions and tasks the software was built for;
- Installation testing: the system is verified upon installation in the target environment. Installation testing can be viewed as system testing conducted once again according to hardware configuration requirements. Installation procedures may also be verified;
- Alpha testing: before releasing the system, it is deployed to some in-house users for exploring the functions and business tasks. Generally, there is no test plan to follow, but the individual tester determines what to do;
- Beta testing: the same as alpha testing but the system is deployed to external users. In this case the amount of detail, the data, and approach taken are entirely up to the individual testers. Each tester is responsible for creating their own environment, selecting their data, and determining what functions, features, or tasks to explore. Each tester is also responsible for identifying their own criteria for whether to accept the system in its current state or not;

- Conformance testing/functional testing: the test cases are aimed at validating that the observed behavior conforms to the specifications. In particular, it checks whether the implemented functions are as intended and provide the required services and methods. This test can be implemented and executed against different test targets, including units, integrated units, and systems;
- Nonfunctional testing: it is specifically aimed at verifying nonfunctional properties of the system such as performance, reliability, usability, and security;
- Regression testing: it is the selective reexecution of test cases to verify that code modifications have not caused unintended effects and that the system or component still complies with requirements. In practice, the objective is to show that a system which previously passed the tests still does. Notice that a trade-off must be made between the assurance given by regression testing every time a change is made and the resources required to do that.

2.3 Test Levels

During the development life cycle of a software product, testing is performed at different levels and can involve the whole system or parts of it. Depending on the process model adopted, then, software testing activities can be articulated in different phases, each one addressing specific needs relative to different portions of a system. Whichever the process adopted, we can at least distinguish in principle between unit, integration, system, and regression test [1]. These are the testing stages of a traditional phased process (see Section 5). However, even considering different, more modern, process models, a distinction between these test levels remains useful to emphasize three logically different moments in the verification of a complex software system. None of these levels is more relevant than another, and more importantly a stage cannot supply for another, because each addresses different typologies of failures.

2.3.1 Unit Test

A unit is the smallest testable piece of software, which may consist of hundreds or even just a few lines of source code, and generally represents the result of the work of one programmer. The unit test purpose is to ensure that the unit satisfies its functional specification and/or that its implemented structure matches the intended design structure [1].

Unit tests can also be applied to check interfaces (parameters passed in correct order, number of parameters equal to number of arguments, parameter, and argument matching), local data structure (improper typing, incorrect variable name, inconsistent data type), or boundary conditions.

2.3.2 Integration Test

Generally speaking, integration is the process by which software pieces or components are aggregated to create a larger component. Integration testing is specifically aimed at exposing the problems that can arise at this stage. Even though the single units are individually acceptable when tested in isolation, in fact, they could still result in incorrect or inconsistent behavior when combined in order to build complex systems. For example, there could be an improper call or return sequence between two or more components [1].

Integration testing thus is aimed at verifying that each component interacts according to its specifications as defined during preliminary design. In particular, it mainly focuses on the communication interfaces among integrated components.

There are not many formalized approaches to integration testing in the literature, and practical methodologies rely essentially on good design sense and the testers intuition. Integration testing of traditional systems was done substantially in either a nonincremental or an incremental approach.

In a nonincremental approach the components are linked together and tested all at once ("big-bang" testing) [1]. In the incremental approach, we find the classical "top-down" strategy, in which the modules are integrated one at a time, from the main program down to the subordinated ones, or "bottom-up," in which the tests are constructed starting from the modules at the lowest hierarchical level and then are progressively linked together upwards, to construct the whole system. Usually in practice, a mixed approach is applied, as determined by external project factors (e.g., availability of modules, release policy, and availability of testers) [1].

In modern Object Oriented, distributed systems, approaches, such as top-down or bottom-up integration and their practical derivatives, are no longer usable, as no "classical" hierarchy between components can be generally identified. Some other criteria for integration testing imply integrating the software components based on identified functional threads [1]. In this case, the test is focused on those classes used in reply to a particular input or system event (thread-based testing); or by testing together those classes that contribute to a particular use of the system. Finally, some authors have used the dependency structure between classes as a reference structure for guiding integration testing, i.e., their static dependencies or even the dynamic relations of inheritance and polymorphism [1].

2.3.3 System Test

System test involves the whole system embedded in its actual hardware environment and is mainly aimed at verifying that the system behaves according to the user requirements. In particular, it attempts to reveal bugs that cannot be attributed to components as such, to the inconsistencies between components, or to the planned interactions of components and other objects (which are the subject of integration testing).

Summarizing, the primary goals of system testing can be [1] (i) discovering the failures that manifest themselves only at system level and hence were not detected during unit or integration testing; (ii) increasing the confidence that the developed product correctly implements the required capabilities; and (iii) collecting information useful for deciding the release of the product.

System testing should therefore ensure that each system function works as expected, failures are exposed and analyzed, and additionally that interfaces for export and import routines behave as required.

Generally, system testing includes testing for performance, security, reliability, stress testing, and recovery [1]. In particular, test and data collected applying system testing can be used for defining an operational profile necessary to support a statistical analysis of system reliability (see Section 3.3).

A further test level, called *Acceptance Test*, is often added to the above subdivision. This is, however, more an extension of system test, rather than a new level. It is in fact a test session conducted over the whole system, which mainly focuses on the usability requirements more than on the compliance of the implementation against some specification. The intent is hence to verify that the effort required from end users to learn to use and fully exploit the system functionalities is acceptable.

2.3.4 Regression Test

Properly speaking, regression test is not a separate level of testing (it is listed among test objectives in Section 2.2), but may refer to the retesting of a unit, a combination of components or a whole system after modification, in order to ascertain that the change has not introduced new faults [1].

As software produced today is constantly in evolution, driven by market forces and technology advances, regression testing takes by far the predominant portion of testing effort in industry. Since both corrective and evolutive modifications may be performed quite often, to rerun after each change all previously executed test cases would be prohibitively expensive [11]. Therefore various types of techniques have been developed to reduce regression testing costs and to make it more effective.

Selective regression test techniques [1] help in selecting a (minimized) subset of the existing test cases by examining the modifications (for instance at code level, using control flow and data flow analysis). Other approaches instead prioritize the test cases according to some specified criterion (for instance maximizing the fault detection power or the structural coverage), so that the test cases judged the most effective with regard to the adopted criterion can be taken first, up to the available budget (see Section 4.4.3 for more details).

3. TEST CASES GENERATION

Test cases generation is among the most important and intensive activities of software testing. A lot of research has been devoted in the last decades to automatic test case generation with the consequent development of different techniques and tools. In this section, we provide an overview of prominent test case generation approaches evidencing the main issues and challenges of each of them.

3.1 Search-Based Testing

Search-based software testing is one of the emerging methodologies for automated test generation. It is based on an heuristic, such as a genetic algorithm [12], to optimize search techniques and automate the test case definition. Such heuristic uses a problem-specific fitness function for on side guiding the search of solutions from a potentially infinite search space, and from the other limiting the required execution time. Search-based software testing is not a recent proposal, indeed the first approach dates back to 1976 [13]. However, in the last years there is an huge increase of proposals due to their flexibility and applicability. Commonly, the application of search-based testing requires that [14]: the solutions for the problem should be encoded so that they can be manipulated by the heuristic algorithm; the fitness function needs to be defined for each specific problem and should be able to guide the search to promising areas of the search space by evaluating candidate solutions.

Due to their versatility, search-based approaches have been adopted in several areas such as [14] functional testing, temporal testing, integration testing, regression testing, stress testing, mutation testing, test prioritization, interaction testing, state machine testing, and exception testing.

Common weaknesses of search-based testing are related to the interaction with external environment, the definition of the fitness function, and the automated oracle. In the former case, the main difficulty is related to the handling interactions with external environment or components with which the system under test could be dependent. In particular, issues could be risen in checking the existence of files or directories. In literature, possible solutions propose to include, in the file or database, test data that can be read back by the program under test, or to use mock objects.

In case of the fitness functions, due to their heuristics nature, there is the possibility they fail to give adequate guidance to the search. A common example is the so-called *flag* problem in structural test data generation [15]: a branch predicate consists of a Boolean value (*the flag*), yielding only two branch distance values one for when the flag is true, and one for when it is false. The *testability transformation* [15] is the most adopted solution to deal with this problem. It consists in the generation of a temporary version of the program under test that can be used to generate the test data and then discarded.

About the automated oracle, notwithstanding the huge interest devoted to this topic, reducing the human activity in the evaluation of the testing results is still an issue. Indeed, available proposals for fitness functions are more focused on maximizing the (structural) coverage, while minimizing the testing effort. The seeding work of [16] tries to alleviate this problem proposing solutions based on knowledge management, such for instance the possibility of explicitly select the starting point of any search-based approach or use the program source of information for setting up the types of inputs that may be expected.

3.2 Model-Based Testing

In recent years, model-based testing (MBT) has become increasingly successful thanks to the emergence of model-centric development paradigms, such as UML and MDA [17]. Models are used for capturing knowledge and specify a system with different levels of accuracy. The main goal of model-based testing is the automatic generation, execution, and evaluation of test cases based on a formal model of the SUT. The work in [18] provides a taxonomy of characteristics, similarities, and differences of MBT techniques and classifies existing tools according to them.

The four main approaches known as model-based testing are [19] (i) generation of test input data from a domain model; (ii) generation of test input data from an environment model; (iii) generation of test cases with oracle from a behavior model; and (iv) generation of executable test scripts from abstract tests.

However, model-based testing suffers from some drawbacks that may prevent its use in some application areas, such as the availability of a system specification, which not always exists in practice. Moreover, when it exists, this specification should be complete enough to ensure some relevance of the derived test suite. Finally, this specification cannot include all the implementation details and is restricted to a given abstraction level. Therefore, to become executable, the derived test cases need to be refined into more concrete interaction sequences. Automating this process still remains an open problem.

Despite the continuous evolving of MBT field, as it could be observed in the increasing number of MBT techniques published at the technical literature, there is still a gap between research related to MBT and its application in the software industry. The authors of [20] present an overview of MBT approaches supporting the selection of MBT techniques for software projects and the risk factors that may influence the use of these techniques in the industry.

3.3 Black-Box vs White-Box Testing

In this section, alternative classifications of test techniques are provided: the first, called black-box testing, is based on the input/output behavior of the system (Section 3.3.1); the second, called white-box testing, is based on the structure and internal data of the software under test (Section 3.3.2); the third called gray-box testing trying to mix the previous one (Section 3.3.3). Generally, the presented test techniques are not alternative approaches but can be used in combination as they use and provide different sources of information [21].

3.3.1 Black-Box Testing

Black-box testing, also called functional testing, relies on the input/output behavior of the system. In particular, the system is subjected to external inputs, so that the corresponding outputs are used to verify the conformance of the system to the specified behavior, with no assumptions of what happens in between. Therefore, in this process we assume knowledge of the (formal or informal) specification of the system under test, which can be used to define a behavioral model of the system (a transaction flowgraph) [1]. A complete black-box test would consist of subjecting the program to all possible input streams and verifying the outcome produced, but as stated in Section 2 this is theoretically impossible. For this, different techniques can be applied such as:

- Testing from formal specifications: in this case it is required that specifications be stated in a formal language, with a precise syntax and semantics. The tests are hence derived automatically from the specification, which are also used for deriving inductive proofs for checking the correct outcome [9].

- Equivalence partitioning: the functional tests are derived from the specifications written in structured, semiformal language. The input domain is partitioned into equivalence classes so that elements in the same class behave similarly. In this context, the category partition is a well-known and quite intuitive method, which provides a systematic, formalized approach to partition testing [22].

- Boundary-values analysis: this is a complementary approach to equivalence partitioning, and concentrates on the errors occurring at boundaries of the input domain. The test cases are thus chosen near the extremes of the class [22].

- Random methods: they consists of generating random test cases based on a uniform distribution over the input domain. It is a low-cost technique because large sets of test patterns can be generated cheaply without requiring any preliminary analysis of software [23, 24].

- Operational profile: test cases are produced by a random process meant to produce different test cases with the same probabilities with which they would arise in actual use of the software [25].

- Decision tables: the decision tables are rules expressed in a structural way used to express the test experts' or design experts' knowledge. Decision tables can be used when the outcome or the logic involved in the program is based on a set of decisions and rules which need to be followed [23].

- Cause–effect graphs: these are combinatorial logic networks that can be used to explore in systematic way the possible combinations of input conditions. By analyzing the specification, the relevant input conditions or causes, and the consequent transformations and output conditions, the effects are identified and modeled into graphs linking the effects to their causes [23].

- Combinatorial testing: in combinatorial testing, test cases are designed to execute combinations of input parameters [23]. Because providing all combinations is usually not feasible in practice, due to their extremely large numbers, combinatorial approaches able to generate smaller test suites for which all combinations of the features are guaranteed, are preferred [24]. Among them, common approach is all-pair testing technique, focusing on all possible discrete combinations of each pair of input parameters. We refer to [26] for a complete overview of the most recent proposals and tools.
- State transition testing: this type of testing is useful for testing state machine and also for navigation of graphical user interface [23]. A wide variety of state transition systems exist, including finite state machines [27], I/O Automata [28], Labeled Transition Systems [29], UML state machines [30], Simulink/Stateflow [31], and PROMELA [32].
- Evidence-based testing: in evidence-based software engineering (EBSE) the *best* solution for a practical problem should be identified based on evidence [33]. The process for solving practical problems based on a rigorous research approach includes the following different steps that can be applied also to testing activities [34]: (i) identify the evidence and formulate a question; (ii) track down the best evidence to answer the question; (iii) critically reflect on the evidence provided with respect to the problem and context that the evidence should help to solve. In software engineering, two approaches that allow to identify and aggregate evidence are systematic mapping studies and systematic reviews [35].

One of the points against the black-box testing is its dependence on the specification's correctness and the necessity of using a large amount of inputs in order to get good confidence of acceptable behavior.

3.3.2 White-Box Testing

The white-box testing, also called structural testing, requires complete access to the object's structure and internal data, which means the visibility of the source code. The tests are derived from the program's structure, which is also used to track which parts of the code have been executed during testing. For this, some of the commonly used techniques for test case selection are

- Control flow-based criteria: these techniques use the control flow graph representation of a program in which nodes correspond to sequentially executed statements while edges represent the flow of control between statements. The aim of white-box testing criteria is to cover as much as possible the control flow graph, limiting the number of selected test cases. In particular, they differentiate in: *statement coverage*, which is based on executable statements; *branch coverage*, which focuses on the blocks and case statements that affect the control flow; *condition coverage*, which relies on subexpressions independently of each other; *path coverage*, which is based on the possible paths exercised through the code [1, 23].

- Data flow coverage: in data flow testing, a data definition of a variable is a location where a value is stored in memory (definition) and a data use is a location where the value of the variable is accessed for computations (c-use) or for predicate uses (p-use). The data flow testing goal is to generate tests that execute program subpaths from definition to use. Traditional data flow analysis techniques work on control flow graphs annotated with specific information on data usage [1, 23].

3.3.3 Gray-Box Testing

The gray-box testing tries to combine the two aforementioned proposals [21]. In this case, the tests are generated exploiting limited knowledge of the internal working and the knowledge of input/output behavior of the system under test. Important aspects of gray-box testing techniques are the possibility of exploiting the interface definition and functional specification rather than source code and the focus on the user's point of view rather than designer one. On the other side, usually gray-box testing does not assure the coverage assessment as the access to source code is completely not available. Indeed, there is the possibility that many program paths remain untested or that the tests could be redundant. The other name of gray-box testing is translucent testing.

3.4 Symbolic Execution

Symbolic execution is a popular program analysis technique, introduced in the 1970s, that has found interest in recent years in the research community, with particular emphasis on applications to test generation [10]. The main idea behind symbolic execution is to use symbolic values, instead of actual data, as input values, and to execute the program by manipulating program expressions involving the symbolic values. The state of a symbolically executed program is represented by the symbolic values of program variables and by a Boolean formula over the symbolic inputs representing constraints which the inputs must satisfy. As a result, the output values computed by a program are expressed as a function of the input symbolic values and a symbolic execution tree represents the execution paths. The tree nodes represent program states and they are connected by program transitions [36].

The main challenges of symbolic execution in processing real-world code are related to [37]:

- Environment interactions: interactions with the file system or the network through libraries and system calls could cause side-effects and affect the execution. Evaluating any possible interaction outcome is generally unfeasible since it could generate a large number of execution states. A typical strategy is to consider popular library and system routines and create models that can help the symbolic engine to analyze only significant outcomes.

- State space explosion and path selection: the existence of real-world looping programs might exponentially increase the number of execution states and prevent the symbolic execution engine to exhaustively explore all the possible states within a reasonable amount of time. In practice, heuristics are used to guide and prioritize states exploration. In addition, efficient mechanisms for evaluating multiple states in parallel are implemented in symbolic engines.

- Constraint solver limitations: constraint solvers suffer from a number of limitations when they have to deal with nonlinear constraints over their elements. Symbolic execution engines normally rely on optimizations and algebraic simplifications.

However, to overcome the limitations of symbolic execution due to path explosion and the time spent in constraint solving, a standard approach consists into combining concrete and symbolic executions of the code under test. This approach is known as *concolic* testing [38] and tightly couples both concrete and symbolic executions that run simultaneously, and each gets feedback from the other. The basic idea is that a *concolic* execution engine uses the concrete execution to drive the symbolic execution, then after choosing an arbitrary input to begin with, it executes the program both concretely and symbolically by simultaneously updating concrete and symbolic stores as well as the path constraints.

Symbolic execution has been included in several recent tools offering the capability to systematically exploring many possible execution paths at the same time without necessarily requiring concrete inputs. The most common ones evidencing the growing impact of symbolic execution in practice are

- Symbolic Java PathFinder (Symbolic JPF): it is a symbolic execution framework that implements a nonstandard bytecode interpreter on top of the Java PathFinder model checking tool [39]. It allows to perform automated generation of test cases and to check properties of code during test case generation. The framework performs a nonstandard bytecode interpretation and uses JPF to systematically generate and execute the symbolic execution tree of the code under analysis. It has been used for testing a prototype NASA flight software component helping discovering a serious bug.

- CUTE and jCUTE: CUTE (a concolic unit testing engine) and jCUTE (CUTE for Java) handle multithreaded programs that manipulate dynamic data structures using pointer operations [40]. They combine concolic execution with dynamic partial order reduction to systematically generate both test inputs and thread schedules. Both tools have been applied to test several Java and C open-source software.
- KLEE: it is a symbolic execution tool, capable of automatically generating tests that achieve high coverage on a diverse set of complex and environmentally intensive programs [41]. An important feature of KLEE is its ability to handle interactions with real-world data read from the file system or over the network. Moreover, it is applied to a variety of areas, including wireless sensor networks, automated debugging, testing of binary device drivers, and online gaming.

3.5 Nonfunctional Testing

An important aspect for software testing is the possibility of determine and evaluate properties of the product such as determine how fast the product responds to a request or how long it takes to do an action. Nonfunctional testing is performed at all test levels and is focused on software attributes such performance, security, scalability, and so on. Nowadays, there exist more than 150 types of nonfunctional testing. Here below, the most common nonfunctional testing types are briefly schematized:

- Performance testing: it is specifically aimed at verifying that the system meets the specified performance requirements. It usually targets validation of response times, throughput, resource-utilization levels, and other specific performance characteristics of the software under test. However, the term performance testing is a general one that many time is associated to various attributes or characteristics of nonfunctional testing. We refer to [42] for a specific definition of this term.
- Load testing: it focuses on assessing the behavior of a system under pressure in order to detect load-related problems where the term load refers to the rate at which different service requests are submitted to the system

under test (SUT) [43]. Possible related problems can be either functional problems that appear only under load (e.g., deadlocks, racing, buffer overflows, and memory leaks) or violations in nonfunctional quality-related requirements under load (e.g., reliability, stability, and robustness). We refer to [43] for more details about the most recent proposals and tools.

- Stress testing: it is the process of putting a system far beyond its capabilities to verify its robustness and/or to detect various load-related problems. Contrary to load testing in which the maximum allowable load is generated, in stress testing, the load generated is more than what the system is expected to handle [43].

- Volume testing: it is the process in which internal program or system limitations, such for instance storage requirements, are tried. It involves the ability of the systems to exchange data and information [42].

- Failover testing: it validates a system's ability to be able to allocate extra resource when it encounters heavy load or unexpected failure so to continue normal operations [42].

- Security testing: it assesses whether security properties related to confidentiality, integrity, availability, authentication, authorization, and non-repudiation are correctly implemented. Security testing demonstrates either the conformance with the security properties or addresses known vulnerabilities by means of malicious, nonexpected input data set [44]. The emerging security testing techniques integrate on different well-known approaches such as model-based testing, code-based testing and static analysis, penetration testing, and dynamic analysis as well as regression testing. We refer to [44] for a complete overview of the most recent testing techniques and tools.

- Reliability testing: testing is used as a means to improve reliability; in such a case, the test cases must be randomly generated according to the operational profile, i.e., they should sample more densely the most frequently used functionalities [45].

- Compatibility testing: it is the process for verifying whether the software can collaborate with different hardware and software facilities or with different versions or releases of the same hardware or software [42].

- Usability testing: it evaluates the ease of using and learning the system and the user documentation, as well as the effectiveness of system functioning in supporting user tasks, and, finally, the ability to recover from user errors. This testing is particularly important when testing GUI [46].

- Scalability testing: it is focused on verifying the ability of the software to increase and scale up on any of its nonfunctionality requirements such as load, number of transactions, and volume of data [45].

4. TEST CHALLENGES

Beyond decades of research activities, techniques, and actors in software testing, there are still many aspects that remain more challenging due to the complexity, pervasiveness, and criticality of continual software development. In this section, we overview several outstanding research challenges that need to be addressed to advance the state of the art in software testing and represent the directions to be followed for coping with the rapid advances of the software market.

4.1 Oracle Problem

An important component of testing is the oracle. Indeed, a test is meaningful only if it is possible to decide about its outcome. The difficulties inherent to this task, often oversimplified, had been early articulated in [47, 48]. Ideally, an oracle is any (human or mechanical) agent that decides whether the program behaved correctly on a given test. The oracle is specified to output a reject verdict if it observes a failure (or even an error, for smarter oracles), and approve otherwise. Not always the oracle can reach a decision; in these cases the test output is classified as inconclusive.

In a scenario in which a limited number of test cases are executed, the oracle can be the tester himself/herself, who can either inspect a posterior the test log or even decide a priori, during test planning, the conditions that make a test successful and code these conditions into the employed test driver [16].

Oracle can be derived by the analysis of textual (natural language) documentation describing the functionalities expected from the SUT to varying degrees. The partial and ambiguous nature of this documentation has often forced the manual oracle decisions. However, as overviewed in [48], some proposals try either to construct a formal specification exploiting user and developer documentations and source code comments or to restrict the natural language to a semiformal one enabling automatic processing.

When the tests cases are automatically derived or also when their number is quite high, in the order of thousands or millions, a manual log inspection or codification of the test results is not thinkable. Automated oracles must then be implemented.

In this case, a possible solution is to use pseudo-oracle [49], i.e., an alternative version of the program produced independently, e.g., by a different programming team or written in an entirely different programming language [48]. Alternatively, regression testing results can be used, i.e., the test outcome is compared with earlier version executions (which, however, in turn had to be judged passed or failed).

However, most of the available proposals for oracle definition are derived from a specification of the expected behavior [47]. According to [48] approaches based on formal specification can be classified into four categories:

- Model-based specification languages: the purpose is to exploit the models and a syntax that defines desired behavior in terms of its effects on the model so to derive the expected oracle. Different proposals discuss how to use the abstract specification to define high-level test oracles [30, 50]. However, the models or the documents describing the specification could be very abstract, quite far from concrete execution output, and consequently, oracle definition could result quite problematic;

- State transition systems: this kind of approaches focuses on the formal modeling of the system thought state transition systems. In particular, they focus on the reaction of a system to stimuli, i.e., *transitions*, and abstract a property of the states. A rigorous empirical evaluation of test oracle construction techniques using state transition systems is provided in [51];

- Assertions and contracts: if assertions could be embedded into the program so to provide run-time checking capability, conditions are instead expressly specified to be used as test oracles. As consequence, the produced execution traces could be logged and analyzed so to derive the oracle verdicts [52];

- Algebraic specifications: the purpose is to define equations over program operations that hold when the program is correct. The software is represented in terms of equational axioms that specify the required properties of the operations. Typically, these languages employ first-order logic to prove properties of the specification, like the correctness of refinements. Abstract data types (ADT), which combine data and operations over that data, are well suited to algebraic specification [47].

The formal specifications have the advantage that they can be used both for test case derivation and oracle specification as well. However, the gap between the abstract level of specifications and the concrete level of executed tests only allows for partial oracles implementations, i.e., only necessary (but not sufficient) conditions for correctness can be derived [47].

In view of these considerations, it should be evident that the *oracle* might not always judge correctly. So the notion of coverage of an oracle is introduced to measure its accuracy. It could be measured for instance by the probability that the oracle rejects a test (on an input chosen at random from a given probability distribution of inputs), given that it should reject it [48], whereby a perfect oracle exhibits a 100% coverage, while a less than perfect oracle may yield different measures of accuracy.

4.2 Full Automation

An important challenge of decades of research in software testing has been to improve the degree of attainable automation, either by developing advanced techniques for generating the test inputs or, beyond test generation, by finding innovative support procedures to automate the testing process [53]. Due to this, the market for tester support tools is in a huge expansion and recent forecasts show that it shall grow up to $34 billion worldwide by 2017, which opens relevant business opportunities for new innovative testing platforms.

Test automation tools promise to increase the number of tests they run and the frequency at which they run them by many orders of magnitude. Test automation can be incredibly effective, giving more coverage and new visibility into the software under test. However, 100% automatic testing remains still a dream [53]. Applicability of test automation is still limited and its adaptation to testing contains practical difficulties in usability.

Practitioners frequently report disastrous failures in the attempt to reduce costs by automating software tests, particularly at the level of system testing [54]. This is due to a gross underestimation of the money and time necessary to automate tests. Automating tests takes time due both to selection, building, installation, and integration of the testing tools and to planning and implementing automated tests. Often this effort is equivalent to that of manual tests. Automated tests may generate a lot of results that can take much more staff involvement and costs for analysis and isolation of discovered faults than manual tests [55].

The trade-off between automated and manual testing is discussed in literature. The authors of [56] present a cost model that uses the concept of opportunity cost to balance benefits, objectives, and risks mitigation of automated and manual testing. The work in [57] presents an extensive review of experiences of test automation detailing the used tools, the application domain, the life cycle development process, the project dimension, as well as the successful or failure results. The challenge would be the full development of a powerful integrated test environment which by itself can automatically take care of possibly instrumenting the deployed software deriving drivers and stubs as well as the most suitable test cases, executing them and finally handling the obtained results providing a test report.

An attempt in this direction is represented by *perpetual testing*, also known as *continuous testing* [58], which uses machine resources to continuously run tests in the background, providing rapid feedback about test failures as source code is edited, reducing in this way wasted time by 92%–98%.

The principle of continuous testing is today a key feature of DevOps which extends agile principles to entire software delivery pipeline [59]. The main idea is to move the testing process to early in software life cycle but also allows the tests to be carried out on production-like system executing the test suite on every software build generated automatically without any user intervention.

Finally, existing techniques aim to reduce automation costs by automating test automation. The main approach consists into a sequence of natural language test steps enabling a sequence of procedure calls with accompanying parameters that can drive testing without human intervention. This technique has been proven effective in reducing the cost of test automation by automating over 82% of the steps contained in a test suite [60].

4.3 Scalability of Testing

The complexity of software systems is not only caused by their functionalities, but it is also due to the complexity of the platform and environment in which they run. Mobile devices, distributed wireless networked and virtualized environment, large-scale clusters, and mobile clouds are just some examples. Additionally, in settings like software as a service (SaaS) [61], big data applications [62], web of systems [63], and cyber-physical systems [64], there is from one hand, the need of computation able to scale up as well as scale out; and from the other hand, an increasing demand of automated testing of these large-scale software systems, as stated in Section 4.2.

Recent studies focus on scalability of mutation testing which requires much time and computational resources to execute the overall test suite against all mutants. An empirical study of scalability of selective mutation testing is proposed in [65]. The authors of [65] show how the program size and the total number of nonequivalent mutants affect the effectiveness of selective mutation testing based on the subsets of selected mutants. In particular, as either the program size or the total number of nonequivalent mutants increases, the number of selected mutants increases slowly, while the proportion of selected mutants decreases, evidencing a good scalability of selective mutation testing.

4.4 Test Effectiveness

Evaluating the program under test, measuring the efficacy of a testing technique, or judging whether testing is sufficient and can be stopped are important aspects of the software testing, which in turn would require having measures of the effectiveness. The following sections provide details about the different aspects of effectiveness.

4.4.1 Measuring the Software

For evaluating the program under test different measurements and approaches can be applied as reported in [66]:

- Linguistic measures: these are based on proprieties of the program or of the specification text. This category includes for instance the measurement of: sources lines of code (LOC), statements, number of unique operands or operators, and function points;
- Structural measures: these are based on structural relations between objects in the program and comprise control flow or data flow complexity. These can include measurements relative to the structuring of program modules, e.g., in terms of the frequency with which modules call each other;
- Hybrid measures: these may result from the combination of structural and linguistic properties;
- Fault density: this is a widely used measure in industrial contexts and foresees the counting of the discovered faults and their classification by their type. For each fault class, fault density is measured by the ratio between the number of faults found and the size of the program.

4.4.2 Measuring the Testing Technique

For evaluating the testing approach, sometimes, coverage/thoroughness measures can be adopted. In this case, adequacy criteria evaluate the testing approach through the percentage of exercised set of elements identified in the program or in the specification by testing. However, the widespread adopted methods focus on the evaluation of the fault detection effectiveness of the considered testing technique. With different degrees of formalization, these proposals measure how much the test cases are able to revealing categories of likely or predefined faults. Among them, mutation analysis [67] is the standard technique to assess the effectiveness of a testing approach. In mutation testing, a mutant is a slightly modified version of the program under test, differing from it by a small, syntactic change. Every test case exercises both the original and all the generated mutants: if a test case is successful in identifying the differences between the program and a mutant, the latter is said to be killed. The underlying assumption of mutation testing is that, by looking for simple syntactic faults, more complex, but real, faults will be found. For the technique to be effective, a high number of mutants must be automatically derived in a systematic way.

Mutation testing can be adopted at any level of testing (unit, integration level, system level) and can be applied both for white-box, black-box, or gray-box testing. In literature, there are plenty of proposals and tools for many programming languages. We refer to [1, 67] for an overview of the most recent proposals about Fortran programs, Ada programs, C programs, Java programs, C# programs, and AspectJ programs. Mutation testing has also been used for interfaces testing [68] or more in general during the design level [67, 69] for mutating finite state machines, statecharts, petri nets, and network protocols. Recently mutation testing has been applied to web services [1] or to security policies [70, 71].

4.4.3 Test Cases Selection and Prioritization

In software testing, to improve fault detection rate at a given test execution time, common solutions rely on the application of proper strategies for test cases selection or prioritization. Test case selection aims to reduce the cardinality of the test suites while keeping the same effectiveness in terms of coverage or fault detection rate; test case prioritization aims at defining a test execution order according to some criteria (e.g., coverage and fault detection rate), so that those tests that have a higher priority are executed before the ones having a lower priority.

Many proposals address test cases selection for regression systems, aiming to speed up testing with a focus on detecting change-related faults [72]. The main idea is to select and run from the full test suite those tests that are relevant to the changes made to different software revisions, guaranteeing that the outcome of the tests that are not selected will not be affected by the changes. The effectiveness of regression test selection is measured by the ratio of the number of tests selected to run over the total number of tests [72].

Some recent proposals try to automate test cases selection techniques using information retrieval [73], others exploit the tracking of dynamic dependencies of tests on Java and JUnit files without require integration with version-control systems [74].

Test case prioritization techniques [75] schedule test cases in order to increase:

- the rate of fault detection, namely, the likelihood of revealing faults earlier in a run of regression tests;
- the likelihood of revealing regression errors related to specific code changes earlier in the regression testing process [76];
- the coverage of code under test;
- the confidence in the reliability of the system under test.

In practice, the test case prioritization problem may be intractable, so test case prioritization techniques rely typically on heuristics. A review of such prioritization techniques is presented in [77]. In [78], several test prioritization techniques are used to increase the fault detection rate of test suites. More recent results [79] still confirm the effectiveness of test case prioritization based on fault detection rate and show the flexibility of the approach for application in different contexts. However, as demonstrated in [80], no prioritization metric is the best one for any system: indeed, the performance of the prioritization approach varies according to the considered application and could depend on the evaluated test suites [81]. Another proposal addresses time-constrained test prioritization in the context of integer programming [82]. An approach that is currently considered very promising is based on the notion of test similarity [83]: the intuition behind similarity-based prioritization is that when resources are limited and only a subset of test cases within a large test suite can be executed, then it is convenient to start from those that are the most dissimilar according to a predefined distance function. In the context of testing of access control systems, some prioritization criteria based on similarity of access control requests are presented in [81].

5. TESTING PROCESS

During recent years, software testing has increased its role in the process of development. It is no longer focused on the defects detection after code completion, but it is now an integrated and significant activity performed during the whole software life cycle. Its critical nature and the importance for the overall quality of the final products led adopting the good practice of starting its management at the early stages of software development during the requirements analysis and proceeding with its organization systematically and continuously during the entire development process up to the code level.

Without referring to any specific software development process, usually the basic development phases can be summarized as in the following [1]:

- Requirements analysis: services, constraints, goals, and features of the overall system are established and organized. The list of requirements specifies the guidelines that the project must adhere to. At the end of this phase a set of documented, actionable, measurable, testable, and traceable requirements should be defined to a level of detail sufficient for system design;
- System and software design: this phase targets both hardware and software requirements and establishes the overall system architecture. The representation of the system is usually provided in terms of functionalities easily transformable into one or more executable applications/programs;
- Implementation: the established design is implemented in order to satisfy the list of requirements identified in the requirement analysis and definition phase. The implemented units or subsystems are integrated into a complete system to be properly tested;
- Validation and testing: quality attributes as well as requirement assessment are calculated and validated. Different test techniques can be included in the process development, each one targeting a different aspect of the proposed system.

Considering in particular the validation and testing phase, even if its management can depend strictly on the development process adopted for delivering the software products, the main phases can be resumed in [1]:

- Planning: as for any other process activity, the testing must be planned and scheduled. Thus, the time and effort needed for performing and completing software testing must be established in advance during the early stages of development. This also includes the specification of the personnel involved, the tasks they must to perform, and the facilities and equipment they may use;
- Test cases generation: according to the test plan constraints a set(s) of the test cases must be generated by using a (several) test strategy(ies);
- Test cases execution: the test cases execution may involve testing engineers, outside personnel or even customers. It is important to document every action performed in order to allow the experiments' duplication and meaningful and truthful evaluations of the results obtained;
- Test results analysis: the collected testing results must be evaluated to determine whether the test was successful (the system performs as

expected, or there are no major unexpected outcomes) and used for deriving measures and values of interest;
- Problem reporting: a test log documents the testing activity performed. This should contain, for example, the date in which a test was conducted, the data of the people who performed the test, the information about the system configuration and any other relevant data. Anomalies or unexpected behaviors should be also reported;
- Postclosure activities: the information relative to failures or defects discovered during testing execution are used for evaluating the performance and the effectiveness of the developed testing strategy(ies) and determining whether the process development adopted needs some improvements.

In the remaining of this section some of the most important development processes are briefly introduced and the role of testing inside them discussed.

5.1 Sequential Models

One of the first models to be designed was the waterfall model (W–model), also known as the cascade model, where the software development life cycle was constructed from a sequential set of stages [1]. It starts with a requirements analysis at the top level and finishes with the testing activity, so that defects are discovered close to the releasing, with a deep impact on the overall development costs. Usually, the waterfall model is the recommended framework for creating software which provides backend functionality, meaning a software whose main scope is to provide a service for other applications.

The V-model is the successive variation of the W–model, where a V-shape folded at the coding level is included. It was important to demonstrate that testing activities should be planned and designed as early as possible in the life cycle. Sometimes the V-model can be used in large projects which can include several subsystems and third-party systems.

The most common alternatives of the sequential approaches are evolutionary ones which are based on iterations. In this case, the system is not defined in advance but evolves through the evaluation of the achieved intermediate results.

5.2 Iterative Models

During the last 10 years, part of software research has been dedicated to the improvement of the development process (SPI initiatives). In this context, the CMM model (Capability Maturity Model) [5], developed at the Software Engineering Institute (SEI) is a de facto reference used by thousands of organizations together with the SPICE framework [6]. We report below a brief description of the commonly used process assessment models referring to [6] for further details.

As summarized in [5], the CMM model is a framework that describes the elements required for an effective software process. In particular, it focuses on an evolutionary improvement path from an ad hoc, immature process to a mature, disciplined process. It presents sets of recommended practices in a number of key process areas that have been shown to enhance software development and maintenance capability. The CMM guides developers in gaining control of their development and maintenance processes, and evolving toward a culture of software engineering and management excellence. CMM is not a framework that advocates magical and revolutionary new ideas, but it is in fact a tailored compilation of the best practices in software testing. Specifically, it improves quality of software delivered; it increases the customer satisfaction; it helps in achieving targeted cost savings; it also ensures stability and consistent high performance.

Other methods for managing the program improvement are the IDEAL framework [84] defined at the SEI, and the Rational Unified Process (RUP) [7].

The IDEAL method is an integrated approach for SPI defined by the SEI which identifies five phases: (i) initiating, which specifies the business goals and objectives that will be realized or supported; (ii) diagnosing, which identifies the organization's current state with respect to a related standard or reference model; (iii) establishing, which develops plans to implement the chosen approach; (iv) acting, which brings together everything available to create a best guess solution specific to organizational needs and puts the solution in place; and (v) leveraging, which summarizes lessons learned regarding processes used to implement IDEAL.

The RUP, which is a detailed refinement of the Unified Process (UP) [7], presents itself as a web-enabled software engineering process useful for: improving team productivity, delivering of software best practices to all team members, guiding the user in applying UML during the process development, and providing an extensive set of guidelines, templates, and examples. It is in particular a customizable framework, adaptable to the different organization exigencies, supported by tools (tightly integrated with Rational tools) which automates a large part of the process development. A central role of this process is represented by the RUP Best Practices, which are mainly guidelines for a well-established process development. RUP identifies six best practices which are develop software iteratively, manage requirements, use component-based architectures, visually model software, verify software quality, and control changes to software.

The RUP structure is characterized by: a static structure that describes the process (who is doing what, how, and when); dynamic structure that details how the process rolls out over time; an architecture-centric process that defines and details the architecture; a use-case driven process which specifies how use cases are used throughout the development cycle.

The RUP encourages testing early by offering a number of mechanisms to integrate testing more closely with the software development effort. In particular, RUP targets to: (i) making test a distinct discipline; (ii) using an iterative development approach; (iii) continuously verifying quality and letting use cases drive development; (iv) scheduling implementation based on risk; (v) managing changes strategically; and (vi) using the right-sized process.

5.3 Agile Development Process

The term *agile software development* was created by the Agile Manifesto [85]. Agile software development is basically an iterative approach that focuses on incremental specification, design, and implementation, while requiring full integration of testing and development. Agile development process has been originated by a precedent development practice, the rapid application development (RAD) methodology [86]. RAD uses a large number of iterations, with every iteration being a complete development cycle. At the end of each iteration, a complete executable product is released. The iteration product is a subset of the complete desired product, and it will be increased from iteration to iteration until the final product is released. Natural evolution of RAD is therefore the nowadays known agile software development approaches.

In the agile environment, testing is a frequent activity as small amounts of code are tested immediately upon being written. According to the Manifesto for agile software development, agile main points are (i) individuals and interactions over processes and tools; (ii) working software over comprehensive documentation; (iii) customer collaboration over contract negotiation; and (iv) responding to change over following a plan.

The intent is to produce high quality software in a cost-effective and timely manner, and in the meantime, meet the changing needs of end users.

In an agile environment, testing is often included within development in the form of test driven development (TDD) [85], which is a programming technique that promotes code development by repeating short cycles. It combines test-first development with refactoring and methods involving four steps: (i) write a test for an unimplemented functionality or behavior; (ii) supply the minimal amount of code to pass the test; (iii) refactor the code; (iv) check that all tests are still passing after the changes were done.

By developing only enough code to pass the tests and then using the refactoring as an improvement method for the design quality, TDD leads to better code quality and improves the confidence in the code as well as increasing the productivity.

Recent applications of agile software development are the XP (eXtreme Programming) and Scrum [87] development methods. In eXtreme Programming (XP), after a short planning stage, development goes through analysis, design, and implementation stages quickly (from 1 to 4 weeks). XP success relies on skilled and well-prepared software developers that are able to improve development quality and productivity.

The Scrum is more focused on the delivering of objected-oriented software [87]. Origin of term Scrum came from the popular sport rugby, because rugby strategies have been used the first time to describe its hyper productive development processes that are (i) a holistic team approach; (ii) constant interaction among team member; and (iii) unchanging core team members.

Basic adjectives characterizing Scrum implementations are transparency (visibility), inspection, and adaptation. Transparency or visibility means that any aspect of the process that affects the outcome must be visible and known to everybody involved in the process. Inspection requires that various aspects of the process must be inspected frequently enough so that unacceptable variances in the process can be detected. Adaptation requires that the inspector should adjust the process if one or more aspects of the process are in an unacceptable range.

6. DOMAIN-SPECIFIC TESTING

Recent years have witnessed the emergence of domain-specific approaches in software testing. These approaches are tailored to the specific needs of the domain and leverage domain knowledge to adapt and customize well-known testing solutions. Specifically, in the last decades software development is driven by emerging trends such as the widespread diffusion of mobile technology, cloud infrastructures adoption, as well as big data analysis and software as a service paradigm that point out new constraints and challenges for the testing activity. In the following of this section, we give a brief overview of different aspects and solutions for software testing applied into several domain-specific environments.

6.1 Cloud-Based Testing

Cloud computing is henceforth an accepted alternative for deploying applications and services. Businesses desire to achieve higher level operational performance and flexibility while keeping the development and deployment cost as lower as possible. Meanwhile, cloud computing has an high impact in software testing with the diffusion of a new testing paradigm known as testing as a service (TaaS). TaaS in a cloud infrastructure is a new business and service model, which provides static or dynamic on-demand testing services in the cloud and delivers them as a service to customers.

The main advantages of cloud-based testing deal with [88, 89]:

- Reducing the testing costs and time. Cloud computing offers unlimited storage as well as virtualized resources and shared cloud infrastructures that can help to eliminate required computer resources and licensed software costs as well as to reduce the execution time of large test suites in a cost-effective manner. This allows large IT companies to support many production lines which require diverse computing resources and test tools;
- Performing on-demand test services to conduct large-scale performance and scalability online validation;
- Performing testing of dynamic, complex, distributed applications such as mobile and web applications, leveraging multiple operating systems and updates, multiple browser platforms and versions, different types of hardware, and a large number of concurrent users.

The authors of [89] provide a systematic survey of cloud-based testing techniques including model-based testing, performance testing, symbolic execution, fault injection testing, random testing, privacy aware testing, and others.

There are three main different forms of testing as service (TaaS) in a cloud environment. Each of them has different focuses and objectives [88, 90]:

- Testing on clouds. In this form, software applications are deployed and executed on a cloud, and validated using the provided test services given by TaaS vendors. The major objective here is to take the advantage of large-scale test simulations and elastic computing resources on a cloud;

- Testing over clouds. In this form, software applications are deployed and validated based on different clouds (such as private clouds, public, or hybrid clouds). A typical software system crossing multiple clouds is structured with components, service software, and servers deployed crossing over several clouds;
- Testing of a cloud. It validates the quality of a cloud infrastructure according to the specified capabilities and service requirements.

Besides the many advantages in using cloud-based systems, there are yet many realistic problems of cloud service testing that need to be solved [88, 90]:

- On-demand test environment setup. There is a lack of supporting solutions to assist engineers to setup a required test environment in a cloud using a cost-effective way. To overcome these limitations, solutions aiming to empower the cloud applications with self-configuration, self-healing, and self-protection capabilities are provided;
- The heterogeneity and lack of standards in test tools and their connectivity and interoperability to support testing services;
- Assuring and assessing user privacy and security of cloud-based applications inside a third-party cloud infrastructure.

Finally, the authors of [91] provide a survey of typical tools for cloud testing. The main requirements for these tools include multilayer testing, SLA-based testing, large-scale simulation, and on-demand test environment.

6.2 SOA Testing

As described in Section 3.3, traditional testing approaches are divided into three major classes: black-box, white-box, and gray-box testing. Considering the specific domain of SOA testing while black-box approaches keep the same target of traditional ones, white-box and gray-box proposals assume a slightly different meaning. Indeed, in SOA white-box testing two different points of view can be identified: coverage measured at the level of a service composition (orchestration or choreography) and coverage of a single service internals. Generally, validation of service orchestrations is based on the Business Process Execution Language description[a] considered as an extended control flow diagram. Classical techniques of white-box coverage (e.g., control flow or dataflow) can be used to guide test generation or to assess test coverage so as to take into consideration the peculiarities of the Business Process Execution Language. Other proposals are instead based on formal specification of the workflows, e.g., Petri nets and finite state

---

[a] WSBPEL available at: https://www.oasis-open.org/committees/tc_home.php?wgabbrev=wsbpel.

processes, used for verifying specific service properties [92]. Considering a service choreography, existing research focuses, among others, on service modeling, process flow modeling, violation detection of properties such as atomicity and resource constraints, and XML-based test derivation.

If, on the one side, there are several approaches for structural testing of service compositions, there are few proposals for deriving structural coverage measures of the invoked services. The reason for this is that independent web services usually provide just an interface, enough to invoke them and develop some general (black-box) tests, but insufficient for a tester to develop an adequate understanding of the integration quality between the application and independent web services.

In the rest of this section, a survey of some proposed approaches and tools for supporting SOA testing is presented. Additionally, since black, white, and gray-box testing, albeit successfully executed, do not prevent security weaknesses, an overview of this important aspect is presented in Section 6.2.4.

6.2.1 SOA Black-Box Testing

As a common guideline, SOA black-box testing relies on the functionality provided by the web services. Commonly, the use of the (formal) specification or the XML Schema datatype[b] available allows for the generation of test cases for boundary-value analysis, equivalence class testing or random testing [92]. The derived test suite could have different purposes, such as to prove the conformance to a user-provided specification, to show the fault detection ability assessed on fault models, or to verify the interoperability by exploiting the invocation syntax defined in an associated WSDL (WS Description Language) document.[c] In this last case the formalized WSDL description of service operations and of their input and output parameters can be taken as a reference for black-box testing at the service interface [93, 94]. More details about SOA black-box testing are provided in [92, 95]. Despite the different proposals, test automation is still an open issue in this context [96]. In the specific case of web services, the use of XML-based syntax of WSDL documents could support fully automated WS test generation by means of traditional syntax-based testing approaches.

---

[b] XML Schema description: https://www.w3.org/XML/Schema#dev.
[c] WSDL description: https://www.w3.org/TR/wsdl20-primer/.

6.2.2 SOA White-Box Testing

Most often white-box testing of SOA applications is devoted to the validation of web service compositions (WSC). Validation of WSCs can be usually addressed performing structural coverage testing of a WSC specification. As overviewed in [92, 95] the WSC specification can be abstracted as an extended control flow diagram or transformed into formal specification so to apply structural coverage criteria (e.g., transition coverage). Similar complexity explosion problems may be encountered in such methods, since the amount of states and transitions of the target model can be very high. Alternative proposals adopt either model checking techniques for conformance testing by generating test cases from counterexamples, or use the formal model of the WSC and the associated properties to verify properties such as reachability. However, the complexity of the involved models and model checking algorithms in some cases could make these approaches hardly applicable to real-world WSCs. In [97, 98] an overview of the SOA testing specifically based on data-related models is provided.

6.2.3 SOA Gray-Box Testing

In the context of service-oriented architectures (SOAs), where independent web services can be composed with other services to provide richer functionality, interoperability testing becomes a major challenge. Independent web services usually provide just an interface, enough to invoke them and develop some general functional (black-box) tests, but insufficient for a tester to develop an adequate understanding of the integration quality between the application and the independent web services. To address this lack, gray-box proposals, trying to "whitening" the SOA testing, can be considered [99, 100]. In such gray-box proposals, the main idea is to try to derive data during the service execution so that produced test traces can be collected and analyzed against the specification paths. Many times, the collection of service execution traces is associated with the possibility to introducing an "Observer" stakeholder into the ESOA (extended SOA) framework. In ESOA, the goal is either to monitor code coverage [99] or to monitor ESOA services (passive testing) against a state model [101].

6.2.4 SOA Security Testing

Security aspects are highly critical in designing and developing web services. It is possible to distinguish at least two kinds of strategies for addressing protective measures of the communication among web services: security at the transport level and security at the message level. Enforcing the security at the transport level means that the authenticity, integrity, and confidentiality of the message (e.g., the SOAP message) are completely delegated to the lower-level protocols that transport the message itself from the sender to the receiver. Such protocols use public key techniques to authenticate both the end points and agree to a symmetric key, which is then used to encrypt packets over the (transport) connection. Since SOAP messages may carry vital business information, their integrity and confidentiality need to be preserved, and exchanging SOAP messages in a meaningful and secured manner remains a challenging part of system integration. Unfortunately, those messages are prone to attacks based on an on-the-fly modification of SOAP messages (XML rewriting attacks or XML injection) that can lead to several consequences such as unauthorized access, disclosure of information, or identity theft. Message-level security within SOAP and web services is addressed in many standards such as WS-Security,[d] which provides mechanisms to ensure end-to-end security and allows to protect some sensitive parts of a SOAP message by means of XML Encryption and XML Signature [44].

The activity of fault detection is an important aspect of (web) service security. Indeed, most breaches are caused when a system component is used in an unexpected manner. Improperly tested code, executed in a way that the developer did not intend, is often the primary culprit for security vulnerability. Robustness and other related attributes of web services can be assessed through the testing phase and are designed first off by analyzing WSDL document to know what faults could affect the robustness quality attribute of web services, and secondly by using the fault-based testing techniques to detect such faults. Focusing in particular on testing aspects, the different strategies and approaches that have been developed over the years can be divided into passive or active mechanisms [44, 92].

Passive mechanisms consist of observing and analyzing messages that the component under test exchanges with its environment and are specially used either for fault management in networks or for checking whether a system respects its security policy. Active testing is based on the generation and the application of specific test cases in order to detect faults.

---

[d] https://www.oasis-open.org/committees/tc_home.php?wg_abbrev=wss.

All the techniques have the purpose of providing evidence in security aspects, i.e., that an application faces its requirements in the presence of hostile and malicious inputs. Like functional testing, security testing relies on what is assumed to be a correct behavior of the system, and on non-functional requirements. However, the complexity of (web) security testing is bigger than functional testing, and the variety of different aspects that should be taken into consideration during a testing phase implies the use of a variety of techniques and tools. An important role in software security is played by negative testing [44, 92], i.e., test executions attempting to show that the application does something that it is not supposed to do. Negative tests can discover significant failures, produce strategic information about the model adopted for test case derivation, and provide overall confidence in the quality and security level of the system. Other common adopted methodologies and techniques include [44, 92]

- Fuzz testing: it involves generating semivalid data and submitting them in defined input fields or parameters (files, network protocols, API calls, and other targets) in an attempt to break the program and find bugs. Semivalid data are correct enough to keep parsers from immediately dismissing them, but still invalid enough to cause problems. Fuzzing covers a significant portion of negative test cases without forcing the tester to deal with each specific test case for a given boundary condition. In the specific area of web service security, fuzzing inputs can often be generated by programmatically analyzing the WSDL or sample SOAP requests and making modifications to the structure and content of valid requests. File fuzzing strategy is also used for detecting XML data vulnerability;
- Injection: in general, web services can interact with a variety of systems and for this reason they must be resistant to injection attacks when external systems are accessed or invoked. The most prevalent injection vulnerabilities include SQL injection, command injection, LDAP injection, XPath injection, and code injection. Recently, SQL Injection attack has become a major threat to web applications. SQL injection occurs when a database is queried with an SQL statement which contains some user-influenced inputs that are outside the intended parameters range;
- Policy-based testing: an important aspect in the security of modern information management systems is the control of accesses. Data and resources must be protected against unauthorized, malicious or improper usage or modification. For this purpose, several standards have been

introduced that guarantee authentication and authorization, such for instance the eXtensible Access Control Markup Language (XACML),[e] and to rule the writing of the access control policies. Thus policy-based testing is the testing process to ensure the correctness of policy specifications and implementations. By observing the execution of a policy implementation with a test input (i.e., access request), the testers may identify faults in policy specifications or implementations, and validate whether the corresponding output (i.e., access decision) is as intended. Although policy testing mechanisms vary because there is no single standard way to specify or implement access control policies, in general, the main goals to conduct policy testing are to ensure the correctness of the policy specifications, and the conformance between the policy specifications and implementations. The recent approaches on XACML policy testing are divided in the following main categories: (i) fault models and mutation testing that are based on a fault model to describe simple faults in XACML policies; (ii) testing criteria that determine whether sufficient testing has been conducted and it can be stopped and measure the degree of adequacy or sufficiency of a test suite. They include structural coverage criteria and fault coverage criteria; and (iii) test generation proposals specifically focused on the access control policies [102, 103].

6.3 GUI-Based Testing

Currently, with the widespread use of graphical user interface (GUI), the GUI testing is receiving a lot of attention. However, the GUI development includes different activities that force to face the GUI testing from different perspectives such as test coverage, test case generation, and test oracle and regression testing. Among these, the test case generation is one of the most important and therefore the one that received most attention in literature. As stated in [104] some proposals for GUI testing can be based on the manual derivation of test cases, i.e., method calls to the instances of the classes under test. Assertions are inserted in the test cases to determine whether the classes/ methods are executed correctly. However, because manual coding of test cases can be tedious, capture/replay techniques can be adopted. These proposals capture sequences of events that testers perform manually on the GUI, define the test case as a sequence of input events and replay derived test cases automatically on the GUI.

---

[e] http://docs.oasis-open.org/xacml/3.0/xacml-3.0-core-spec-os-en.html.

Referring to the classification provided into [105] the available proposals can be grouped as in the following:

- Based on finite state machines: this kind of approach is the most adopted for the GUI testing. Usually the input and output sequences are used for generating hierarchal models based on them. FSM can be then exploited for different testing approaches such as coverage for all-paths and all-transitions. Peculiarities of such kind of testing are [106] (i) GUI has an enormous number of states and every state should be tested; (ii) the GUI input space is very large because of permutations of inputs and events which effect the GUI; (iii) the complex dependencies cannot be avoided in the GUI system; and (iv) the GUI system can cause external effects at any time;
- Goal-driven approaches: these proposals exploit the definition of predefined goals that can be used both as input and as output, by generating sequences of actions, which reach these goals. These sequences of actions are used as test cases for GUIs. This technique allows the analysis of user interactions using model checking, and the synthesis of user interactions to executable GUI applications;
- Based on abstractions: in these proposals, abstractions may be based on structural features of GUI applications, e.g., the enabledness of a button (enabled or disabled) using a Boolean value, or the current value of slider control using an integer value. Abstraction can be either manually inferred or automatically derived by static analysis of the code [107];
- Model-based approaches: all these techniques require the creation of a model of the software or its GUI, and algorithms to use the model to generate test cases. In some cases, the models are created manually; in others, they are derived in an automated manner. Also the test case generation for some techniques is manual, but for most is automated. In [104] an exhaustive overview of the most recent proposals is provided.

6.4 Mobile Testing

Rapid advances of mobile technology and wireless networking push the wide adoption of mobile applications in different critical domains, from banking to mobile cyber-physical systems, to patients monitoring. According to the recent studies, the market for cloud-based mobile applications will grow exponentially in the next years. To guarantee the reliability and security of such mobile applications, their testing is required on many different configurations of devices and operating systems. Approaches to make such testing automated, systematic, and measurable are needed. According to recent studies [108], the revenues from mobile application testing tools will reach $800 million by the end of 2017 due to the test automation demand for mobile applications. What makes testing mobile software more demanding than testing computer software is the great complexity of the environment. Specifically, the diversity of relevant devices, operating systems, and networks as well as the fast updates of mobile browsers and mobile platforms and technologies make the testing of mobile web applications very expensive.

Recent research addresses mobile testing as a service (MTaaS) [109] as a possible solution to reduce the complexity of mobile testing by leveraging a cloud-based scalable mobile testing environment to assure predefined quality of service requirements and service-level agreements. The authors of [109] discuss about issues and solutions of MTaaS proposing a mobile testing as a service infrastructure aiming at reducing high costs in current mobile testing practice and environments and supporting mobile scalability test. The work in [110] presents a set of existing cloud-based services and tools for mobile application testing whereas [111] proposes a framework, called automated mobile testing as a service (AM-TaaS), which offers automated test for mobile applications including emulation of mobile devices with different characteristics under virtual machines and cloud infrastructure. However, well-defined test models and processes that address the distinct needs of mobile testing are yet lacking. In particular, quality of mobile applications has different meaning and criticality for different users and in different contexts. Nowadays, the main challenge about mobile testing is improving its cost effectiveness by combining:

- Use of test automation: traditional testing of mobile applications, often done by manual execution of test cases and visual verification of the results, is an effort-intensive and time-consuming process. Automating the testing activities produces a great reduction of costs and effort in the whole mobile applications software development life cycle;
- Use of emulators and actual devices: emulators can be useful for testing features of the application that are device independent. However, due to diversity in mobile hardware, actual devices will be used for validating the results;
- Testing the mobile environment for application complexity: this is needed in order to address the varieties of mobile and wireless technologies, like Wi-Fi, WiMax, 3G, 4G, etc., as well as to validate performance and robustness of the mobile system in real-world network conditions;

- Testing on heterogeneous mobile platforms: the applications will be tested on different platforms for checking compatibility with different mobile operating systems;
- Large-scale on-demand mobile test services: they allow to implement testing techniques and strategies responding to on-demand requests of test generation, execution, and control.

6.5 Big Data Testing

Big data provides not only large amounts of data but also various data types such as images, videos, interactive maps, time stamps, and associated metadata. The main four characteristics of big data are volume, variety, velocity, and value. Data is continuously being generated from machines, sensors from Internet of Things, mobile devices, network data traffic, and application logs. This requires advances in data storage, mining and business intelligence technologies making it possible to preserve increasing amounts of data generated directly or indirectly by users and analyze it to yield valuable new insights [112].

Big data also present new challenges with respect to their maintenance, testing, and benchmarking. While big data systems have advanced their capabilities of data analysis, scalability, processing and fault tolerance, database testing, and benchmarking have not moved forward to provide data generators, data sets, and workloads. The authors of [113] provide a comparative analysis of techniques for big data testing, including genetic algorithms, clustering techniques, performance, and regression test approaches. In particular, the emerging challenges in the field of testing big data relate to the need of quickly generating huge, realistic and scalable data sets, as well as the need for well-defined workloads that capture the nature of novel, modern analysis tasks [114].

Experimental studies often reuse data sets from well-known standardized benchmarks for performance evaluation of database systems, like TPC-C [115], and XMLGen [116]. They typically provide open-source tools for data and workload generation, which can be easily adapted and used by third parties reducing the overall effort required to prepare and execute the experiment [114]. An alternative approach deals with the implementation of custom data generators for the comparison of approaches for large-scale data analysis.

The main limitations of existing approaches of benchmarking and testing of big data systems are about the lack of realistic data sets. Existing data sets are build on simplistic assumptions such as uniform distributions or over-simplified schema that often are not representative for real-world data. An attempt to overcome these limitations is to automatically extract the domain information from a ground truth data set, which is often available in practice. This information is then integrated into a data generator specification for a specific target environment and then used to create a concrete data generator instance that is able to mimic the original data set [114].

Different solutions to generate representative data sets in the big data context and to control the size of the test set, deal with input space partitioning testing associated to input domain model. The tester partitions the input domain model, selects test values from partitioned blocks, and applies combinatorial coverage criteria to generate tests. Parallel computing and Hadoop are used to speed up data generation [117].

Finally, the evaluation and testing of big data analytic systems requires more complex, realistic, and universally useful workload specifications that involve machine learning algorithms, information extraction, and graph analysis/mining.

6.6 Automotive Testing

In automotive domain, one of the most important aspects is security. Currently, existing solutions for security assessment of software systems basically perform either static analysis (i.e., the software is not executed) or dynamic analysis (performed by executing the software on specific inputs), although any life cycle will likely apply a combination of both. A 2009 survey carried out within the NIST SAMATE project[f] provides an analysis of more than 70 tools either specifically conceived for security assessment or more generally for software correctness related to security, and catalogs them under the following categories:

- Static analysis: aids analysts in locating security-related issues;
- Source code fault injection: the source code is instrumented by inserting changes and then executed to observe the changes in state and behavior that emerge;
- Dynamic analysis: refers generically to security testing approaches, such as coverage-based or profiling;
- Architectural analysis: aims at identifying flaws in the software architecture and determining resulting risks to information assets;

---

[f] http://samate.nist.gov.

- Pedigree analysis: identifies software coming from an external source (e.g., open-source software);
- Binary code analysis and disassembler analysis: both categories review binary code. Since source code is not needed, they can be applied to commercial off-the-shelf (COTS) components;
- Binary fault injection: focuses on likely faults in the real-time operation of the software (e.g., memory faults or other error conditions provided by the processor);
- Fuzzing: a form of negative testing, in which the software is exercised under random invalid data, generally specific to a particular type of input;
- Malicious code detectors: search for malicious logic embedded in programs;
- Bytecode analysis: bytecode contains more semantic information about the program execution than an equivalent binary. Bytecode analysis tools are actively investigated.

Many works try to transfer existing security assessment methods to the automotive domain, addressing specific risks and challenges of automotive software. Nowadays, several works [118] concur that security cannot be addressed separately from safety in the software engineering process of automotive systems, thus assessment approaches should converge and complement each other.

The widespread utilization of model-based design and production code generation in the automotive development process enables the application of automatic verification and validation techniques earlier in the software life cycle, and at an higher level of abstraction using for instance Simulink models [31].

However, many testing methodologies and tools are customized to address specific challenges of the automotive domain. In [119] a detailed survey of the most recent proposals is provided.

Other security testing activities are aimed at finding implementation errors that could be exploited by an outside attacker to cause potential functionality problems. Such activities are also used to establish to what extent the target system can resist an attack and generally consist of: (i) functional automotive security testing able to ensure general compliance with the specifications and standards for the security functionality implemented, encryption algorithms and authentication protocols of a vehicular IT system; (ii) vulnerability scanning applied to all relevant applications, source codes, networks, and backend infrastructures of an automotive system in order to test the system for common security vulnerabilities such as security loopholes or security configurations with known weaknesses taken from a continuously updated database of automotive security vulnerabilities; (iii) fuzzing that is used to expose the implementation to unexpected, invalid, or random input in the hope that the target will react in an unexpected way and, as a result, uncover new vulnerabilities; (iv) penetration tests applied in a final step to test security of the whole system by applying attacking methods conducted by trusted individuals to check whether ECUs are vulnerable and could allow unauthorized users access.

7. DISCUSSION AND CONCLUSIONS

This chapter presented a journey through the world of software testing and its emerging techniques, ranging over many fields from definition to organization, from its applicability and analysis of effectiveness, to the challenges and specific issues of some of the most important application domains. However, due to the vast and articulated research discipline, the covering into one chapter of all ongoing and foreseen research directions and emerging technologies is impossible. Indeed, this chapter proposes broadness against depth overview and it is an attempt to depict a comprehensive and extensible roadmap of the most important topics, challenges, and future directions of testing activity. Fig. 1 provides a graphical overview of the main chapter contents.

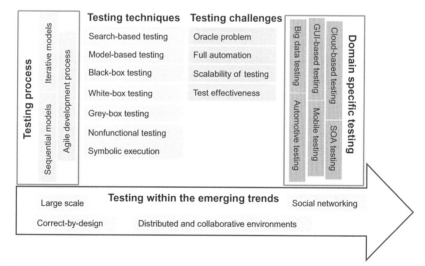

Fig. 1 Software testing roadmap.

This chapter would be a reference of the most recent testing techniques and presented an almost complete overview of the new methods, approaches, and tools useful to the reader for managing, controlling, and evaluating software testing development. It targets students, researchers, software developers, and practitioners looking for solutions for their problems and improvements in the different activities of the testing process. As emerged from this chapter, software testing is not an isolated activity; it has many fruitful relations between different areas of software engineering and many times it is the bridge between different disciplines.

Due to the increasing complexity of software systems and the variety of execution environments, software testing will continue to be a key activity in the software development process and quality assurance.

Over the years software testing has had to keep the path of the trends, innovations, and modifications provided by the new software development paradigms. In order to trigger always updated and practical solutions, software testing needs to learn from the state of practice and to elaborate the future vision useful to predict the problems, market exigencies, and quality issues that could rise. In particular, the growth of cloud computing brings the need of benchmarks and on-demand test environment construction to measure the performance and scalability metrics of new applications considering the special features of cloud such as dynamic scalability, scalable testing environments, SLA-based requirements, and cost-models. New testing methods will be developed to assess compatibility, interoperability, and multitenancy ability of cloud applications that must be able to work across multiple environments, various cloud platforms, client technologies, and browsers.

Possible suggestions for more in-depth and effective methodologies and approaches can be the use of realistic settings and large-scale systems as well as the simulation or involvement of real stakeholders working simultaneously in distributed environments. In particular, large-scale and software intensive IT systems are created by integrating and orchestrating independently controlled and managed systems. The most limiting factor of their development is software validation, which typically requires very costly and complex testing processes. Existing approaches for testing these systems focus on orchestrating the different phases of the testing process including test operation, test injection, monitoring and reporting, sometimes according to response time, bandwidth usage, throughput, and adaptability, with the aim of continuous integration and deployment. New testing techniques able to deal with the inherent complexity of these systems will be based on the *divide-and-conquer* principle, which is commonly used for architecting complex software. They will leverage a novel test orchestration theory and toolbox enabling the creation of complex test suites for large systems as the composition of simple testing units. These new solutions will allow the reusability of testing knowledge, architectures, and code, making testing activity more effective and less expensive. Moreover, these new testing techniques, based on the orchestration topology of simple test cases, will also target the important problem of automated partial oracle derivation of tests of large systems by inferring test oracle specification from the composition of expected and obtained outputs of the executed testing units.

An important role is given by the possibility of leveraging the software testing proposals from the specific programming language and execution environment so that high-level, application-independent attributes such as trust, security, and performance can be easily and automatically verified with a drastic reduction of testing time and effort.

Considering in particular the new collaborative and distributed software development processes such as Agile, Scum, and DevOp, the testing techniques should make easier the simultaneous verification and the automatic alignment of test activity with decomposition and distribution of the target systems. Additionally, due to the short development time, new proposals drastically decreasing learning time and addressing the multiple levels of IT personnel are becoming a real pressing so improve software testing and productivity.

Moreover, due to the increasing popularity of social networking services and their massive integration into the users' everyday life, another future target of testing activity is represented by the specific requirements of these services. Besides, performance and scalability validation of the decentralized online social networks, new testing techniques will be devoted to assess security and privacy of users' personal data in order to prevent local attacks. These techniques will specifically target assessment of privacy and sharing information policies as well validation of decentralized management of users' social profiles and data storage solutions.

Finally, the recent trend of adopting correct or secure-by-design development process forces testing activity to propose tools and methodology useful for testing and analyzing design level artifacts. In particular, simulation, model checking, and model-based approaches should be the key success for improving the quality of the developed systems.

REFERENCES

[1] P. Ammann, J. Offutt, Introduction to Software Testing, Cambridge University Press, 2016.

[2] E.W. Dijkstra, Notes on Structured Programming, Technische Hogeschool Eindhoven, Eindhoven, 1970.

[3] A. Bertolino, Software testing, in: A. Abran, P. Bourque, R. Dupuis, J.W. Moore (Eds.), Guide to the Software Engineering Body of Knowledge—SWEBOK, IEEE Press, Piscataway, NJ, 2001, p. 69, ISBN: 0769510000.

[4] R.H. Thayer, R. Hunter, Software Process Improvement, IEEE Computer Society, 2001.

[5] M. Paulk, Capability Maturity Model for software, in: Encyclopedia of Software Engineering, Wiley Online Library, 1993.

[6] K.E. Emam, W. Melo, J.-N. Drouin, SPICE: The Theory and Practice of Software Process Improvement and Capability Determination, IEEE Computer Society Press, 1997.

[7] P. Kruchten, The Rational Unified Process: An Introduction, Addison-Wesley Professional, 2004.

[8] A. Bertolino, E. Marchetti, A brief essay on software testing, in: R.H. Thayer, M.J. Christensen (Eds.), Software Engineering, The Development Process, vol. 3, Wiley-IEEE Computer Society Press, 2005.

[9] J.V. Guttag, J.J. Horning, Larch: Languages and Tools for Formal Specification, Springer Science & Business Media, 2012.

[10] C. Cadar, K. Sen, Symbolic execution for software testing: three decades later, Comm. ACM 56 (2) (2013) 82–90.

[11] M. Gligoric, S. Negara, O. Legunsen, D. Marinov, An empirical evaluation and comparison of manual and automated test selection, in: Proc. of the 29th ACM/IEEE Int. Conf. on Automated Software Engineering, ACM, 2014, pp. 361–372.

[12] S. Xanthakis, C. Ellis, C. Skourlas, A. Le Gall, S. Katsikas, K. Karapoulios, Application of genetic algorithms to software testing, in: Proceedings of the 5th International Conference on Software Engineering and Applications, 1992, pp. 625–636.

[13] W. Miller, D.L. Spooner, Automatic generation of floating-point test data, IEEE Trans. Softw. Eng. (3) (1976) 223–226.

[14] P. McMinn, Search-based software testing: past, present and future, in: 2011 IEEE Fourth International Conference on Software Testing, Verification and Validation Workshops (ICSTW), IEEE, 2011, pp. 153–163.

[15] M. Harman, L. Hu, R. Hierons, J. Wegener, H. Sthamer, A. Baresel, M. Roper, Testability transformation, IEEE Trans. Softw. Eng. 30 (1) (2004) 3–16.

[16] P. McMinn, M. Stevenson, M. Harman, Reducing qualitative human oracle costs associated with automatically generated test data, in: Proceedings of the First International Workshop on Software Test Output Validation, ACM, 2010, pp. 1–4.

[17] K. Lano, Model-Driven Software Development with UML and Java, Course Technology Press, 2009.

[18] M. Utting, A. Pretschner, B. Legeard, A taxonomy of model-based testing approaches, Soft. Test. Verification Reliab. 22 (5) (2012) 297–312.

[19] M. Utting, B. Legeard, Practical Model-Based Testing: A Tools Approach, Morgan Kaufmann, 2010.

[20] A.C. Dias-Neto, G.H. Travassos, A picture from the model-based testing area: concepts, techniques, and challenges, Adv. Comput. 80 (2010) 45–120, https://doi.org/10.1016/S0065-2458(10)80002-6. http://www.sciencedirect.com/science/article/pii/S0065245810800026.

[21] M.E. Khan, F. Khan, et al., A comparative study of white box, black box and grey box testing techniques, Int. J. Adv. Comput. Sci. Appl. 3 (6) (2012).

[22] A. Bhat, S.M.K. Quadri, Equivalence class partitioning and boundary value analysis–a review, in: 2015 2nd International Conference on Computing for Sustainable Global Development (INDIACom), 2015, pp. 1557–1562.

[23] G.J. Myers, C. Sandler, T. Badgett, The Art of Software Testing, John Wiley & Sons, 2011.

[24] A. Arcuri, L. Briand, Formal analysis of the probability of interaction fault detection using random testing, IEEE Trans. Softw. Eng. 38 (5) (2012) 1088–1099, ISSN 0098-5589, https://doi.org/10.1109/TSE.2011.85.

[25] M.R. Lyu, et al., Handbook of Software Reliability Engineering, IEEE Computer Society Press, California, 1996.

[26] C. Nie, H. Leung, A survey of combinatorial testing, ACM Comput. Surv. 43 (2) (2011) 11:1–11:29, ISSN 0360-0300, https://doi.org/10.1145/1883612.1883618.

[27] D. Lee, M. Yannakakis, Principles and methods of testing finite state machines—a survey, Proc. IEEE 84 (8) (1996) 1090–1123.

[28] N.A. Lynch, M.R. Tuttle, An Introduction to Input/Output Automata, Massachusetts Institute of Technology. Laboratory for Computer Science, 1988.

[29] J. Tretmans, R.M. Hierons, Model based testing with labelled transition systems, in: J.P. Bowen, M. Harman (Eds.), Formal Methods and Testing, Springer, 2008, pp. 1–38.

[30] E. Börger, R. Stärk, Abstract State Machines: A Method for High-level System Design and Analysis, Springer Science & Business Media, 2012.

[31] A. Tiwari, Formal Semantics and Analysis Methods for Simulink Stateflow Models, Technical report, Citeseer, 2002.

[32] C. Baier, J.-P. Katoen, K.G. Larsen, Principles of Model Checking, MIT Press, 2008.

[33] A. Kasoju, K. Petersen, M.V. Mäntylä, Analyzing an automotive testing process with evidence-based software engineering, Inf. Softw. Technol. 55 (7) (2013) 1237–1259, ISSN 0950-5849, https://doi.org/10.1016/j.infsof.2013.01.005. http://www.sciencedirect.com/science/article/pii/S0950584913000165.

[34] B.A. Kitchenham, T. Dyba, M. Jorgensen, Evidence-based software engineering, in: Proceedings. 26th International Conference on Software Engineering, 2004, pp. 273–281, https://doi.org/10.1109/ICSE.2004.1317449.

[35] R.E. Santos, C.V. de Magalhães, F.Q. da Silva, The use of systematic reviews in evidence based software engineering: a systematic mapping study, in: Proceedings of the 8th ACM/IEEE International Symposium on Empirical Software Engineering and Measurement, ACM, 2014, p. 53.

[36] C.S. Păsăreanu, W. Visser, A survey of new trends in symbolic execution for software testing and analysis, Int. J. Softw. Tools Technol. Transfer 11 (4) (2009) 339–353.

[37] R. Baldoni, E. Coppa, D.C. D'Elia, C. Demetrescu, I. Finocchi, A Survey of Symbolic Execution Techniques, 2016. arXiv preprint arXiv:1610.00502.

[38] K. Sen, Concolic testing, in: Proceedings of the Twenty-second IEEE/ACM International Conference on Automated Software Engineering, ASE '07, ACM, New York, NY, 2007, pp. 571–572, https://doi.org/10.1145/1321631.1321746. 978-1-59593-882-4.

[39] C.S. Pasareanu, P.C. Mehlitz, D.H. Bushnell, K. Gundy-Burlet, M. Lowry, S. Person, M. Pape, Combining unit-level symbolic execution and system-level concrete execution for testing NASA software, in: Proceedings of the 2008 International Symposium on Software Testing and Analysis, ACM, 2008, pp. 15–26.

[40] K. Sen, D. Marinov, G. Agha, CUTE: a concolic unit testing engine for C, SIGSOFT Softw. Eng. Notes 30 (5) (2005) 263–272, ISSN 0163-5948, https://doi.org/10.1145/1095430.1081750.

[41] C. Cadar, D. Dunbar, D. Engler, KLEE: unassisted and automatic generation of high-coverage tests for complex systems programs, in: Proc. of the 8th USENIX Conf. on Operating Systems Design and Implementation, OSDI'08, 2008, pp. 209–224, http://dl.acm.org/citation.cfm?id=1855741.1855756.

[42] W.E. Lewis, Software Testing and Continuous Quality Improvement, CRC Press, 2016.

[43] Z.M. Jiang, A.E. Hassan, A survey on load testing of large-scale software systems, IEEE Trans. Softw. Eng. 41 (11) (2015) 1091–1118, ISSN 0098-5589, https://doi.org/10.1109/TSE.2015.2445340.

[44] M. Felderer, M. Büchler, M. Johns, A.D. Brucker, R. Breu, A. Pretschner, Chapter One-Security Testing: A Survey, Adv. Comput. 101 (2016) 1–51.

[45] S.R. Dalal, M.R. Lyu, C.L. Mallows, Software Reliability, Wiley Online Library, 2014.

[46] E. Geisen, J.R. Bergstrom, Usability Testing for Survey Research, Morgan Kaufmann, 2017.

[47] M. Pezze, C. Zhang, Automated test oracles: a survey, Adv. Comput. 95 (2015) 1–48.

[48] E.T. Barr, M. Harman, P. McMinn, M. Shahbaz, S. Yoo, The oracle problem in software testing: a survey, IEEE Trans. Softw. Eng. 41 (5) (2015) 507–525.

[49] M.D. Davis, E.J. Weyuker, Pseudo-oracles for non-testable programs, in: Proceedings of the ACM'81 Conference, ACM, 1981, pp. 254–257.

[50] D.K. Peters, D.L. Parnas, Using test oracles generated from program documentation, IEEE Trans. Softw. Eng. 24 (3) (1998) 161–173.

[51] S. Mouchawrab, L.C. Briand, Y. Labiche, M. Di Penta, Assessing, comparing, and combining state machine-based testing and structural testing: a series of experiments, IEEE Trans. Softw. Eng. 37 (2) (2011) 161–187.

[52] P. Loyola, M. Staats, I.-Y. Ko, G. Rothermel, Dodona: automated Oracle data set selection, in: Proceedings of the 2014 International Symposium on Software Testing and Analysis, ISSTA 2014, ACM, New York, NY, 2014, pp. 193–203, https://doi.org/10.1145/2610384.2610408. 978-1-4503-2645-2.

[53] A. Bertolino, Software testing research: achievements, challenges, dreams, in: 2007 Future of Software Engineering, IEEE Computer Society, 2007, pp. 85–103.

[54] C. Persson, N. Yilmazturk, Establishment of automated regression testing at ABB: Industrial experience report on 'avoiding the pitfalls', in: Proceedings of the 19th IEEE International Conference on Automated Software Engineering, IEEE Computer Society, 2004, pp. 112–121.

[55] D. Hoffman, Cost benefits analysis of test automation, in: STAR West, vol. 99, 1999.

[56] R. Ramler, K. Wolfmaier, Economic perspectives in test automation: balancing automated and manual testing with opportunity cost, in: Proceedings of the 2006 International Workshop on Automation of Software Test, ACM, 2006, pp. 85–91.

[57] D. Graham, M. Fewster, Experiences of Test Automation: Case Studies of Software Test Automation, Addison-Wesley Professional, 2012.

[58] D. Saff, M.D. Ernst, Reducing wasted development time via continuous testing, in: 14th International Symposium on Software Reliability Engineering, IEEE, 2003, pp. 281–292.

[59] M. Virmani, Understanding DevOps & bridging the gap from continuous integration to continuous delivery, in: Fifth International Conference on Innovative Computing Technology (INTECH), IEEE, 2015, pp. 78–82.

[60] S. Thummalapenta, S. Sinha, N. Singhania, S. Chandra, Automating test automation, in: 2012 34th International Conference on Software Engineering (ICSE), 2012, pp. 881–891, https://doi.org/10.1109/ICSE.2012.6227131.

[61] P. Buxmann, T. Hess, S. Lehmann, Software as a service, Wirtschaftsinformatik 50 (6) (2008) 500–503.

[62] P. Zikopoulos, C. Eaton, et al., Understanding Big Data: Analytics for Enterprise Class Hadoop and Streaming Data, McGraw-Hill Osborne Media, 2011.

[63] T. Isakowitz, M. Bieber, F. Vitali, Web information systems, Commun. ACM 41 (7) (1998) 78–80.

[64] R. Baheti, H. Gill, Cyber-physical systems, in: The Impact of Control Technology, vol. 12, IEEE Control Systems Society, 2011, pp. 161–166.

[65] J. Zhang, M. Zhu, D. Hao, L. Zhang, An empirical study on the scalability of selective mutation testing, in: IEEE 25th International Symposium on Software Reliability Engineering (ISSRE), IEEE, 2014, pp. 277–287.

[66] N. Fenton, J. Bieman, Software Metrics: A Rigorous and Practical Approach, CRC Press, 2014.

[67] Y. Jia, M. Harman, An analysis and survey of the development of mutation testing, IEEE Trans. Softw. Eng. 37 (5) (2011) 649–678, ISSN 0098-5589, https://doi.org/10.1109/TSE.2010.62.

[68] M.E. Delamaro, J.C. Maidonado, A.P. Mathur, Interface mutation: an approach for integration testing, IEEE Trans. Softw. Eng. 27 (3) (2001) 228–247.

[69] F. Belli, C.J. Budnik, A. Hollmann, T. Tuglular, W.E. Wong, Model-based mutation testing: approach and case studies, Sci. Comput. Program. 120 (2016) 25–48.

[70] E. Martin, T. Xie, A fault model and mutation testing of access control policies, in: Proc. of WWW, 2007, pp. 667–676.

[71] A. Bertolino, S. Daoudagh, F. Lonetti, E. Marchetti, XAMUT: XACML 2.0 mutants generator, in: IEEE Sixth International Conference on Software Testing, Verification and Validation Workshops (ICSTW), IEEE, 2013, pp. 28–33.

[72] A. Shi, T. Yung, A. Gyori, D. Marinov, Comparing and combining test-suite reduction and regression test selection, in: Proceedings of the 2015 10th Joint Meeting on Foundations of Software Engineering, ACM, 2015, pp. 237–247.

[73] C. Magalhães, F. Barros, A. Mota, E. Maia, Automatic selection of test cases for regression testing, in: Proc. of the 1st Brazilian Symp. on Systematic and Automated Software Testing, ACM, 2016, p. 8.

[74] M. Gligoric, L. Eloussi, D. Marinov, Practical regression test selection with dynamic file dependencies, in: Proc. of the 2015 Int. Symp.on Software Testing and Analysis, ACM, 2015, pp. 211–222.

[75] P.R. Srivastava, Test case prioritization, J. Theoretical Appl. Inf. Technol. 4 (3) (2008) 178–181.

[76] R.K. Saha, L. Zhang, S. Khurshid, D.E. Perry, An information retrieval approach for regression test prioritization based on program changes, in: IEEE/ACM 37th IEEE International Conference on Software Engineering (ICSE), vol. 1, IEEE, 2015, pp. 268–279.

[77] G. Rothermel, R.H. Untch, C. Chu, M.J. Harrold, Test case prioritization: an empirical study, in: IEEE International Conference on Software Maintenance (ICSM '99), 1999, pp. 179–188, https://doi.org/10.1109/ICSM.1999.792604.

[78] G. Rothermel, R.H. Untch, C. Chu, M.J. Harrold, Prioritizing test cases for regression testing, IEEE Trans. Softw. Eng. 27 (10) (2001) 929–948.

[79] H. Do, G. Rothermel, On the use of mutation faults in empirical assessments of test case prioritization techniques, IEEE Trans. Softw. Eng. 32 (9) (2006) 733–752.

[80] S. Elbaum, G. Rothermel, S. Kanduri, A.G. Malishevsky, Selecting a cost-effective test case prioritization technique, Softw. Qual. J. 12 (3) (2004) 185–210.

[81] A. Bertolino, S. Daoudagh, D. El Kateb, C. Henard, Y. Le Traon, F. Lonetti, E. Marchetti, T. Mouelhi, M. Papadakis, Similarity testing for access control, Inf. Softw. Technol. 58 (2015) 355–372.

[82] L. Zhang, S.-S. Hou, C. Guo, T. Xie, H. Mei, Time-aware test-case prioritization using integer linear programming, in: Proceedings of the Eighteenth International Symposium on Software Testing and Analysis, ACM, 2009, pp. 213–224.

[83] K. Wu, C. Fang, Z. Chen, Z. Zhao, Test case prioritization incorporating ordered sequence of program elements, in: Proceedings of the 7th International Workshop on Automation of Software Test, IEEE Press, 2012, pp. 124–130.

[84] B. McFeeley, IDEAL: A User's Guide for Software Process Improvement, Technical report, DTIC Document, 1996.

[85] M. Fowler, J. Highsmith, The Agile Manifesto, Softw. Dev. 9 (8) (2001) 28–35.

[86] P. Beynon-Davies, C. Carne, H. Mackay, D. Tudhope, Rapid application development (RAD): an empirical review, Eur. J. Inf. Sys. 8 (3) (1999) 211–223.

[87] H. Kniberg, Scrum and XP from the Trenches, Lulu.com, 2015.

[88] J. Gao, X. Bai, W.-T. Tsai, Cloud testing-issues, challenges, needs and practice, Soft. Eng. Int. J. 1 (1) (2011) 9–23.

[89] I. Chana, A. Rana, et al., Empirical evaluation of cloud-based testing techniques: a systematic review, ACM SIGSOFT Soft. Eng. Notes 37 (3) (2012) 1–9.

[90] J. Gao, X. Bai, W.-T. Tsai, T. Uehara, Testing as a service (TaaS) on clouds, in: IEEE 7th Int. Symp. on Service Oriented System Engineering (SOSE), IEEE, 2013, pp. 212–223.

[91] X. Bai, M. Li, B. Chen, W.-T. Tsai, J. Gao, Cloud testing tools, in: IEEE 6th Int. Symp. on Service Oriented System Engineering (SOSE), 2011, pp. 1–12.

[92] C. Bartolini, A. Bertolino, F. Lonetti, E. Marchetti, Approaches to functional, structural and security SOA testing, in: Performance and Dependability in Service Computing: Concepts, Techniques and Research Directions, IGI Global, 2012, pp. 381–401.

[93] C. Bartolini, A. Bertolino, E. Marchetti, A. Polini, WS-TAXI: a WSDL-based testing tool for web services, in: International Conference on Software Testing Verification and Validation (ICST), IEEE, 2009, pp. 326–335.

[94] D. Petrova-Antonova, K. Kuncheva, S. Ilieva, Automatic generation of test data for XML schema-based testing of web services, in: 10th International Joint Conference on Software Technologies (ICSOFT), vol. 1, IEEE, 2015, pp. 1–8.

[95] M. Bozkurt, M. Harman, Y. Hassoun, Testing and verification in service-oriented architecture: a survey, Softw. Test. Verification Reliab. 23 (4) (2013) 261–313.

[96] L.M. Hillah, A.-P. Maesano, L. Maesano, F. De Rosa, F. Kordon, P.-H. Wuillemin, Service Functional Testing Automation with Intelligent Scheduling and Planning, in: Proc. of the 31st Annual ACM Symp. on Applied Computing, ACM, New York, NY, 2016, pp. 1605–1610, https://doi.org/10.1145/2851613.2851807. 978-1-4503-3739-7.

[97] C. Bartolini, A. Bertolino, E. Marchetti, I. Parissis, Data flow-based validation of web services compositions: perspectives and examples, in: R. de Lemos, F. Di Giandomenico, C. Gacek, H. Muccini, M. Vieira (Eds.), Architecting Dependable Systems V, Springer, 2008, pp. 298–325.

[98] W. Hummer, O. Raz, O. Shehory, P. Leitner, S. Dustdar, Testing of data-centric and event-based dynamic service compositions, Softw. Test. Verification Reliab. 23 (6) (2013) 465–497.

[99] C. Bartolini, A. Bertolino, S. Elbaum, E. Marchetti, Whitening SOA testing, in: Proceedings of the 7th Int. Conf. ESEC-FSE, ACM, 2009, pp. 161–170.

[100] C. Ye, H.-A. Jacobsen, Whitening SOA testing via event exposure, IEEE Trans. Softw. Eng. 39 (10) (2013) 1444–1465.

[101] A. Benharref, R. Dssouli, M.A. Serhani, R. Glitho, Efficient traces' collection mechanisms for passive testing of web services, Inf. Softw. Technol. 51 (2) (2009) 362–374.

[102] E. Martin, Automated test generation for access control policies, in: Companion to the 21st ACM SIGPLAN Symposium on Object-Oriented Programming Systems, Languages, and Applications, ACM, 2006, pp. 752–753.

[103] A. Bertolino, S. Daoudagh, F. Lonetti, E. Marchetti, The X-CREATE framework: a comparison of XACML policy testing strategies, in: WEBIST, 2012, pp. 155–160.

[104] A.M. Memon, B.N. Nguyen, Advances in automated model-based system testing of software applications with a GUI front-end, Adv. Comput. 80 (2010) 121–162.

[105] I.A. Qureshi, A. Nadeem, GUI testing techniques: a survey, Int. J. Futur. Comput. Commun. 2 (2) (2013) 142.

[106] I. Banerjee, B. Nguyen, V. Garousi, A. Memon, Graphical user interface (GUI) testing: Systematic mapping and repository, Inf. Softw. Technol. 55 (10) (2013) 1679–1694.

[107] S. Arlt, E. Ermis, S. Feo-Arenis, A. Podelski, Verification of GUI Applications: A Black-Box Approach, in: Springer, Berlin, Heidelberg, 2014, pp. 236–252, https://doi.org/10.1007/978-3-662-45234-9_17. 978-3-662-45234-9.

[108] ABI Research's Mobile Application Technologies, $200 Million Mobile Application Testing Market Boosted by Growing Demand for Automation, https://www.abiresearch.com/press/200-millionmobile-application-testing-market-boos.

[109] J. Gao, W.-T. Tsai, R. Paul, X. Bai, T. Uehara, Mobile testing-as-a-service (MTaaS)—infrastructures, issues, solutions and needs, in: IEEE 15th Int. Symp. on High-Assurance Systems Engineering (HASE), IEEE, 2014, pp. 158–167.

[110] O. Starov, S. Vilkomir, A. Gorbenko, V. Kharchenko, Testing-as-a-service for mobile applications: state-of-the-art survey, in: W. Zamojski, J. Sugier (Eds.), Dependability Problems of Complex Information Systems, Springer, 2015, pp. 55–71.

[111] I.K. Villanes, E.A.B. Costa, A.C. Dias-Neto, Automated Mobile Testing as a Service (AM-TaaS), in: 2015 IEEE World Congress on Services, 2015, pp. 79–86, https://doi.org/10.1109/SERVICES.2015.20.

[112] K. Michael, K.W. Miller, Big Data: New Opportunities and New Challenges [Guest editors' introduction], Computer 46 (6) (2013) 22–24, ISSN 0018-9162, https://doi.org/10.1109/MC.2013.196.

[113] A. Abidin, D. Lal, N. Garg, V. Deep, Comparative analysis on techniques for big data testing, in: 2016 Int. Conf. on Information Technology (InCITe)—The Next Generation IT Summit on the Theme—Internet of Things: Connect your Worlds, 2016, pp. 219–223, https://doi.org/10.1109/INCITE.2016.7857620.

[114] A. Alexandrov, C. Brücke, V. Markl, Issues in big data testing and benchmarking, in: Proc. of the Sixth Int. Workshop on Testing Database Systems, ACM, 2013, p. 1.

[115] TPC Benchmarks & Benchmark Results, http://www.tpc.org/ (accessed April 2017).

[116] XMark—An XML Benchmark Project, http://www.xml-benchmark.org/ (accessed April 2017).

[117] N. Li, A. Escalona, Y. Guo, J. Offutt, A scalable big data test framework, in: 2015 IEEE 8th International Conference on Software Testing, Verification and Validation (ICST), 2015, pp. 1–2, https://doi.org/10.1109/ICST.2015.7102619.

[118] C. Robinson-Mallett, Coordinating security and safety engineering processes in automotive electronics development, in: Proceedings of the 9th Annual Cyber and Information Security Research Conference, 2014, pp. 45–48.

[119] B. Antonia, C. Antonello, D.G. Felicita, L. Giuseppe, L. Francesca, M. Eda, M. Fabio, M. Ilaria, M. Paolo, Secure Software Engineering for Connected Vehicles: A Research Agenda, Technical reports no. IIT TR-18/2015, http://www.iit.cnr.it/en/node/36711.

ABOUT THE AUTHORS

Francesca Lonetti is a researcher at CNR–ISTI. She received her Ph.D. in Computer Science from the University of Pisa in 2007, Italy. Her current research focuses on monitoring and testing of software systems. In particular, she is interested in testing of security systems (access control and usage control systems), methodologies and tools for robustness testing of web services, modeling, and validation approaches of non-functional properties and business process modeling and assessment. Her expertise on these research topics has been applied in the context of several national and European research projects including Learn PAd, CHOReOS, TAS3, NESSoS, CONNECT, and D-ASAP. She is and has been part of the program committee of several international conferences and workshops in the field, such as ICST, MODELSWARD, AST, and QUATIC. She has been a publications chair of ICST 2012 and proceedings cochair of ICSE 2015. She is/has been a cochair of the 11th IEEE/ACM International Workshop on Automation of Software Test (AST 2016) and the International Workshop on domAin-specific Model-based AppRoaches to vErification and validation (Amaretto 2016).

Eda Marchetti is a researcher at CNR–ISTI. She graduated summa cum laude in Computer Science from the University of Pisa (1997) and got a Ph.D. from the same University (2003). Her research activity focuses on software testing and in particular security testing, testing of access control systems, model-based testing, SOA, and component-based testing. She has taking the role of WP coordinator in recently concluded projects including TAS3, Presto4u. She has been involved in several EU project:

TELCERT, CONNECT, CHOReOS, NESSoS, and TAROT. She is currently involved in the monitoring and testing process of Learn PAd. She has served as a reviewer for several international conferences and journals, and she has been part of the organizing and program committee of several international workshops and conferences. She has (co)authored over 50 papers in international journals and conferences.

CHAPTER FOUR

Optimizing the Symbolic Execution of Evolving Rhapsody Statecharts

Amal Khalil, Juergen Dingel

School of Computing, Queen's University, Kingston, ON, Canada

Contents

Advances in Computers, Volume 108
ISSN 0065-2458
https://doi.org/10.1016/bs.adcom.2017.09.003

Abstract

Model-driven engineering (MDE) is an iterative and incremental software development process. Supporting the analysis and the verification of software systems developed following the MDE paradigm requires to adopt incrementality when carrying out these crucial tasks in a more optimized way.

Communicating state machines are one of the various formalisms used in MDE tools to model and describe the behavior of distributed, concurrent, and real-time reactive systems (e.g., automotive and avionics systems). Modeling the overall behavior of such systems is carried out in a modular way and on different levels of abstraction (i.e., it starts with modeling the behavior of the individual objects in the system first then modeling the interaction between these objects). Similarly, analyzing and verifying the correctness of the developed models to ensure their quality and their integrity is performed on two main levels. The intralevel is used to analyze the correctness of the individual models in isolation of the others, while the interlevel is used to analyze the overall interoperability of those that are communicating with each other.

One way to facilitate the analysis of the overall behavior of a system of communicating state machines is to build the global state space (also known as the global reachability tree) of the system. This process is very expensive and in some cases it may suffer from the state explosion problem. Symbolic execution is a technique that can be used to construct an abstract and a bounded version of the system global state space that is known as a symbolic execution tree (SET), yet the size of the generated trees can be very large especially with big and complex systems that are composed of multiple objects. As the system evolves, one way to avoid regenerating the entire SET and repeating any SET-based analyses that have been already conducted is to utilize the previous SET and its analysis results in optimizing the process of generating the SET of the system after the change. In this chapter, we propose two optimization techniques to direct the successive runs of the symbolic execution technique toward the impacted parts of an evolving state machine model using memoization (MSE) and dependency analysis (DSE), respectively. The evaluation results of both techniques showed significant reduction in some cases compared with the standard symbolic execution technique.

1. INTRODUCTION

Model-driven engineering (MDE) is a *model-centric* software engineering approach that aims at improving the productivity and the quality of software artifacts by focusing on models as first-class artifacts in place of code. MDE has been widely used for over a decade in many domains such as automotive and telecommunication industries. *Iterative-incremental* development and *model-based analysis* are central to MDE in which artifacts typically undergo several iterations and refinements during their lifetime that may require changes to their initial design versions. As these models evolve, it is necessary to assess their quality by repeating the analysis and the verification of these models after every iteration or refinement. This process, if not optimized, can be very tedious and time consuming.

State machines are a visual state-based formalism used to describe the behavior of reactive systems and they are found, with some variation, in every MDE tool (e.g., IBM Rational Rhapsody's Statecharts, IBM Rational Software Architect's UML-RT, or MathWorks Simulink's Stateflow).

Symbolic execution is a well-known analysis technique that is used to analyze the execution paths of behavioral software artifacts (e.g., programs [1] and state-based models [2, 3]). The output of the analysis is a symbolic execution tree (SET) which provides the basis for various types of analysis and verification. One of the key challenges of symbolic execution is scalability, especially when applied to big, complex artifacts where the size of the output SET becomes very large. Repeating the entire analysis even after small changes is not the best solution.

This chapter describes research investigating two complementary optimization techniques that leverage the similarities between state machine versions to reduce the cost of symbolic execution of the evolved version.

This research is motivated by a number of facts. First, symbolic execution has been shown to be a very powerful method for the analysis of programs and there are already several commercial code analysis tools built based on it (e.g., CodeSonar [4]). Similarly, the technique has been adopted and applied in the context of state-based models (e.g., the IAR visualSTATE—Verificator [5]). Second, there is an interest from our industrial partner,

General Motors Corporation (GM), to improve the model-level analysis capabilities of the IBM Rhapsody tool. Third, research on optimizing the symbolic execution of evolving programs has been recently addressed [6, 7], however, to the best of our knowledge we are the first to consider such optimizations for the symbolic execution of evolving state machines.

1.1 Research Statement and Scope

The main goal of this research is to prove that the efficiency and cost of symbolic execution of evolving state machines can be measurably improved using incremental analysis.

Since we aim to use IBM Rhapsody Statecharts as the input state-based models to build a proof of concept implementation of our proposed techniques, we will only use IBM related tools (e.g., the IBM Rational Rhapsody Developer RulesComposer Add-On and the IBM Rational Rhapsody DiffMerge) to develop the different components of our proposed architectures as presented in Section 4 and 5.

Rhapsody Statecharts offer a full set of advanced features for specifying reactive behavior. Supporting the entire set of these features is beyond the scope of this work, however, we have selected a subset that contains the most frequently used features. This includes composite and orthogonal states, interlevel transitions (also known as group transitions), and condition and junction connectors. History connectors are an example of an advanced feature that is not supported.

The prototype implementation of our proposed techniques is based on the KLEE symbolic execution engine [8]. Since KLEE supports only C code, we consider only a subset of Java or C++ action code that has the same syntax as C code, including basic assignment statements, conditional statements, and iterative statements. The data types supported are the basic data types including characters, Booleans, and integers. We also consider the statements used in action code for sending events.

The Statechart models to be used in the evaluation were selected in accordance with the aforementioned assumptions.

1.2 Contributions

The proposed research aims to improve the current state of the art in the area of model-based analysis in an evolutionary software development environment. Our contributions specifically include (1) the development, formalization, and proof-of-concept implementation of the proposed optimization techniques mentioned in Section 5, (2) an evaluation that will provide the results showing the benefits of our research methodology, and (3) a concrete example of research that can enrich the analysis capabilities of an existing MDE tool.

1.3 Chapter Organization

Section 2 gives an overview of the background. Section 3 discusses the related work. Section 4 presents our standard symbolic execution technique for individuals and communicating Rhapsody Statecharts. Section 5 demonstrates the optimization ideas employed in this research. Section 6 explains the implementation details of the proposed work. Section 7 outlines the evaluation procedure and presents and discusses the evaluation results. Section 8 concludes the work presented in this chapter and outlines future work.

2. BACKGROUND

In this section we provide the background information of the main technologies used in this research. First, we introduce the modeling context used and the types of models considered. Then, we present an overview of symbolic execution and dependence analysis. Finally, we specify the tools we use to realize model differencing and model transformation tasks.

2.1 Modeling and Analysis in the IBM Rational Rhapsody Developer MDE Tool

The IBM Rational Rhapsody Developer framework [9] is one of the commercial tools that support model-driven development, where models are the primary artifacts for software development and are iteratively refined until code can be automatically generated by the tool. The tool is heavily used in practice (e.g., in the automotive industry). It has a number of specialized models (or diagrams) used to describe a software system at different levels of abstraction. We refer only to the two types of models we use in this work: object model diagram (OMD) and Statecharts. A detailed description of all features supported by these models is outside the scope of this work. For simplicity, only a limited set of these features is used in the illustrative example shown in Fig. 1.

2.1.1 Rhapsody Object Model Diagrams

The OMDs are used to specify the structure and the static relationships of the classes and objects in a software system. Rhapsody's OMDs serve, to some extent, the same purpose as the UML's composite structure diagrams, class diagrams and object diagrams. The top-level object in an OMD is usually a composite class that represents the root class of the system and which is also considered as the system container. The IBM Rational Rhapsody code generator directly translates the elements and relationships modeled in an OMD into source code in a specified high-level language (e.g., C, C++, C#, Java, or Ada). Fig. 1A shows a system AB that is composed of two communicating objects: itsA and itsB. Object itsA is of class A that has an integer attribute x that is initialized to 0 and a protocol of a set of three messages: e1, e2, and e3. Similarly, object itsB is of class B that has an integer attribute y that is initialized to 5 and a protocol of three messages: e4, e5, and e6. Message e1 has an integer argument e1Arg1 and message e5 has an integer argument e5Arg1. The behavior of each object is modeled by the Rhapsody Statecharts shown in Figs. 1B and C, respectively.

2.1.2 Rhapsody Statecharts

Rhapsody Statecharts[a] (also known as Harel's Statecharts [10]) are a visual state-based formalism implemented in the IBM Rational Rhapsody framework to describe the behavior of reactive systems. They extend conventional Mealy machines—a type of finite state machines (FSMs) that perform their action only on firing transitions—with advanced features such as hierarchical states, orthogonal regions, and action code in states and transitions. An action is a reactive behavior associated with a state or a transition that can be carried out when, for instance, entering a state, leaving a state, or when firing a transition. Examples of these actions include updating the values of the machine variables or producing an output to be sent to the environment via events. Rhapsody allows the use of C, C++, or Java to express these actions. A transition is defined in the general format: event [guard]/action, that specifies the firing event of the transition, an optional guard condition and an optional action code. Events are stimuli delivered to the Statechart from the environment or other objects in the system and they can have optional user-defined parameters (also called arguments). Timeout events are internal events which can be generated using timers within the

---

[a] The term "Statechart" is used exchangeably with the term "State Machine" in this document.

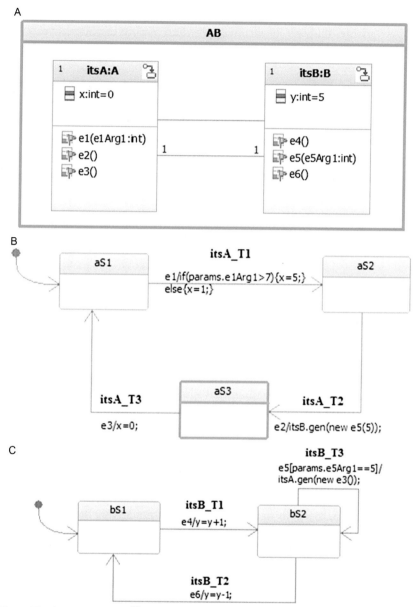

Fig. 1 The base version AB^{V0} of an example model AB of two communicating objects. (A) The Object Model Diagram (OMD) of our simple model AB. (B) The Statechart of object itsA in (A). (C) The Statechart of object itsB in (A).

Statecharts. In contrast with the machine variables, which are considered as global variables that can be accessed anywhere within the Statechart, event arguments are considered as local variables that have a limited scope that is restricted only to the guard and the action code of the transitions that receive their event. A guard is an optional Boolean expression that can be defined over machine variables and events arguments. A transition is fired when its source state is active, its corresponding event is received, its guard evaluates to true and no higher priority transition is enabled. When a transition is fired, its action is executed and, consequently, the system goes to the transition's target state. Besides the standard assignment and conditional statements that can be used in describing the action code of a transition, there is a domain-specific send event statement which is used to generate an output event to be sent to a specified object. An example of such statement is shown in the action Java code of transition `itsA_T2` in Fig. 1B, which is `itsB.gen(new e5(5))`; that is used to generate event `e5` with the event's argument `e5Arg1` set to 5 to object `itsB`. Accordingly, the argument value of the event `e5` is stored in a variable called `params` that can be accessed using the statement `params.e5Arg1`.

Apart from whether the given Statecharts satisfy their specification, there are a list of basic and intuitive design rules any given Statechart model should satisfy. This includes the reachability of states, the satisfiability of guards and the availability of input events. The task of verifying whether these rules are satisfied looks very simple for Statecharts like the ones in Fig. 1B and C but it gets more complex as these models get larger and when more advanced modeling features are used, which is the typical case when building industrial real-time systems.

2.1.3 Analysis and Verification of State-Based Models

Existing analysis methods to analyze state-based models are divided into two categories: *model-level analysis* and *code-level analysis*. The motivation of the code-level analysis approaches stems from the fact that these models are executable and we can easily generate their corresponding code and use some of the model checkers for C/C++ or Java code or one of the static analysis tools would help in finding possible problems at the code-level. However, it is difficult for a general static analysis tool to figure out the high-level intent of any piece of code we feed to it and therefore they are typically incapable of revealing model-level problems (e.g., unreachable states and dead ends). Also, it is not always the case that we would be able to generate the code, especially with incomplete specifications. Additionally, the results provided from a code-level analysis are typically more detailed which usually makes them less intuitive for designers and modelers. Other challenges include the difficulty of tracing back the errors generated in the code to its source at the model level and to adapt existing analysis tools (e.g., model checkers, symbolic execution engines, or static analysis tools) to handle domain-specific statements found in the generated code. For these reasons, we believe that model-level analysis is the most appropriate way to handle this task. Therefore, the motivation of our research is to highlight the importance of the types of analysis or verification to be carried out on the model level.

2.2 Symbolic Execution

Symbolic execution is a program analysis technique that allows the execution of programs in a parametric way using symbolic inputs to derive precise characterizations of their properties and their execution paths. It has first introduced in the 1970s by Clarke [11] and King [1] for program testing. Since 2003, much research effort has been devoted to improve the effectiveness, the efficiency, and the applicability of the traditional technique [12–14].

The main idea is to substitute program inputs with symbolic values and then execute the program parametrically such that (1) the values of all program variables are computed as symbolic expressions over the symbolic input values, and (2) all possible execution paths are explored. The result from the symbolic execution of a program is a tree-based structure called SET. The nodes of a SET represent the symbolic program states and the edges represent the transitions between these states. Each program symbolic state consists of a program location, the set of program variables and their symbolic valuations, and a path constraint (PC) which is the conjunction of all the logical constraints collected over the program variables to reach that program location. The feasible paths of a SET characterize all the distinct execution paths of a program. Decision procedures and SMT solvers are used to check the satisfiability of each path constraint (PC).

The resulting SET forms the basis for many program analysis, verification, and testing activities (e.g., bug finding, dead code detection, invariants checking, program equivalence checking, regression analysis, and test case generation). For instance, it can be used to optimize the testing process by generating the minimum number of test cases required to inspect all feasible program paths. Fig. 2B (resp., 2D) shows the SET of the sample code in Fig. 2A (resp., 2C). The exploration is carried out in a depth-first manner. Examples of other exploration strategies are breadth first,

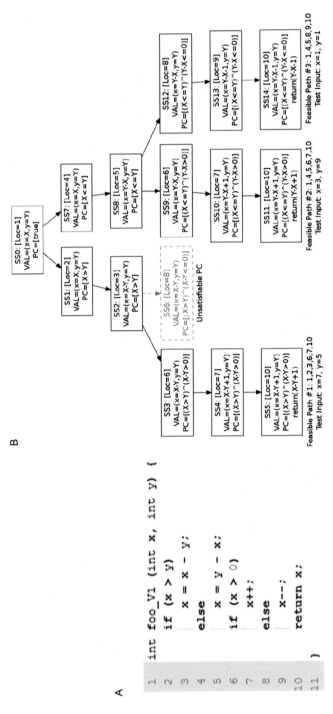

Fig. 2 Symbolic execution of programs—Example 1. (A) Code Snippet of `foo_V1`. (B) SET of `foo_V1`. (C) Code Snippet of `foo_V2`. (D) SET of `foo_V2`.

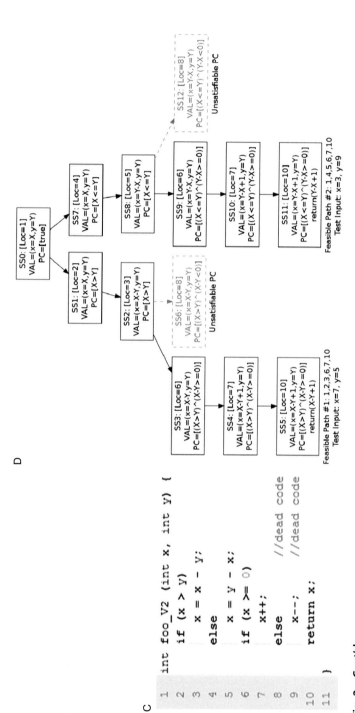

C

```
   int foo_V2 (int x, int y) {
1
2      if (x > y)
3         x = x - y;
4      else
5         x = y - x;
6      if (x >= 0)
7         x++;
8      else            //dead code
9         x--;         //dead code
10     return x;
11  }
```

D

SS0: [Loc=1]
VAL=(x=X,y=Y)
PC=[true]

SS1: [Loc=2]
VAL=(x=X,y=Y)
PC=[X>Y]

SS7: [Loc=4]
VAL=(x=X,y=Y)
PC=[X<=Y]

SS2: [Loc=3]
VAL=(x=X-Y,y=Y)
PC=[X>Y]

SS6: [Loc=8]
VAL=(x=X-Y,y=Y)
PC=[(X>Y)^(X-Y<0)]

Unsatisfiable PC

SS8: [Loc=5]
VAL=(x=Y-X,y=Y)
PC=[X<=Y]

SS12: [Loc=8]
VAL=(x=Y-X,y=Y)
PC=[(X<=Y)^(Y-X<0)]

Unsatisfiable PC

SS3: [Loc=6]
VAL=(x=X-Y,y=Y)
PC=[(X>Y)^(X-Y>=0)]

SS4: [Loc=7]
VAL=(x=X-Y+1,y=Y)
PC=[(X>Y)^(X-Y>=0)]

SS5: [Loc=10]
VAL=(x=X-Y+1,y=Y)
PC=[(X>Y)^(X-Y>=0)]
return(X-Y+1)

Feasible Path #1: 1,2,3,6,7,10
Test Input: x=7, y=5

SS9: [Loc=6]
VAL=(x=Y-X,y=Y)
PC=[(X<=Y)^(Y-X>=0)]

SS10: [Loc=7]
VAL=(x=Y-X+1,y=Y)
PC=[(X<=Y)^(Y-X>=0)]

SS11: [Loc=10]
VAL=(x=Y-X+1,y=Y)
PC=[(X<=Y)^(Y-X>=0)]
return(Y-X+1)

Feasible Path #2: 1,4,5,6,7,10
Test Input: x=3, y=9

Fig. 2—Cont'd

random, or heuristic-based exploration. Solving the path constraints of the leaf nodes of each SET will provide us with the sample input values that can be used to test each feasible path. Program locations that do not have any corresponding symbolic states in the program SET indicate dead code. For example, program locations 8–9 in the code snippet foo_V2 are dead code.

One of the biggest challenges in symbolic execution is the path explosion problem especially for large programs and programs with unbounded loops and recursion. Examples of proposed solutions include setting upper bounds for loops, summarizing loop effects, employing some state matching criteria (e.g., state subsumption) for pruning redundant paths and reducing the state space, using heuristics for path finding to achieve some user-defined coverage criteria, or dividing a program into independent parts and run the symbolic execution for each part in parallel [14]. For example, Fig. 3B shows a finite version of the SET of the code snippet in Fig. 3A [12]. This finitized SET version is obtained by performing a state matching technique called subsumption checking to determine when a symbolic state is revisited and hence pruning the exploration at this point. For any given two symbolic

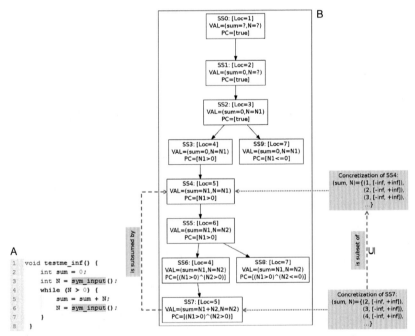

Fig. 3 Symbolic execution of programs—Example 2. (A) Code Snippet of testme_inf [12]. (B) A finite SET of testme_inf.

states SS_i and SS_j, SS_j is subsumed by SS_i if both symbolic states represent the same program location and if the concretization of SS_j is included in the concretization of SS_i. For example, in Fig. 3B, the symbolic state $SS7$ is subsumed by the symbolic state $SS4$ and hence the exploration of the symbolic state $SS7$ can be avoided.

Other challenges include (1) the limitation of constraint solvers in solving very complex and nonlinear constraints and (2) the inability to handle external library calls. Examples of proposed solutions include concolic symbolic execution (an variant of the traditional technique that enables the use of concrete values for unmanageable symbolic expressions) [15, 16], performing constraints simplification and providing models to simulate or abstract the behavior of external modules.

Examples of existing symbolic execution tools are jCUTE and Java Path-Finder (JPF) for Java, EXE and its successor KLEE for C and C++, and Pex for the .NET Framework.

Symbolic execution is not restricted only to programs; it has been also applied to a variety of state-based models including Input Output Symbolic Transition Systems [2], Statecharts [17], UML State Machines [18], and recently UML-RT State Machines [3].

2.3 Dependence Analysis

Dependence analysis has been extensively studied in research areas such as program refactoring, parallelization, and optimization. It also supports many activities in software maintenance such as comprehension, slicing, impact analysis, and regression test reduction. Dependence analysis has been primarily studied in the context of programming languages where dependences are defined between statements and variables using their control flow graphs (CFGs). Eventually, the same concept has been adapted for other types of software artifacts to capture potential interactions within such artifacts, e.g., Extended Finite State Machines (EFSMs) [19–21] where interactions are defined between transitions since they are the "active" elements of these types of models. Examples of dependences are control dependence, data dependence, and communication dependence.

In our setting, we transform Rhapsody Statecharts into an EFSMs-like representation called Mealy-like machines (MLMs) (the formal definition on an MLM is presented in Section 4.1.1), therefore we can adopt the definitions of dependencies from conventional EFSMs. In the sequel, we will refer to a set of definitions for computing these dependences as adopted from [22] based on the work presented in [19–21].

2.3.1 Control Dependence

Classical definitions of control dependence in sequential programs state that a statement s_j is control dependent on a statement s_i if statement s_i is a conditional that affects the execution of statement s_j. For example, in an `if-then-else` construct, statements in the two branches of the conditional statement are control dependent on the predicate. Since state-based formalisms differ from sequential programs, some adapted definitions are given to capture the control dependence in such formalisms. The one that applies to state machines with possible nontermination is called nontermination sensitive control dependence (NTSCD) (see Definition 2), which is given in terms of maximal paths (see Definition 1).

Definition 1 (Maximal path).

A path in a state machine is defined as a finite sequence of consecutive transitions that are organized such that no state is visited more than once. A path is called maximal if it terminates in an end state (state with no outgoing transitions) or it terminates in a cycle or it reaches the initial state again. A cycle is a set of transitions that forms a strongly connected component.

Definition 2 (Nontermination sensitive control dependence).

$t_i \xrightarrow{\text{NTSCD}} t_j$ means that t_j is nontermination sensitive control dependent on a transition t_i iff:

(1) for all paths $\pi \in$ `MaximalPaths(targetState(`t_i`))`, the `sourceState(`t_j`)` belongs to π;

(2) t_i has at least one sibling t_k and there exists a path $\pi' \in$ `MaximalPaths(sourceState(`t_k`))` such that `sourceState(`t_j`)` does not belong to π';

where `MaximalPaths(s)` is the set of paths that have s as the source state of the first transition on each path and t_k is said to be a sibling transition of t_i if `sourceState(`t_k`) = sourceState(`t_i`)`.

2.3.2 Data Dependence

Typical data dependence definitions are given in terms of variable definitions and uses. In the context of state machines, a variable is used on a transition if its value appears in the guard of the transition or appears on the right side of an assignment-statement or in the Boolean expression of an if-statement, or in an output-statement in the action code of the transition. A variable is defined on a transition if it appears on the left-hand side of an assignment-statement in the action code of the transition. Based on this, the following definition is given [19, 22].

Definition 3 (Data Dependence (DD)).

$t_j \xrightarrow{DD} t_k$ means that t_k is data dependent on t_j with respect to variable v iff:

(1) $v \in Def(t_j)$, where $Def(t_j)$ is the set of variables defined by the action code of transition t_j;

(2) $v \in Use(t_k)$, where $Use(t_k)$ is the set of variables used in the guard or the action code of transition t_k;

(3) there exists a path in the state machine from t_j to the targetState(t_k) along which v is not modified.

2.3.3 Communication Dependence

Two transitions are called communication dependent iff: (1) each transition belongs to a different MLM (the formal definition on an MLM is presented in Section 4.1.1) and (2) one transition is triggered by an event that is generated by the action code of the second transition [19, 22].

Definition 4 (Communication Dependence (COMD)).

Given two transitions t_j and t_k, $t_j \xrightarrow{COMD} t_k$ means that t_k is communication dependent on t_j iff:

(1) $t_j \in T(MLM_j)$ and $t_k \in T(MLM_k)$, where $j \neq k$;

(2) $e(t_k)$—the event (or trigger) of t_k is generated by the action code of t_j.

The detailed steps for computing the maximal paths, the nonterminating sensitive control dependence (NTSCD), the data dependence (DD) in an MLM, as well as the communication dependence (COMD) in a model of communicating MLMs are shown in Algorithm 1, Algorithm 2, Algorithm 3, and Algorithm 4, respectively. In our work we consider only data and communication dependences, however, we listed the others for completeness. Fig. 4 depicts the data and communication dependences that exist between the Statecharts describing the behavior of the two objects itsA and itsB of System AB shown in Fig. 1.

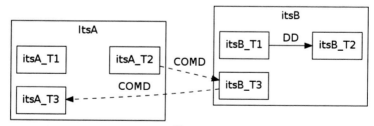

Fig. 4 Dependency graph of system AB^{v0}.

Algorithm 1 Maximal Paths Computation

Inputs:

1: An MLM representation of a Rhapsody Statechart model $MLM = (S, T, V, s_0)$ that is defined in terms of states S, transitions T, variables V and an initial state $s_0 \in S$

2: A mapping, $outTrans : S \rightarrow \mathcal{P}(T)$, of states to their outgoing transitions

Output:

1: $maximalPaths(MLM)$

Steps:

1: $endStates(MLM) \leftarrow \emptyset$

2: $cycles(MLM) \leftarrow \emptyset$

3: $maximalPaths(MLM) \leftarrow \emptyset$

4: **for all** $(s \in S) \wedge (outTrans(s) = \emptyset)$ **do**

5:    $endStates(MLM) \leftarrow endStates(MLM) \cup \{s\}$

6: **end for**

7: //Explore the Statechart in depth-first-search manner

8: Push the initial state s_0 into a stack st

9: Mark $outTrans(s_0)$ as unexplored

10: **while** st is not empty **do**

11:    $currentStateToExplore \leftarrow topElement(st)$

12:    **for all** $(t_o \in T) \wedge (t_o \in outTrans(currentStateToProcess)) \wedge (t_o$ is still unexplored) **do**

13:       **if** $(targetState(t_0) = s_0) \vee (targetState(t_0) \in endStates(MLM)) \vee (targetState(t_0) \in cycles(MLM))$ **then**

14:          Report a maximal path $\pi \leftarrow st + targetState(t_o)$

15:          $maximalPaths(MLM) \leftarrow maximalPaths(MLM) \cup \{\pi\}$

16:          Mark t_o as explored

17:       **else if** $targetState(t_0) \in st$ **then**

18:          Report a cycle $\circlearrowleft \leftarrow st.subList(indexOf(targetState(t_0)), size(st)) + targetState(t_0)$

19:          $cycles(MLM) \leftarrow cycles(MLM) \cup \{\circlearrowleft\}$

20:          Report a maximal path $\pi \leftarrow st.subList(0, size(st)) + targetState(t_0)$

21:          $maximalPaths(MLM) \leftarrow maximalPaths(MLM) \cup \{\pi\}$

22:          Mark t_o as explored

23:       **else**

24:          Push $targetState(t_0)$ into st

25:          Mark $outTrans(targetState(t_0))$ as unexplored

26:          Mark t_o as explored

27:          Break

28:       **end if**

29:    **end for**

30:    Pop $currentStateToProcess$ from st

31: **end while**

Algorithm 2 Nonterminating Sensitive Control Dependence (NTSCD) Computation

Inputs:

  1: An MLM representation of a Rhapsody Statechart model $MLM = (S, T, V, s_0)$ that is defined in terms of states S, transitions T, variables V and an initial state $s_0 \in S$

  2: *maximalPaths(MLM)*

Output:

  1: A mapping, $CD: T \to \mathcal{P}(T)$, of transitions to their control dependent transitions

Steps:

  1: // Finding *maximalPaths(T)*
  2: **for all** $t \in T$ **do**
  3:     **for all** $\pi \in maximalPaths(MLM)$ **do**
  4:       **if** $\pi.indexOf(targetState(t)) = \pi.indexOf(sourceState(t) + 1$ **then**
  5:         *maximalPaths(t)* $\leftarrow$ *maximalPaths(t)* $\cup \{\pi\}$
  6:       **end if**
  7:     **end for**
  8: **end for**
  9: // Finding $CD(T)$
10: **for all** $(t_i \in T) \wedge (siblingTransitions(t_i).size() \geq 1)$ **do**
11:     $CD(t_i) \leftarrow \emptyset$
12:     **for all** $(s \in S) \wedge (s \in \forall maximalPaths(t_i))$ **do**
13:       **if** $(t_k \in T) \wedge (t_k \in siblingTransitions(t_i)) \wedge (\exists \pi \in maximalPaths(t_k)) \wedge (s \notin \pi)$ **then**
14:         **for all** $(t_o \in T) \wedge (t_o \in outTrans(s))$ **do**
15:           $CD(t_i) \leftarrow CD(t_i) \cup \{t_o\}$
16:         **end for**
17:       **end if**
18:     **end for**
19: **end for**

Algorithm 3 Data Dependence (DD) Computation in an MLM

Inputs:

  1: An MLM representation of a Rhapsody Statechart model $MLM = (S, T, V, s_0)$ that is defined in terms of states S, transitions T, variables V and an initial state $s_0 \in S$

  2: A mapping, $Use: T \to \mathcal{P}(V)$, of transitions to the set of variables each transition uses

  3: A mapping, $Def: T \to \mathcal{P}(V)$, of transitions to the set of variables each transition defines

Output:

1: A mapping, $DD: T \rightarrow \mathcal{P}(T)$, of transitions to their data dependent transitions with respect to V

Steps:

1: **for all** $t \in T$ such that: $(\exists v \in V) \wedge (v \in Use(t))$ // i.e., for all transitions that use at least one variable **do**

2:   **for all** $(v \in V) \wedge (v \in Use(t))$ // i.e., for all the variables that are used in t **do**

3:     // Apply backward depth-first-search to find the first transitions defining the variable v

4:     Push the $sourceState(t)$ into a stack st

5:     **while** st is not empty **do**

6:       $currentStateToProcess \leftarrow topElement(st)$

7:       **for all** $(t_i \in T) \wedge (t_i \in incomingTransition(currentStateToProcess))$ $\wedge (t_i$ has not yet been explored)) **do**

8:         $DD(t_i) \leftarrow \emptyset$

9:         **if** $v \in Def(t_i)$ // i.e., v is defined in t_i **then**

10:           $DD(t_i) \leftarrow DD(t_i) \cup \{t\}w.r.t.v$

11:           Mark t_i as explored

12:         **else**

13:           Mark t_i as explored

14:           Push the $sourceState(t_i)$ into st

15:           Break

16:         **end if**

17:       **end for**

18:     **end while**

19:   **end for**

20: **end for**

Algorithm 4 Communication Dependence (COMD) Computation

Inputs:

1: A global model/system of n asynchronously communicating MLMs $M = (MLM^M, Q^M, s_0^M, V^M, V_0^M)$ with $MLM^M = (MLM_1, MLM_2, ..., MLM_n)$ and $MLM_i = (S_i, V_i, E_i, EA_i, T_i, s_{0i}, V_{0i})$ is the ith MLM in M

2: A mapping, $Send: T \rightarrow \mathcal{P}(E^o)$, of all transitions in MLM^M to the set of output events each transition may generate as a result of evaluating/executing their action code

Output:

1: A mapping, $COMD: T \rightarrow \mathcal{P}(T)$, of transitions to their communication dependent transitions

Steps:
 1: **for all** $MLM_i \in MLM^M$ **do**
 2:   **for all** $t_x \in T_i$ **do**
 3:     $COMD(t_x) \leftarrow \emptyset$
 4:     **for all** $t_y \in T$ **do**
 5:       **if** $e(t_x) \in Send(t_y) \wedge MLM(t_x) \neq MLM(t_y)$ **then**
 6:         $COMD(t_x) \leftarrow COMD(t_x) \cup \{t_y\}$
 7:       **end if**
 8:     **end for**
 9:   **end for**
10: **end for**

2.4 Model Differencing

Model differencing is the practice of identifying and locating the changes between different versions of a model. As models evolve, keeping track of their changes is essential for their maintainability during their lifetime. Existing approaches and tools for model differencing depend on some similarity-based matching criteria to guide the search process [23]. Some of them consider the syntactic similarities between models elements and others try to search for the semantic similarities as well [24]. In our work, we use the IBM Rational Rhapsody DiffMerge tool to find the differences between Rhapsody Statecharts. The tool compares Rhapsody projects that belong to the same ancestor (i.e., they are all models originating from the same base version) and generates a difference report with the differences between the matched elements in each project. As we notice, the tool depends on some globally unique identifiers which are assigned internally to model elements to match between the elements in each model and then to find the differences between the matched ones.

2.5 Model Transformation

Model transformation is the technology that is used in the area of MDE to convert models to other software artifacts (e.g., code) [23]. When the transformation is carried out to convert a source model to a target model and both models conform to the same meta-model, we call it an *endogenous transformation*. This type of model transformation is used to perform tasks such as model refactoring and optimization in general. On the other hand when the transformation is carried out to convert a source model to a target model and both models conform to different meta–models, we call this an *exogenous transformation*. This type of transformation is used handle tasks such as code generation, reverse engineering, and migration. In our setting, we need to perform exogenous transformations to transform Rhapsody Statecharts to our internal MLM representation such that we can symbolically execute the model. Therefore, we use the model transformation technologies offered by the IBM Rational Rhapsody Developer RulesComposer Add-On. The add-on is built based on the Eclipse Modeling Framework (EMF) and it provides the infrastructure required to perform common MDE services such as model and meta–model manipulation, model-to-model transformation, and model–to–text generation. The language used to define model transformations is called Model Query Language (MQL) and the language used to define text templates is called Text Generation Language (TGL) and they are both imperative and proprietary languages. MQL is also used to launch text generations and to define the expressions used within TGL.

2.6 Summary

In this section we have presented the background information. First, we have introduced the modeling context used and briefly described the two types of models considered. Namely, the OMDs for describing the structural aspects of the system and the Statecharts for describing the behavioral aspects of the system. We also have showed the basics of two analysis techniques: symbolic execution and dependence analysis. We have introduced symbolic execution of programs with examples showing some of the most common applications of the technique and explaining the state matching criteria used to reduce the state space of programs with loops. Additionally, we have provided the definitions and the algorithms used to compute the three types of dependences that exist between transitions in EFSMs-like representations. Last, we have defined and specified the model differencing and the model transformation tools used in the context of our work.

3. RELATED WORK

In this section we first introduce a general description of our classification of the two most common ways used in the literature to reconcile analysis and evolution efficiently in software development: reused-based (artifact-oriented) reconciliation and reduction-based (analysis-oriented) reconciliation [25]. Second, we discuss related work in the context of the proposed research and based on the two classified reconciliation methods. We review studies that (1) propose approaches for the analysis and the verification of Statecharts-like models and highlight those that are based on symbolic execution, (2) propose incremental analysis or verification techniques for evolving programs and highlight those which are built specifically to improve the efficiency of symbolic execution, and (3) motivate and realize the concept of incremental verification in the context of model-based development.

3.1 Two Ways to Reconcile Analysis and Evolution Efficiently

Analysis and verification are crucial and expensive activities in the development of real-time and safety-critical software systems. The prerequisite preparation required to run these activities is not trivial in terms of the time it costs and the technologies it uses. The output from such prerequisite preparation process is a special type of artifact that represents one or more views of the system under development in the input language of the analysis or the verification method to be used. The generation of these analysis-related artifacts is usually done using some transformation or translation technologies that allow the semantic mapping between the artifacts used to build the system and those that are used to analyze and verify it. The output from running the analysis or verification process on these auxiliary or intermediate artifacts is used to verify or disprove the correctness of the analyzed properties. Classical analysis methods were focused on analyzing source code related artifacts (both statically and dynamically) and on the intralevel as well as the interlevel. The intralevel is used to analyze the correctness of the individual artifacts in isolation of the others, while the interlevel is used to analyze the overall interoperability of those models that are communicating with others. With the emergence of MDE and agile methods, analysis techniques have been extended to analyze the more abstract architecture artifacts (both structural and behavioral). Examples of static analysis-related artifacts and their usages include (1) CFGs for static analysis and optimization [26], (2) dependency graphs for change impact analysis and slicing [19] and for regression test selection [27], (3) theorem prover formalization for formal verification [28], and (4) SETs for reachability analysis, invariant checking and test case generation [2, 29]. Examples of dynamic analysis-related artifacts include run-time and simulation execution traces which can be used for performance and configuration analysis and run-time errors detection [30].

With the iterative and incremental development of modern development life-cycles, software systems are continuously evolving. Maintaining and analyzing an ever-changing system is essential for its quality and continuity, yet it is a very expensive process and requires considerable efforts. Looking for effective techniques to optimize the analysis activity of these systems as they evolve is vital and it starts with a better identification and understanding of the subject change (i.e., the type of the change and its location) as well as the type of the analysis to be carried out and of course taking into consideration whether a previous analysis of the same type has been conducted or not. For instance, some refactoring changes such as consistent renaming should not invalidate a previous analysis result but would require a refactoring of the previously created analysis-related artifacts to maintain the correctness of the result. On the other hand, some corrective, modification, addition or deletion changes may or may not invalidate a previous analysis result and hence require a complete or partial regeneration of the analysis-related artifacts and rerunning the analysis on the newly generated analysis-related artifacts.

In the sequel we discuss two different scenarios where optimization opportunities (or reconciliations) may exist.

(1) *Reused-based (artifact-oriented) reconciliation*: The first scenario is the case where analysis-related artifacts are regularly retained (i.e., not discarded) and they can be used to run different types of analyses (e.g., SETs and dependency graphs). In this case there is an opportunity to optimize the process of regenerating these analysis-related artifacts after a change by reusing the unaffected parts of the already existing ones and regenerating only the change-related parts. The output from this process are complete updated analysis-related artifacts which are consistent with the system in its most recent state (i.e., after the change) and can be used to run any of its related analyses. By distinguishing the newly generated parts from the reused parts, we can also optimize the process of rerunning any previously conducted analyses such that only the newly generated parts of these analysis-related artifacts are considered. Incremental model transformation [31], extreme model checking [32], memoized symbolic execution [7], and regression model checking [33] are all examples of existing approaches that are based on this reconciliation strategy.

(2) *Reduction-based (analysis-oriented) reconciliation:* The second scenario is the case where analysis-related artifacts are discarded but there is certainty about the validity of the system in its previous state (i.e., before the change) with respect to a specific type of analysis and the goal is to find the most effective way to rerun the same type of analysis on the system in its current state (i.e., after the change). In this case, one possible opportunity for optimization is to rerun the analysis in a reduced mode or context taking into consideration only the changed parts and their dependencies. The output from this process are partial, yet sufficient, analysis-related artifacts that can be used to run only the same type of analysis and to conclude the overall correctness of the system after the change even though the new analysis does not cover the whole system. Existing work for incremental equivalence checking [34–39], directed incremental symbolic execution [6], and incremental property checking [40] employ this idea.

3.2 Analysis and Verification of Statechart-Like Models

Model checking-based approaches: The majority of existing analysis methods for Statechart-like models (e.g., UML state machines, UML-RT, STATEMATE, and Stateflow) require a translation to the input language of existing formal verification tools, mainly model checkers such as SPIN [41] or SMV [42]. Most of the time, the translation process is nontrivial and it restricts the support of the key features of the model to those supported by the model checker to be used and hence makes integration and use with modern MDE tools difficult. Another category of approaches analyze the C or Java code generated from these models and use one of the model checkers for these languages (e.g., Java Pathfinder [43] or CBMC—a Bounded Model Checker for C and C++ [44]). On the other hand, there are fewer approaches that intend to provide domain-specific analysis capabilities by developing domain-specific model checkers that are tailored to the modeling language of interest and hence support more sophisticated features and reduce the semantic gap between the modeling language and the specification language of a general-purpose model checker (e.g., the verification methods implemented within the FUJABA Real-Time tool [45] or the VIS model checker in a very early version of the Rhapsody tool [46]).

Symbolic execution-based approaches: Besides the model checking-based approaches, there are several approaches that adopt the symbolic execution analysis technique of programs and use it in the context of state-based models including:

- Modechart specifications (a variation of Harel Statecharts that incorporates timing constraints in the models) for test sequences generation [47];

- Labeled Transition Systems (LTSs) for test case generation [48] and test case selection [49];
- Input Output Symbolic Transition Systems (IOSTSs) [2] for test purpose definition;
- STATEMATE Statecharts [17] and UML State Machines [18] for verifying temporal properties of the subject models;
- Simulink/Stateflows [50] for analysis and test case generation of flight software using Java PathFinder and Symbolic PathFinder;
- UML-RT State Machines [3] to support a variety of analyses for this type of models.

Our research belongs to this category of approaches. We use symbolic execution to support the analysis of a slightly different type of state-based models which are Rhapsody Statecharts. Our technique differs from others in that (1) it reuses the off-the-shelf symbolic execution engine KLEE [8] and (2) it supports the incremental analysis of the subject models as they evolve.

3.3 Incremental Analysis and Verification Techniques for Evolving Programs

A comprehensive survey of existing work in this area of research is outside the scope of our study. Therefore, we only provide examples of two categories of work that use similar ideas as ours in building their incrementality approaches, namely, reuse-based (or artifact-oriented) incremental approaches and reduction-based (or analysis–oriented) incremental approaches. We noticed that the majority of existing work belongs to the first category.

Reused-based (artifact-oriented) incremental approaches: In the first category of approaches incrementality is achieved by reusing analysis-related artifacts. Examples of analysis-related artifacts that have been reused in the literature are state-space representations (e.g., abstract reachability trees and SETs) [32, 33, 51, 52], CFGs and dependency graphs [52, 53], constraint solving results [7, 8, 54–56], counter example traces [40, 57], abstraction precisions [58], some compressed representation in the form of hash codes of verified properties [59], and function summaries [60].

In analogy to the idea of extreme programming that advocates the use of regression testing side-by-side with stepwise software development, the idea of extreme model checking is proposed in [32] to support regression verification. The proposed approach extended the lazy abstraction algorithm implemented in the unbounded model checker BLAST for C programs to support incremental checking of safety properties. This is done by caching and reusing the abstract reachability trees resulting from a previous model checking run. A similar approach is presented in [51] to perform an incremental state-space exploration for evolving Java programs by recording and reusing the state-space resulting from a previous model checking run to speed up the exploration process of subsequent program versions. The realization of this approach is built in the context of the Java PathFinder (JPF) symbolic execution framework. Inspired by this latter work, a technique of regression model checking is presented in [33]. Two modes of operations are defined recording mode and pruning mode. Recording mode is used to compute and store the state space and state coverage resulting from exploring a new program location. When a change is made to a program whose state space before the change has previously been computed and recorded, the model checker operates in the pruning mode where a change impact analysis is used to identify the impacted elements in the old state space that need to be updated. Only program locations related to these impacted elements are going to be reexplored during the recording operation mode of the model checker.

The idea of memoizing and reusing the path constraints found in the SET of a Java program to optimize the symbolic execution of the program's next revision is introduced in [7]. Reusing a previous SET will avoid the reexecution of common paths between the new version of a program and its previous one and hence speed up the symbolic execution of the new version.

The idea of caching and reusing constraint solving proofs in the context of symbolic execution is first used in the implementation of the symbolic execution engine KLEE [8]. The goal is to reduce the number of calls to the constraint solver while running the symbolic execution and hence mitigating some of the scalability issues related to this technique. The same idea has been further explored and applied in the context of other symbolic execution engines such as JPF [54, 55], Crest, and JBSE [56]. In all such work, a repository of constraint proofs is created that is augmented with a number of powerful simplification techniques to facilitate the optimal reuse of constraints including slicing, normalization and canonicalization. This constraint reusing mechanism can be applied to optimize the symbolic execution of similar or different programs.

The idea of reusing previous model checking outcomes including the counter-examples of failed properties (also called verification witnesses) is presented in [57] to optimize the following reverification runs and in [40] to perform incremental property checking. A key goal in such approaches is to reduce the number of properties to be rechecked after an update has been made to the set of properties required for a given program.

Different types of analysis can be performed in the context on model checking (e.g., explicit-value analysis and predicate analysis). Each type needs to be explicitly defined and supplied to the model checker to determine the scope of the analysis (e.g., location-scoped, function-scoped, or global-scoped) and to guide the abstraction process followed by the model checker to create a proper and sufficient abstraction of the model state space. The definitions of such analysis-based requirements are called program precisions which can be recorded and reused for speeding up the subsequent analysis of an evolving program as presented in [58]. Intuitively, the effectiveness of such an approach depends on the class of program changes introduced in a program's next revision which may or may not require refinement of old precisions.

The idea of saving and reusing a compressed representation (hash codes) of the parts of the model parse tree relevant to a given property is presented in [59] for efficient regression verification. When a change is committed in the system, a new hash code is computed and is compared with the hash code value generated from the previous verification run. If both values are identical then there is no need to reverify the given property, otherwise a reverification is needed.

In [60], an incremental upgrade checking approach is presented and implemented in the context of bounded model checking (BMC). It reuses the function summaries generated from model checking a previous program version to check the validity of their corresponding modified functions in the new program version. If such validity check fails for a given modified function, an upgrade check is performed to create a new valid summary to replace the invalid one.

A whole program path profiling technique is presented in [61] that uses execution traces collected from running a system test suite to generate a compact graphical representation in a form of directed acyclic graph (DAG) of the control flow of the entire system including interprocedural paths. The generated DAG can be used to perform more precise program analysis such as dynamic impact analysis [62]. An incremental version of such dynamic impact analysis is presented in [53] which reuses the DAG of a previous release of a system to incrementally maintain the DAG of the system as it evolves (i.e., a change is made to its behavior or its test suite).

In [52], an incremental approach for interprocedural analysis of safety properties is presented. The main idea depends on reusing previous derivation graphs resulting from the synchronous product of program CFGs and a checking automaton describing some safety properties to speed up the reanalysis of the program after a change has been made to its specification or to its safety properties.

Reduction-based (analysis-oriented) incremental approaches: On the other hand, in the second category of approaches incrementality is achieved by reducing the scope of the new analysis to focus mainly on the parts of the system that are impacted by the change given that (1) some analysis has already been conducted on the system before the change and (2) this analysis proved the system's correctness with respect to some properties. Examples of these approaches are built on the assumption that the previous version of the program is verified and is correct and hence the task of reverifying the program after the change is either limited to the impacted parts [6, 40, 63] or is reduced to an equivalence check between the impacted parts of both program versions [34–39] to conclude the correctness of the modified program version. Program refactoring is an example of an application that can benefit from equivalence checking approaches.

In [6], an intraprocedural level analysis technique called DiSE (directed incremental symbolic execution) is proposed which uses static analysis and change impact analysis to determine the differences between program versions and the impact of these differences on other locations in the program, and uses this information to direct the symbolic execution to only explore these impacted locations. Analyzing the symbolic paths that characterize the behavior of impacted locations helps in discovering bugs and regression testing (test case selection and augmentation). Another important and practical application of this technique is presented in [64] for monitoring safety-critical systems. Extending DiSE with interprocedural analysis capabilities is proposed in [63]. A DiSE-based strategy is used in [40] in the cases where code changes occur along with the update of properties. In such cases, only the impacted parts of the program are considered when analyzing the set of updated properties.

In order to reduce the overhead of performing a complete semantics equivalence check of two compared programs, the notion of uninterpreted functions is employed in [36, 37, 39] to abstract the parts of the programs that are syntactically equivalent and have same inputs. In this case a semantics equivalence check is only applied between the corresponding parts that have not been abstracted. An obvious challenge in this technique is to determine which parts correspond to each other.

The notion of uninterpreted functions is also used in [38] in place of common parts of two program versions which are then analyzed by an extended form of symbolic execution to generate a set of symbolic summaries of behavior that reflects the differences between the two program versions. Analyzing these summaries will reveal equivalency, partial equivalency and deltas in the two versions which can be used for refactoring assurance, change characterization and test suite evolution.

In [34], a change impact analysis is applied to two related program versions to compute the sets of impacted statements in each version and the set of impacted symbolic summaries (constraints) resulting from the symbolic execution of each program version. Checking the equivalence of program versions depends on whether their impacted symbolic summaries are equivalent or not.

A partition-based regression verification is proposed in [65] where the input space of two program versions is partitioned, using a form of symbolic execution in conjunction with random testing, into equivalence-revealing partitions (those that generate the same output in both versions) and difference-revealing partitions. Proving the absence of regression errors is carried out by checking the difference-revealing partitions.

The idea of reducing the equivalence checking of similar programs to a comparison of their deduced predicates summaries is presented in [35]. Proving behavior equivalence of two programs is inferred from the level of coupling found between their predicates.

3.4 Incremental Analysis and Verification in Model-Based Development

In his statement paper "Evolution, Adaptation, and the Quest for Incrementality", Ghezzi [66] argues that supporting software evolution requires building incremental methods and tools to speed up the maintenance process with the focus on the analysis and the verification activities. An incremental approach in such contexts would try to characterize exactly what has been changed and reuse (as much as possible) the results of previous processing steps in the steps that must be rerun after the change. The motivation for this is twofold: time efficiency and scalability. Given the iterative development approach suggested by MDD, we believe that this vision needs to be employed by modern analysis and verification methods.

Surveying the literature for existing work on model-based incremental analysis and verification methods shows two categories of work: change-based incremental analysis and verification methods and compositional incremental analysis and verification methods.

Change-based incremental analysis or verification methods: We notice that the majority of existing work in this category focuses on incremental model-based test case generation and regression testing [67–73]. In [67], a global state reduction technique based on dead and live variables has been used to construct and to incrementally update (upon a change), on-the-fly, the global reachable state space of communication protocols modeled as a set of extended communicating finite state machines (ECFSMs). The generated global state space is then used to verify the correctness of such systems. The idea of incremental test case generation is presented in [69] for testing communication protocols modeled as FSMs. The goal is to test only the modified parts (i.e., transitions) of an evolving FSM specification. The experimental results provided showed significant gains, especially when the percentage of the modifications is not higher than 20% of the original specification. The same idea has also been explored in the context of software product lines [71] using SAT-based analysis and UML-RT models [72] using symbolic execution. In [71], properties of features are defined as first-order logic formulas in Alloy which are then automatically mapped to a transformation that defines an incremental refinement of test suites. In [72], SETs of the subject models (i.e., a model and its evolved versions) are manipulated to incrementally generate their test suites. Techniques for model-based regression test reduction and regression test generation of extended finite state machines (EFSMs) are presented in [68, 70] using an incremental manipulation of the control and data dependence graphs of these models.

The idea of incremental dependency graph adaptation and incremental slice computation is proposed in [74] for software product lines based on the concept of delta modeling [75] where a family of product variants are defined in terms of a core product and a set of deltas describing the changes to be made to the core product such that a product variant is obtained. In the same context, a framework for incremental regression-based testing is proposed in [73] to efficiently test the selected product variants.

Incremental techniques for consistency checking of UML models are presented in [76] using an automatic scope validation of consistency rules instantiation and in [77] using a change impact analysis on the dependencies between model elements and inconsistency rules. Both approaches automatically detect the model changes that violate user-defined consistency rules during their validation.

Our research shares the same motivation as this category of existing work, however it differs in the analysis technique employed and type of models involved.

Compositional incremental analysis or verification methods: In contrast to the change-based incremental verification methods, compositionality-based incremental analysis or verification methods focus on optimizing the analysis or verification process of large systems using different strategies including assume-guarantee reasoning and compositional minimization (or compositional abstraction) [78]. A key aspect of these strategies is to alleviate some of the scalability issues (e.g., state explosion problem) encountered when analyzing large systems. The main idea of the assume-guarantee reasoning strategy is to make use of the modular structure of the system by breaking down the verification process into subprocesses, each of which is concerned with the verification of one component. A global conclusion about the verification of the whole system should be incrementally derived from the verification of its components without the need to construct the global state space of the system. On the other hand, the compositional minimization or abstraction strategy depends on building a reduced version of the global state space of the system by composing a minimized or abstracted version of the local state space of the individual system components.

Examples of compositional reasoning approaches are presented in:

- [79] for UML-RT models using domain-specific communication patterns and the RAVEN real-time model checker;
- [80] for software product lines using an automata-based model checking algorithm;
- [81] for component-based software systems based on an assume-guarantee model checker.

Examples of compositional minimization or abstraction approaches are presented in:

- [82] for integration testing of component-based applications whose behavior is modeled as UML Statecharts using some heuristic reduction rules;
- [83] for verifying systems of asynchronously communicating extended finite state machines (CEFSMs) using a partial product algorithm (or a partial composition technique);
- [84] for verifying systems of distributed programs modeled as synchronously communicating FSMs using a hierarchy-based compositional minimization algorithm;
- [85] for communicating state machines with infinite data space by clustering bisimilar states that have different data values into a finite set of equivalent classes and hence reducing the system state space;

- [86, 87] for communication protocols modeled by FSMs communicating through FIFO queues using symbolic reachability analysis and loop-first search analysis, respectively;
- [88, 89] for labeled transition systems (LTS) by leveraging precise information about the context and interface constraints for the different components in the system and hence eliminating the presence of possible unreachable states in the resulting compositional reachability analysis of the whole system;
- [90] for communicating UML-RT state machines by computing the SETs of individual state machines and then composing them to construct a composite symbolic execution tree which can be used to run multiple system-level analyses.

Our research adopts the notion of abstraction presented above to construct an abstracted version of the global state space of the system to be analyzed using symbolic execution. This construction is performed on-the-fly for the whole system and not incrementally as in the work presented in this category.

3.5 Summary

In this section we have outlined the categories of research work that are relevant to our research. First, we have shown examples of approaches for the analysis and the verification of Statecharts-like models. Next, we have presented incremental analysis and verification techniques for evolving programs. Finally, we have reviewed some of the techniques that implement incremental analysis and verification of behavioral models.

Our work is inspired by the category of work presented in Section 3.3 that is built specifically to improve the efficiency of symbolic execution when it is reapplied to an evolving version of a program, e.g., the work on incremental state-space exploration (ISSE) [51], regression model checking [33], memoized symbolic execution [7], and directed incremental symbolic execution (DiSE) [6]. In contrast to the work in [7, 33, 51] which reuses previous model checking results and constraint solving proofs generated from analyzing a predecessor version of a Java program, we reuse SETs generated from analyzing a predecessor version of a Rhapsody Statechart. Also in contrast to the work in [6] which performs an intraprocedural regression analysis of an evolved Java code, we perform a regression symbolic analysis of an evolved version of a Rhapsody Statechart.

4. STANDARD SYMBOLIC EXECUTION OF RHAPSODY STATECHARTS

In this section, we first explain how we build the SET of individual Rhapsody Statecharts. Then, we explain how we extend it to consider a model of communicating Statecharts. Fig. 5 shows the architecture of an implementation of both techniques.

4.1 Standard Symbolic Execution of Individual Rhapsody Statecharts

In Section 3, we referred to some existing work that adopts the idea of symbolic execution of programs and transfers it to state-based models. We found the approach of Zurowska and Dingel [3] is the most relevant to our work as it has been applied to a type of models, namely, UML-RT state machines, that shares many features with Rhapsody Statecharts. Therefore, we followed a similar procedure to build ours. The two core components of our standard symbolic execution technique of individual Rhapsody Statecharts are shown in Fig. 5. The first component, "SC2MLM Transformation", is a transformation that transforms an individual Statechart into our custom MLM

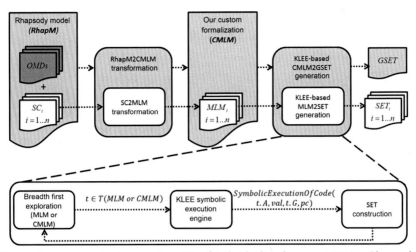

Fig. 5 Standard symbolic execution (SE) of individual and communicating Rhapsody Statecharts.

representation. The second component, "KLEE-based MLM2SET Generation", is a symbolic executor that traverses the MLM model and symbolically executes the action code encountered in each transition to build its symbolic state space. In contrast to the work in [3], our intermediate machine representation of Rhapsody Statecharts takes the form of MLMs instead of their functional finite state machines (FFSMs). Additionally, our symbolic execution module is based on an off-the-shelf symbolic execution engine, KLEE [8], to symbolically execute action code encountered in the Statecharts, whereas an in-house symbolic execution engine is used in [3]. The reason for choosing the MLM formalism is to have actions associated only with transitions, and we do this in such a way that it preserves the behavior of Statecharts.

4.1.1 Statecharts-to-Mealy-Like-Machines Transformation: SC2MLM

We first define our custom MLM formalism that is used to represent a semantically equivalent flattened version of a Rhapsody Statechart. The basic structure of an MLM consists of a set of global variables (sometimes called attributes), a set of simple states with one of them marked as an initial state, and a set of transitions between these states. Simple states in MLMs do not have entry or exit actions. Transitions are characterized in the same way as in Rhapsody Statecharts by the event that triggered them, an optional guard and an action that occurs upon firing them. More advanced features that are found in Rhapsody Statecharts such as composite states, concurrent states, states with entry, and exit actions and choice points (also called condition connectors or OR-connectors, which allow a transition to branch depending on the value of a guard) need to be mapped to fit the structure of MLMs. Our current transformation supports the mapping of these specific features. For example:

- Composite and concurrent states and their outgoing group transitions (formally called group transitions or high-level transitions) are flattened into simple ones;
- Choice points along with their incoming and outgoing branches are replaced by newly created transitions connecting the source state of each choice point with its target states;
- Entry actions of each state are added at the end of the action code of all its incoming transitions; and
- Exit actions of each state are added at the beginning of the action code of all its outgoing transitions.

We followed the semantics of the Rhapsody Statecharts that is used by the code generator of the IBM Rational Rhapsody tool to develop our transformation. We performed exhaustive manual checking and inspection for the set of models that we used to validate the correctness of our transformation. Conducting a formal testing or verification study for this task is outside the scope of this research and it is subject for future work.

Fig. 6 shows an example of a Rhapsody Statechart with the subset of features that we consider in this work and its MLM representation.

Definition 5. (Mealy-like machine)

Formally, an MLM is a tuple $MLM=(S, V, E, EA, T, s_0, V_0)$, where:

- S is a nonempty finite set of *states*;
- V is a set of typed global *variables* (these are the attributes of the class of objects whose behavior is modeled by the MLM);
- E is a nonempty finite set of *events* (also called triggers or signals) by which the machine communicates with its environment; they are divided into disjoint sets of input events E^{inp}, output events E^{out}, and internal events E^{int}, and can optionally have arguments (also called parameters), EA;
- EA is a finite set of event arguments that is disjoint with V;
- T is a set of transitions connecting the states in S. A transition is a tuple $t=(s, e, eA, G, A, s', DD)$, where:
  - s and s' are the source state and the target state, respectively;
  - e is an input or internal event in E^{inp} or E^{int};
  - eA is a subset of EA representing the arguments of e;
  - G is a Boolean expression (condition) defined over a subset of $V \cup eA$ and is called the *guard* of t;
  - A is a fragment of action code written in C, C++, or Java and is called the *action* of t; statements in this fragment can be assignment expressions, conditional statements, iterations or a special type of statement used to generate some output events to be sent to the environment; the execution of this fragment may result in updating the values of a subset of V, constraining some of the variables in V, or the parameters in eA, or it may cause a sequence of output events in E^{out} to be sent;
  - DD is a subset of the transitions in T that have a *data dependency* with t with respect to the variables in V (the initialization of this variable occurs as a result of computing the data dependency between the transitions in T);
- $s_0 \in S$ is the *initial state*.
- V_0 is an *initial valuation* (or initialization) of the variables in V.

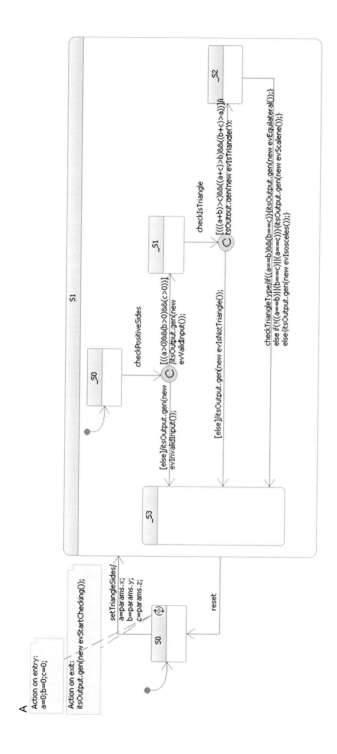

Fig. 6 An example of a Rhapsody Statechart and its corresponding MLM representation. (A) An example of a Rhapsody Statechart as adapted from [91]. (B) The MLM representation of the Rhapsody Statechart in (A).

B

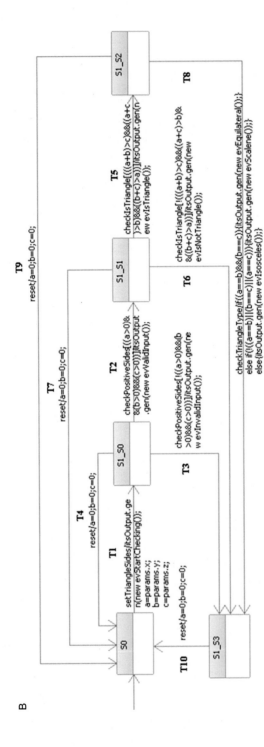

Fig. 6—Cont'd

Example 1. For the example of the MLM model presented in Fig. 6B, we can identify the following:

- $S = \{S0, S1\_S0, S1\_S1, S1\_S2, S1\_S3\}$;
- $V = \{a, b, c\}$;
- $E^{inp} = \{setTriangleSides(x, y, z), checkPositiveSides(), checkIsTriangle(), checkTriangleType(), reset()\}$;
- $E^{out} = \{evStartChecking(), evValidInput(), evInvalidInput(), evIsTriangle(), evIsNotTriangle(), evEquilateral(), evScalene(), evIsoscelees()\}$;
- $E^{int} = \emptyset$;
- $EA = \{x, y, z\}$;
- $s_0 = S0$;
- $V_0 = \{a=0, b=0, c=0\}$;
- $T1 = (S0, setTriangleSides, \{x, y, z\}, \emptyset, \{itsOutput.gen(new evStartChecking()); a=params.x; b=params.y; c=params.z;\}, S1\_S0, \emptyset)$;
- $T2 = (S1\_S0, checkPositiveSides, \emptyset, \{(x>0)\&\&(b>0)\&\&(c>0)\}, \{itsOutput.gen(new evValidOutput());\}, S1\_S1, \{T1\})$;
- ...
- $T10 = (S1\_S3, reset, \emptyset, \emptyset, \{a=0;b=0;c=0;\}, S0, \emptyset)$.

4.1.2 Symbolic Execution of MLM:MLM2SET

The main idea is the same as for the symbolic execution of programs with some variations that reflect the different features of MLMs. For instance, states in MLMs are similar to program locations (i.e., program counters), transitions in MLMs are similar to program statements, and event arguments in MLMs are similar to program arguments (which both represent the input data passed from the environment). Just as symbolic execution of programs replaces the concrete values of all input variables to a program by symbolic values, so too will symbolic execution of MLMs replace the concrete values of all event arguments received by a MLM by a unique set of symbolic values. Consequently, both symbolic execution techniques are able to symbolically trace the given artifact (either a program or an MLM) and compute the constraints and the variable updates associated with each execution path. In this case, both constraints and variable updates are defined over symbolic values. The output from symbolic execution is represented by a symbolic execution tree (SET). The nodes in this tree represent program locations (or MLM states) with the symbolic valuations of program variables (or MLM variables) at these locations and the path constraints collected to reach these locations; therefore, we call them symbolic. The edges in this tree are links between symbolic locations, and they reflect the control flow of the execution. For the symbolic execution of programs, these edges usually have no labels; however, in the symbolic execution of MLMs they are labeled with the event causing the transition (with the symbolic values substituting their arguments if any) and also the sequence of events resulting from the execution of the transition action code (along with their arguments, if any). We call these edges *symbolic transitions*.[b] The root of the tree is the node representing the initial state of the MLM with the MLM variables set to their initial values and the path constraint set to "true" or empty.

Definition 6. (Symbolic Execution Tree)

Formally, a SET of an MLM, $MLM=(S, V, E, EA, T, s_0, V_0)$, is a tuple $SET= (SS, L, ST, ss_0)$ where:

- SS is a finite set of *symbolic states*; a symbolic state is a tuple $ss=(s, val, pc)$ where,
  - s is a *state* in S;
  - val is the *symbolic valuation* of machine variables V;
  - pc is the *path constraint* collected to reach state s.
- L is a finite set of labels; a label is a tuple $l=(e, V^s, \sigma, out)$ where,
  - e is an input or internal event in E^{inp} or E^{int};
  - V^s is a set of symbolic variables, different from V and EA;
  - σ is a mapping from the original event argument names $eA \subset EA$ to the new set of variables names in V^s;
  - out is a sequence of some output events $e_1^{out}, e_2^{out}, ..., e_n^{out}$ in E^{out} and the symbolic valuations of their arguments $e_1^{out}A, e_2^{out}A, ..., e_n^{out}A$ in EA (if any).
- ST is a finite set of symbolic transitions defining the transition relation between symbolic states; a symbolic transition is a tuple $st=(ss, l, ss')$ where,
  - ss is the source symbolic state;
  - l is the label of the symbolic transition;
  - ss' is the target symbolic state.
- ss_0 is the *initial symbolic state*.

---

[b] Please note that we sometimes use the terms "symbolic transition", "symbolic path", and "execution path" to mean the same thing, however, more generally, the term "execution path" refers to a sequence of one or more consecutive symbolic transitions in a given SET.

In the remainder of this section, we provide the details of our algorithm to symbolically execute MLMs as listed in Algorithm 5.

Algorithm 5 Symbolic Execution (SE) of MLMs

Input:
1: An MLM $MLM = (S, V, E, EA, T, s_0, V_0)$

Output:
1: A SET $SET = (SS, L, ST, ss_0)$

Steps:
1: $ss_0 \leftarrow (s_0, v_0, \text{true})$
2: $SS \leftarrow \{ss_0\}$
3: Create a queue q
4: $q \leftarrow [ss_0]$
5: //Explore the MLM in Breadth-First-Search fashion
6: **while** q is not empty **do**
7:     Remove the first element $ss = (s, val, pc)$ from q
8:     **for all** outgoing transitions $t = (s, e, eA, G, A, s')$ of s **do**
9:         Create a unique set V^s of new variables in one-to-one relation with the transition's event arguments eA
10:        Create a map σ between eA and V^s
11:        Substitute every occurrence of a variable $x \in eA$ in the guard G and the action code A by $\sigma(x)$ to obtain $G^s \leftarrow \sigma(G)$ and $A^s \leftarrow \sigma(A)$
12:        **if** G^s is satisfiable given val and pc **then**
13:            $feasiblePaths \leftarrow SymbolicExecutionOfCode(A^s, val, G^s, pc)$
14:            **for all** $feasiblePath = (val', pc', out') \in feasiblePaths$ **do**
15:                Create a new symbolic state, $ss' = (s', val', pc')$
16:                **if** ss' is not subsumed by any previously generated symbolic state in SS **then**
17:                    $SS \leftarrow SS \cup \{ss'\}$
18:                    Create a new label, $l' = (e, V^s, \sigma, out')$
19:                    $L \leftarrow L \cup \{l'\}$
20:                    Create a new symbolic transition, $st' = (ss, l', ss')$
21:                    $ST \leftarrow ST \cup \{st'\}$
22:                    $q \leftarrow enqueue(q, ss')$
23:                **end if**
24:            **end for**
25:        **end if**
26:    **end for**
27: **end while**

The first task in this algorithm concerns the exploration of the MLM, which is done in a breadth-first-search fashion,[c] starting from the initial state and proceeding through each of its outgoing transitions, and so on, until all states have been visited. Since some MLMs may have loops, we need to limit the exploration by some criterion. The criterion that we consider here is to check if a symbolic state is *subsumed by* some other previously explored symbolic state; if so, we should not explore it again.

Definition 7. (Symbolic states subsumption [92])

Formally, a symbolic state ss = (s, val, pc) is subsumed by another symbolic state ss' = (s', val', pc') iff:

(1) s = s';

(2) val = val';

(3) pc is included in pc' (i.e., pc ⇒ pc' or pc is at least as constraining as pc'). The second task is the symbolic execution of transitions. In this task, we perform the following steps for each transition: (1) we create and assign a unique set of symbolic variables to replace the event arguments of the transition, if any (Lines 9–10); (2) we substitute the occurrences of these event arguments in the guard expression and the action code statements by their assigned symbolic values (Line 11); (3) we check if the guard is satisfiable with respect to the variable updates and the path constraint of the current symbolic state (Line 12); if so, (4) we use symbolic execution of programs to execute the updated action code of the transition; the result from the symbolic execution engine[d] is a set of variable valuation and path constraints that trigger different feasible paths in that code (Line 13); (5) We use the results from the previous step to both create a new set of symbolic states representing the target state of this transition and also to label the edges to these new symbolic states (Lines 15–19).

Example 2. The complete SET of the MLM in Fig. 6B is shown in Fig. 7. As we notice, the SET has depth 6, and it has 14 symbolic execution paths identified by the leaves in the tree. In the following, we show the detailed steps for generating the first three levels of the tree, based on Algorithm 5.

As it is shown, the root of the tree is the symbolic state SS1 = (S0, (a=0, b=0, c=0), []). Exploring state S0 of SS1 will symbolically execute its outgoing transition T1, which creates new symbolic values X, Y, and Z, assigns them to the arguments x, y, and z of the event setTriangleSides, updates

---

[c] Depth-first-search would be another possible alternative.
[d] We use the popular open-source symbolic execution engine KLEE [8].

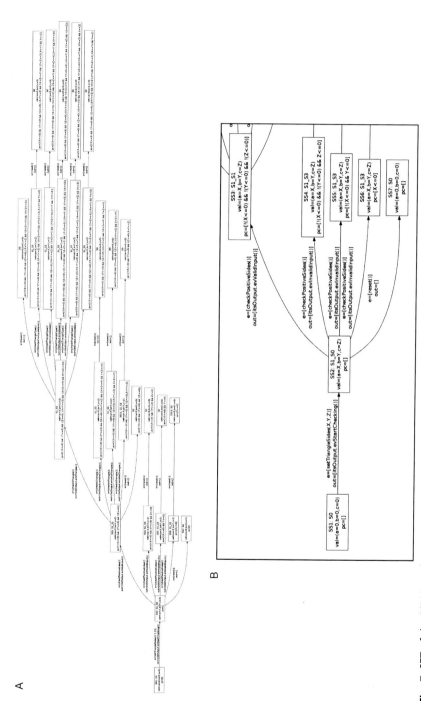

A

B

Fig. 7 SET of the MLM model in Fig. 6B (highlighted Levels: 1, 2, and 3).

the machine variables a, b, and c to have the new symbolic values of the setTriangleSides event arguments, creates a new symbolic state SS2 = (S1_S0, (a=X, b=Y, c=Z), []), and finally forms the label for the symbolic transition (i.e., the edge) connecting SS1 and SS2 with the transition event e = [setTriangleSides(X, Y, Z)] and the output out = [itsOutput.evStart-Checking()]. Because there is no guard associated with the transition T1 and there are no conditional statements in its action code, there is no update to be made to the path constraint of the symbolic state SS2. The next step is exploring state S0_S1 of SS2 which will symbolically execute its outgoing transitions, T2, T3, and T4, in sequence. Symbolically executing T2 creates a new symbolic state SS3 = (S1_S1, (a=X, b=Y, c=Z), [!(X<=0)&&!(Y<=0) &&!(Z<=0)]) and forms the label for the edge between SS2 and SS3with the transition event e = [checkPositiveSides()] and the output out = [itsOutput.evValidInput()]. As we notice, the value of the path constraint of SS3 reflects the guard condition of transition T2. In contrast with T2, symbolically executing T3 creates three symbolic states representing state S1_S2, which are SS4, SS5, and SS6. All have the same variable valuations val = (a=X, b=Y, c=Z) but have different path constraints representing all feasible cases for the guard condition of T3. The edges connecting SS3 with SS4, SS5, and SS6 are all labeled with the input event e = [checkPositiveSide()] and the output out = [itsOutput.evInvalidInput()]. Finally, symbolically executing T4 creates a new symbolic state SS7 = (S0, (a=0, b=0, c=0), []) and forms the label for the edge between symbolic state SS2 and symbolic state SS7 with the transition input event e = [reset()] and the output out = []. The difference between the symbolic state SS7 and the other symbolic states SS4, SS5, and SS6 is that we find that there is another symbolic state, SS1, that has already been explored before, and it subsumes SS7, therefore no further exploration is needed for SS7.

We can notice that the symbolic execution of a transition may result in zero, one or many symbolic edges depending on the satisfiability of the guard condition and action code of the transition. Therefore, we refer to transitions with more than one symbolic edge in the resulting SET as multipath transitions (e.g., transitions T3 and T4 are multipath transitions).

The resulting SET can be used to run many *intralevel* types of analyses (e.g., reachability analysis and guard analysis) to ensure the correctness of the individual objects in isolation of the others.

4.2 Symbolic Execution of a Model of Communicating Rhapsody Statecharts

In the previous subsection, we showed how to symbolically execute individual Statecharts. In this section we present how we extended our standard symbolic execution technique to consider a model of communicating Statecharts. The goal here is to facilitate the analysis of the overall behavior of the system and the interoperability between its communicating objects. Our approach is based on an on-the-fly exploration of all interleaving Statecharts in a given model to construct the global state space of the system represented by a more general SET called a *global symbolic execution tree* (GSET).

The two components of our standard symbolic execution technique of a model of communicating Statecharts are shown in Fig. 5, which are the "RhapM2CMLM Transformation" and the "KLEE-based CMLM2GSET Generation". Each component generalizes the functionality provided by its corresponding one for individual Statecharts. This generalization incorporates the extra features present in the communication setting. For instance, we need to access the OMDs of a Rhapsody model to identify the set of communicating objects in the model and to determine the communication topologies between these objects. This requires the extension of our previous MLM formalism to represent a collection of communicating MLMs and to represent their communication topologies. We also need to extend our previous SET definition to represent, instead, a global SET. In such global SETs, symbolic model states are used in place of simple symbolic states.

4.2.1 Rhapsody Model of Communicating Statecharts-to-Communicating MLMs Transformation: RhapM2CMLM

The input to this component is a Rhapsody model represented by its OMDs and its Statecharts and the output is a Communicating MLMs representation. Our formalism for Communicating MLMs is shown below and it uses the MLM formalism presented in Section 4.1.1 (please see Definition 5). For illustration, we refer to the model example shown in Section 2, Fig. 1. For simplicity, we consider only simple states and simple transitions in the Statecharts used in this example, yet our implementation supports Statecharts with more complex features such as hierarchical states, orthogonal regions, choice points, and group transitions.

Definition 8. (Communicating MLMs)

Formally, a global model of n asynchronously communicating MLMs is a tuple $M = (MLM^M, Q^M, s_0^M, V^M, V_0^M)$ where:

- $MLM^M = (MLM_1, MLM_2, \ldots, MLM_n)$, where $MLM_i = (S_i, V_i, E_i, EA_i, T_i, S_{0i}, V_{0i})$ is the i-th MLM in M that is defined as follows:
  - S_i is a nonempty finite set of *states* and $S_1 \cap S_2 \cap \ldots \cap S_n = \emptyset$;
  - V_i is a set of typed global *variables* (these are the attributes of the class of objects whose behavior is modeled by the MLM);
  - E_i is a nonempty finite set of *events* (also called triggers or signals) by which the machine communicates with its environment. Events can have arguments EA_i (also called parameters) and are partitioned into input events, E_i^{inp}, output events, E_i^{out}, and internal events, E_i^{int} and $E_1^{inp} \cap E_2^{inp} \cap \ldots \cap E_n^{inp} = E_1^{out} \cap E_2^{out} \cap \ldots \cap E_n^{out} = E_1^{int} \cap E_2^{int} \cap \ldots \cap E_n^{int} = \emptyset$;
  - EA_i is a finite set of event arguments that is disjoint with V_i and $EA_1 \cap EA_2 \cap \ldots \cap EA_n = \emptyset$;
  - T_i is a set of transitions connecting the states in S_i. A transition is a tuple $t = (s, e, eA, G, A, s', DD, COMD)$, where:
    - s and s' are the source state and the target state, respectively;
    - e is one of the input or internal events in E_i^{inp} and E_i^{int};
    - eA is a subset of EA_i representing the arguments of e;
    - G is a Boolean expression (condition) defined over a subset of $V_i \cup eA$ and is called the *guard* of t;
    - A is a fragment of action code written in C, C++, or Java and it is called the action of t; statements in this fragment can be assignment expressions, conditional statements, iterations, or a special type of statements used to generate some output events to be sent to the environment; the execution of this fragment may result in updating the values of a subset of V_i, constraining some of the variables in V_i, or the parameters in eA, or it may cause a sequence of output events in E_i^{out} to be sent;
    - DD is a subset of the transitions in T_i that have a *data dependency* with t with respect to the variables in V_i;
    - $COMD$ is a subset of the transitions in T_j of $MLM_j \in M$, where $j = 1, \ldots, n$ and $j \neq i$ that have a *communication dependency* with t.
  - $S_{0i} \in S_i$ is the *initial state*.
  - V_{0i} is an *initial valuation* (or initialization) of the variables in V_i.
- $Q^M = (q_1, q_2, \ldots, q_n)$ is the set of communication queues in M (these queues allow the exchange of messages between MLMs);
- $S_0^M = (S_{01}, S_{02}, \ldots, S_{0n})$, where S_{0i} is the initial state for MLM_i;

- $V^M = V_1 \cup V_2 \cup \ldots \cup V_n$, where V_i is the set of variables in MLM_i and $V_1 \cap V_2 \cap \ldots \cap V_n = \emptyset$;
- $V_0^M = V_{01} \cup V_{02} \cup \ldots \cup V_{0n}$, where V_{0i} is the initial valuation of the variables V_i in MLM_i.

Example 3. Based on this formalism, we can represent the Rhapsody model shown in Section 2, Fig. 1 as the following CMLM:

$M^{AB} = ((\text{MLM}_{itsA}, \text{MLM}_{itsB}), (q_{itsA}, q_{itsB}), (aS1, bS1), (x, y), (0, 5))$ where:

- $\text{MLM}_{itsA} = (\{aS1, aS2, aS3\}, \{x\}, \{e1, e2, e3\}, \{e1Arg1\}, \{itsA\_T1, itsA\_T2, itsA\_T3\}, aS1, \{0\})$;
    - $itsA\_T1 = (aS1, e1, e1Arg1, \emptyset, (if(params.e1Arg1>7)\{x=5;\}else \{x=1;\}), aS2, \emptyset, \emptyset)$;
    - $itsA\_T2 = (aS2, e2, \emptyset, \emptyset, (itsB.gen(newe5(5));), aS3, \emptyset, \emptyset)$;
    - $itsA\_T3 = (aS3, e3, \emptyset, \emptyset, (x=0;), aS1, \emptyset, \{itsB\_T3\})$;
- $\text{MLM}_{itsB} = (\{bS1, bS2\}, \{y\}, \{e4, e5, e6\}, \{e5Arg1\}, \{itsB\_T1, itsB\_T2, itsB\_T3\}, bS1, \{5\})$;
    - $itsB\_T1 = (bS1, e4, \emptyset, \emptyset, (y=y+1;), bS2, \emptyset, \emptyset)$;
    - $itsB\_T2 = (bS2, e5, e5Arg1, (params.e5Arg1==5), (itsA.gen(newe3());), bS1, \{itsB\_T1\}, \emptyset)$;
    - $itsB\_T3 = (bS2, e6, \emptyset, \emptyset, (y=y-1;), bS2, \emptyset, \{itsA\_T2\})$.

Additionally, we can recognize that (1) events e1, e2, e4, and e6 are external events, while events e3 and e5 are internal events and (2) transition itsA_T1 is a multipath transition due to the if-else conditional statement in the transition's action code.

4.2.2 Symbolic Execution of Communicating MLMs:CMLM2GSET

The input to this component is a set of Communicating MLMs representing a Rhapsody model and the output is a GSET. The formal definition of a GSET extends our previous definition in Section 1.1 for the SET of an individual MLM. It replaces the definition for the symbolic states in the tree to consider instead symbolic model states and it adds an extra element to represent the queues' contents in a given model.

Definition 9. (Global SET)

Formally, a global SET of a model of n asynchronously communicating MLMs, $M = (\text{MLM}^M, Q^M, s_0^M, V^M, V_0^M)$ where $\text{MLM}_i = (S_i, V_i, E_i, EA_i, T_i, s_{0i}, V_{0i})$ is the ith MLM in M, is a tuple $\text{GSET}^M = (SS^M, L^M, ST, ss_0^M)$ where:

- SS^M is a finite set of *symbolic model states*; a symbolic model state is a tuple $ss^M = (s^M, val, pc, q^M)$ where,
  - s^M is a *model state* in $S^M=(s_1, s_2, \ldots, s_n)$ where $s_i \in S_i$ of MLM_i;
  - val is the *symbolic valuation* of the model variables V^M;
  - pc is the *path constraint* collected to reach the model state s^M;
  - q^M is the content (or update) of the model queues Q^M;
- L is a finite set of labels; a label is a tuple $l=(e, V^s, \sigma, out)$ where,
  - e is an input or internal event in MLM_i for some $i=1, \ldots, n$;
  - V^s is a set of symbolic variables, different from V^M and $EA_i(MLM_i)$ for all $i=1, \ldots, n$;
  - σ is a mapping from the original event argument names $eA_i \subset EA_i(MLM_i)$ for all $i=1, \ldots, n$ to the new set of variables names in V^s;
  - out is a sequence of some output events $e_1^{out}, e_2^{out}, \ldots, e_n^{out}$ in E_i^o and the symbolic valuations of their arguments $e_1^{out}A, e_2^{out}A, \ldots, e_n^{out}A$ in $EA_i(MLM_i)$ (if any).
- ST is a finite set of symbolic transitions defining the transition relation between symbolic states; a symbolic transition is a tuple $st=(ss^M, l, ss'^M)$ where:
  - ss^M is the source symbolic model state;
  - l is the label of the symbolic transition;
  - ss'^M is the target symbolic model state.
- ss_0^M is the *initial symbolic model state*.

Now we explain the steps to symbolically execute a model of communicating MLMs as listed in Algorithm 6.

The first step in this algorithm is an on-the-fly breadth-first-search exploration for each MLM in the model M concurrently, starting from the initial model state and proceeding through the outgoing transitions of the initial state in each MLM, and so on, until all model states have been visited or until we reach a certain depth limit in case of unbounded loops. Checking if a symbolic model state has been visited is done using the subsumption criterion definition listed below that extends Definition 7 to consider the queues' contents of symbolic model states.

Definition 10. (Symbolic Model States Subsumption [92])

Formally, a symbolic model state $ss^M = (s^M, val, pc, q^M)$ is subsumed by another symbolic mode state $ss'^M = (s'^M, val', pc', q'^M)$ iff:

(1) $s^M = s'^M$;

(2) $val = val'$;

(3) $q^M = q'^M$;

(4) pc is included in pc' (i.e., $pc \Rightarrow pc'$ or pc is more constrained than pc').

Symbolic model states that are marked "visited" do not require further exploration and therefore they form the leaves in a GSET. In Fig. 8, we identified the subsumed nodes for all the leaves of the GSET(AB^{V0}). For example, symbolic model state SS21, which is a leaf node in the GSET (AB^{V0}), is subsumed by symbolic model state SS11, which is a nonleaf node in the GSET(AB^{V0}).

A common problem when building the global state space of asynchronous systems are unbounded queues which are continuously filled with messages (sent either from the environment or from the peers in the system) that cannot be consumed at the same rate. In this case the exploration will exhibit an infinite state behavior and will not end. Possible solutions for this problem are addressed in [93] using bounded queues or assuming a "slow" environment in which external events (i.e., events that are sent by the environment to the system) are offered only when the system reaches a stable state (i.e., when the execution of all internal events are completed and all system queues have become empty). In such an environment, a higher priority is given to internal events over external ones.

In our approach, we adopt the "slow" environment assumption used in [93] by allowing the execution of transitions with external events only if the system is in a stable state.

The second step is the symbolic execution of transitions which is done using the open-source symbolic execution engine KLEE [8]. For each such transition that has an external event or an available internal event (i.e., the event exists in the head of the queue of the MLM to which the transition belongs to) (Line 11) we do the following:

(1) we create and assign a unique set of symbolic variables to replace the event arguments of the transition, if any (Lines 12–13);

(2) we substitute the occurrences of these event arguments in the guard expression and the action code statements, by their assigned symbolic values (Line 14);

(3) we check if the guard is satisfiable with respect to the variable updates and the path constraint of the current symbolic state (Line 15); if so,

(4) we use the symbolic execution engine KLEE [8] to execute the updated action code of the transition; the results are a set of variable assignments and path constraints that trigger different feasible paths in that code (Line 13);

(5) we use the results from the previous step to both create a new set of symbolic model states representing the target state of this transition and also to label the edges to these new symbolic model states (Lines 17–28).

Algorithm 6 Symbolic Execution (SE) of CMLMs

Input:
1: A system/model of n communicating MLMs $M = (MLM^M, Q^M, s_0^M, V^M, V_0^M)$

Output:
1: A Global SET $GSET^M = (SS^M, L, ST, ss_0^M)$

Steps:
1: $ss_0^M \leftarrow (s_0^M, V_0^M, \text{true}, q_0^M)$
2: $SS^M \leftarrow \{ss_0^M\}$
3: Create a queue q
4: $q \leftarrow [ss_0^M]$
5: //Breadth-First-Search fashion exploration for each MLM in M
6: **while** q is not empty **do**
7:    Remove the first element $ss^M = (s^M, val, pc, q^M)$ from q
8:    **for all** MLM_i in MLM^M **do**
9:      $s_i \leftarrow s^M(i)$ where $s_i \in S_i(MLM_i)$
10:      **for all** outgoing transitions $t_i = (s_i, e, eA, G, A, s_i)'$ of s_i **do**
11:        **if** $((q^M = \phi) \wedge (e \in E_i^{inp})) \vee ((head(q_i^M) = e) \wedge (e \in E_i^{int}))$ **then**
12:          Create a unique set V^s of new variables in one-to-one relation with the transition's event arguments eA
13:          Create a map σ between eA and V^s
14:          Substitute every occurrence of a variable $x \in eA$ in the guard G and the action code A by $\sigma(x)$ to obtain $G^s \leftarrow \sigma(G)$ and $A^s \leftarrow \sigma(A)$
15:          **if** G^s is satisfiable given val and pc **then**
16:             *feasiblePaths* $\leftarrow$ *SymbolicExecutionOfCode*(A^s, val, G^s, pc)
17:             **for all** *feasiblePath* $= (val', pc', out') \in$ *feasiblePaths* **do**
18:                $q'^M \leftarrow out'$
19:                Create a new symbolic state, $ss'^M = (s'^M, val', pc', q'^M)$ where $s'^M = s^M.replace(s_i, s_i')$
20:                **if** ss'^M is not subsumed by any previously generated symbolic model state in SS^M **then**
21:                   $SS^M \leftarrow SS^M \cup \{ss'^M\}$
22:                   Create a new label, $l' = (e, V^s, \sigma, out')$
23:                   $L \leftarrow L \cup \{l'\}$
24:                   Create a new symbolic transition, $st' = (ss^M, l', ss'^M)$
25:                   $ST \leftarrow ST \cup \{st'\}$
26:                   $q \leftarrow enqueue(q, ss'^M)$
27:                **end if**
28:             **end for**
29:          **end if**
30:        **end if**
31:      **end for**
32:    **end for**
33: **end while**

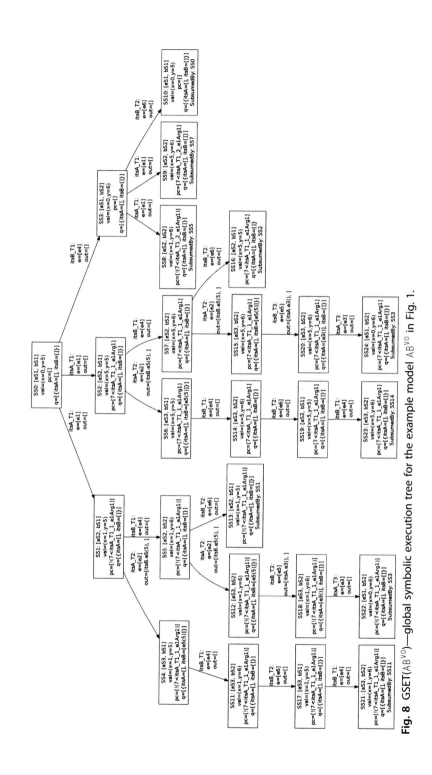

Fig. 8 GSET(AB^{V0})—global symbolic execution tree for the example model AB^{V0} in Fig. 1.

Example 4. Applying this algorithm on the example model in Fig. 1 results in the GSET shown in Fig. 8 which consists of 25 symbolic model states and 9 symbolic execution paths given by the number of leaves in the tree.

4.3 Summary

In this section, we have shown examples of individual and communicating Rhapsody Statecharts. We have presented the formalisms we use to represent a semantically equivalent flattened version of Rhapsody Statecharts and a model of communicating Rhapsody Statecharts such that we can symbolically execute them. The proposed formalisms are called MLMs and Communicating MLMs (CMLMs). We also have provided the formal definitions for the structures of the SET and the GSET resulting from each symbolic execution technique.

5. OPTIMIZING SYMBOLIC EXECUTION OF EVOLVING RHAPSODY STATECHARTS

In this section, we first illustrate with a running example the optimization ideas employed in this research for the symbolic execution of evolving Statecharts. Next, we provide the details of our two proposed techniques, namely, memoization-based symbolic execution (MSE) and dependency-based symbolic execution (DSE), for both individual and communicating Statecharts.

5.1 Motivating Example

In this section we explain using a running example the optimization ideas employed in the next section. For illustration, we use the models in Fig. 9 and their symbolic execution trees (SETs) as shown in Figs. 10 and 11 to show how our optimization techniques work for evolving individual Statecharts.

In Fig. 9, we have two versions of an example of individual Rhapsody Statechart: V0 (in Fig. 9A) is the base version and V1 (in Fig. 9B) is a modified version of the base version where we modified the conditional statement in the action code of transition T2. In Fig. 10, we show the SET of V0, SET(V0), however, in Fig. 11A and B, we show two different illustrations for the symbolic execution tree of V1, SET(V1).

Alternatively, we can use the model version AB^{V1} shown in Fig. 12 and the GSETs in Figs. 13 and 14 to illustrate how the same concepts are applied

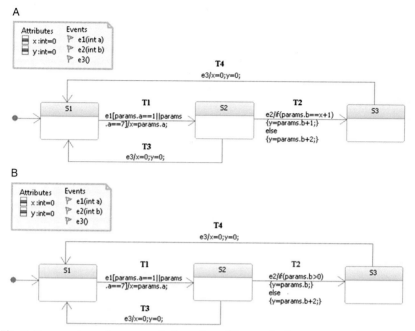

Fig. 9 Two versions of an example Statechart. (A) `V0`—the base version of an example Statechart. (B) `V1`—a modified version of the Statechart in (A).

for evolving communicating Statecharts. The model version AB^{V1} is the successive version of the model example AB^{V0} shown in Fig. 1 with a change in transition `itsA_T2` of the Statechart of object `itsA`.

5.1.1 *Optimization Via Reuse*—An Artifact-Oriented Evolution Support Approach

By comparing the unhighlighted parts (colored black) of the symbolic execution tree of `V1`, SET(`V1`), in Fig. 11A with the SET of `V0`, SET(`V0`), in Fig. 10, we notice that these parts are identical. However, by looking at the highlighted parts (colored red) of the SET(`V1`), we find that:

(1) they are slightly different from their corresponding parts in the SET(`V0`);

(2) they represent the symbolic execution of transition `T2` (the changed transition in `V1`) and subsequent transitions; and

(3) they reflect the parts of the SET(`V0`) that are impacted by the changes made on `V1`.

Therefore, if we already have the SET(`V0`) and we manage to identify the parts of it that need to be updated in order to account for the changes made on `V1`, then we can direct the symbolic execution of `V1` to generate only the

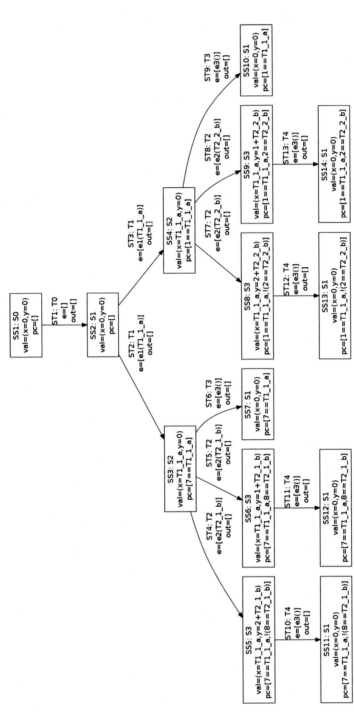

Fig. 10 SET($\mathbb{V}0$)—symbolic execution tree for the Statechart in Fig. 9A.

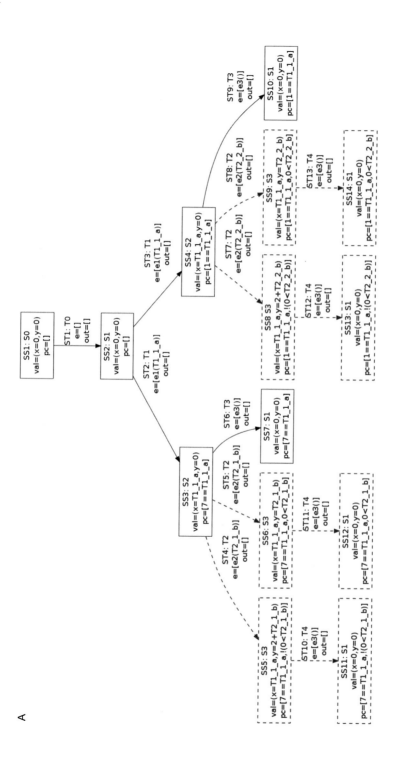

Fig. 11 SET(V1)—symbolic execution tree for the Statechart in Fig. 9B. (A) The parts that are different from SET(V0) are highlighted (in *red*). (B) The pruned parts resulting from the partial exploration of transitions: T1, T3, and T4 are *grayed* out.

B

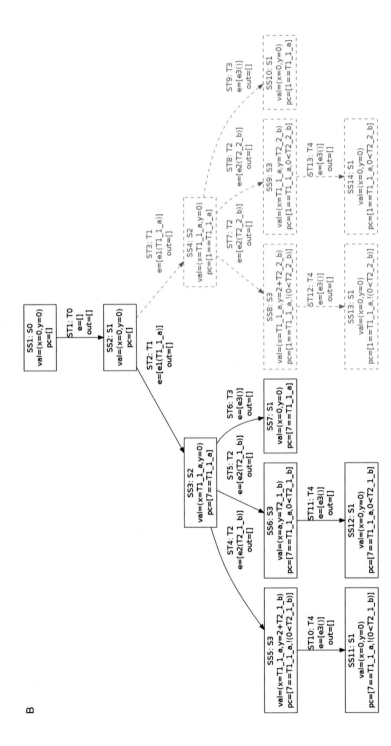

Fig. 11—Cont'd

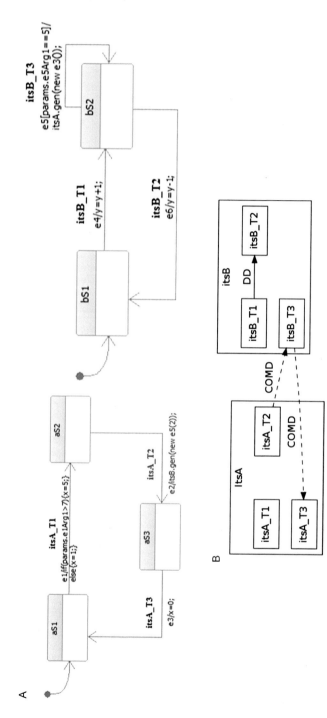

Fig. 12 A modified model version AB^{V1} and its dependency graph. (A) AB^{V1}—a successive version of the example model AB^{V0} in Fig. 1. (B) Dependency graph of the model version AB^{V1}.

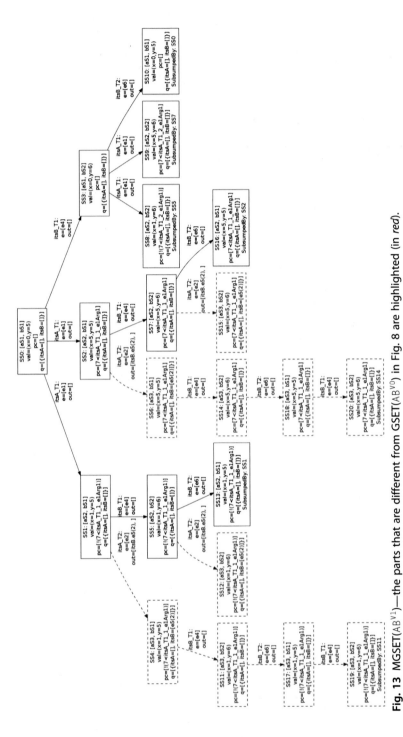

Fig. 13 MGSET(AB^{V1})—the parts that are different from GSET(AB^{V0}) in Fig. 8 are highlighted (in *red*).

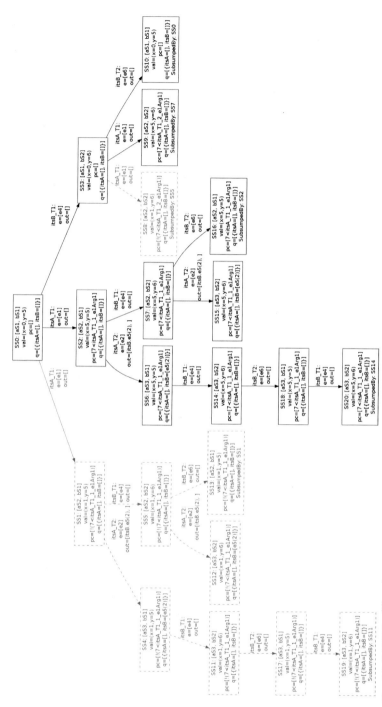

Fig. 14 DGSET(AB^{V1})—the pruned parts resulting from the partial exploration of transition itsA_T1 are *grayed* out.

new updated parts to replace the old ones. The applicability and utility of running this process depends, first, on the existence of the SET(V0) and, second, on the amount of savings gained from reusing it which can be very large for big complex models that undergo some minor changes (i.e., changes with limited impact). Possible measures for quantifying the amount of savings are (1) the time gained from not generating the SET(V1) from scratch; (2) the percentage of nodes (or symbolic states) in the SET(V0) that can be safely reused; or (3) the percentage of execution paths in the SET(V0) that can be safely reused. A very important benefit from this latter measure is that it provides us with a good estimate of how many of the test cases generated from the SET(V0), to test V0, can be reused for testing V1. We notice that the location of the change in the model may suggest the effectiveness of reusing the SET(V0). We believe that, in general, the closer the change is to the initial state of the model, the smaller the number of nodes in the SET(S0) that can be reused. Additionally, the more model paths go through a changed state or transition, the more updates of the SET(V0) are needed and the smaller the effectiveness of reusing SET(V0). For example, having a change in state S1 or transition T1 in V1 will require an update to all the descendants of the symbolic state SS2 in the SET(V0), leaving only two symbolic states to be reused.

5.1.2 *Optimization Via Reduction*—An Analysis-Oriented Evolution Support Approach

First, we assume that the base version of a model (i.e., the version before the change) has already been analyzed successfully to, e.g., determine reachability of states or generate test cases. This assumption enables the optimization described below.

Second, as we discussed in Section 4.1, the symbolic execution of a transition, t_i, in an MLM model may result in zero, one or many symbolic transitions of t_i in the SET of that model, depending on: (1) the path constraints and the variable valuation of the symbolic state representing the source state of t_i; (2) the satisfiability of the guard condition of t_i; and (3) the number of execution paths of the action code of t_i. For example, the symbolic execution of transition T1 in the two versions of the model in Fig. 9 results in two symbolic transitions connecting the symbolic state SS2 of state S1 (the source state of T1), with the two symbolic states SS3 and SS4 of state S2 (the target state of T1). The first symbolic transition is taken if a=7, while the second takes place if a=1 (these are the two cases that satisfy the guard condition of T1). Similarly, the symbolic execution of T2 results in having two symbolic transitions originating from each of the symbolic states SS3 and SS4 and ending at the symbolic states SS5 and SS6 (resp., SS8 and SS9) as a result of having a conditional statement in the action code of T2. On the other hand, the symbolic execution of transition T3, which does not have a guard or any conditional statements in its action code, results in having only one symbolic transition originating from each of the symbolic states SS3 and SS4 and ending at the symbolic states SS7 and SS10, respectively. In the same way, the symbolic execution of transition T4 results in having one symbolic transition originating from each of the symbolic states SS5, SS6, SS8, and SS9 and ending at the symbolic states SS11, SS12, SS13, and SS14, respectively.

Third, as discussed in Section 2.3, applying a dependency analysis on a modified version of a model given the list of the changes made on its previous version allows us to identify all the parts of the model that have an impact on or are impacted by the changes.

Now, with all that said, our question is can we direct the symbolic exploration of a modified version of a model to exhaustively explore all changed transitions and all other transitions that either impact or are impacted by one or more changed transitions and to reduce the exploration of the remaining transitions (i.e., the transitions that neither impact nor are impacted by the change) to a minimum (i.e., to consider only one symbolic transition per such transition)? Our goal is twofold: (1) to explore the minimal set of symbolic transitions to perform the analysis on the evolved state machine model (to, e.g., reveal states that have become unreachable by the change or to generate test cases for the new executions introduced by the change); and (2) to reduce the time of exploration as well as the size of the resulting SETs.

For example, applying a dependency analysis on V1 given that transition T2 has been changed with respect to its base version V0 shows that there is no dependency between T2 and any other transitions as none of them defines a variable that transition T2 uses (and vice versa). Now, with this information in mind, we can direct the symbolic execution of V1 to *"fully"* explore the changed transition T2 (i.e., to explore all its symbolic transitions) and to only *"partially"* explore the rest of the transitions in the model (i.e., to explore only one of its symbolic transitions). Following this process to symbolically execute V1 results in the SET shown in Fig. 11B, which has six symbolic states less than the complete SET(V1). Although the new SET is not complete, it is sufficient to run regression types of analysis. For instance, the SET in Fig. 11B can be used to determine if the change introduced any unreachable states or which test cases must be run to test the execution paths introduced by the change.

As the amount of savings to be gained here depends on the number of symbolic transitions to be pruned from the partial exploration of unimpacted transitions, this technique is most beneficial if applied to Statecharts that have transitions with disjunct guards (i.e., "||") or with action code that has conditional statements, which both result in more than one symbolic transition in the SET. The more symbolic transitions we have for an unimpacted transition, the more savings we gain if it is partially explored.

5.2 Proposed Symbolic Execution Optimizations for Evolving Rhapsody Statecharts

Symbolic execution is an expensive approach especially when applied to big and complex artifacts (programs or models) where the size of the generated SETs can be very large. Software artifacts can undergo several iterations and refinements and repeating the symbolic execution of these artifacts from scratch after every iteration or refinement step can be very tedious and time consuming. The new version of an artifact can be very similar to the previous one, so excluding the unchanged parts from successive runs of the symbolic execution technique reduces the time required for any symbolic execution-based types of analyses. Alternatively, directing the successive runs of the symbolic execution technique away from execution paths that are not impacted and toward execution paths that are impacted also reduces the time required for any symbolic execution-based types of analyses. Inspired by the work in [7] and [6] for optimizing the symbolic execution of evolving programs, we propose two techniques for optimizing the symbolic execution of evolving Rhapsody Statecharts. The two techniques are applied for both individual Statecharts and communicating Statecharts.

5.2.1 Memoization-Based Symbolic Execution

In this section we present our MSE technique for both individual and communicating Statecharts. The architectures of implementations of the proposed approaches are shown in Fig. 15A and B, respectively.

5.2.1.1 Technique Description

The basic idea of this technique is to reuse the results from a previous run of the symbolic execution technique on the model version before the change

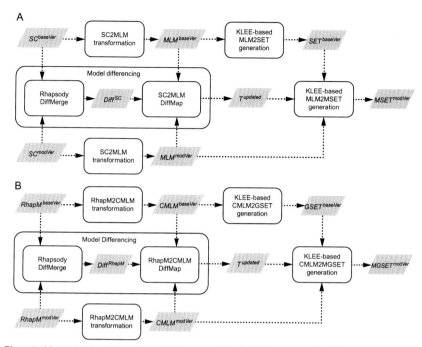

Fig. 15 Memoization-based symbolic execution (MSE) proposed architecture. (A) MSE architecture for evolving individual Statecharts. (B) MSE architecture for evolving communicating Statecharts.

to optimize the next run of the technique on the model version after the change such that only the parts of the SET (resp., GSET) that are affected by the change will be regenerated. It is obvious that a key requirement for this technique is to have a stored version of the SET (resp., GSET) of the model before the change. Another requirement is to have a list of the changes made to the previous version of the model which will be used to remove the parts of the previous SET (resp., GSET) that are affected by the changes and to mark their corresponding source nodes (i.e., symbolic states or symbolic model states) as "*needing an update*". Rerunning the SE technique on the modified model version will be directed to generate the required updates for these marked nodes.

To cope with the requirements of this technique, we needed to add to the definition of the symbolic state (resp., symbolic model state) presented earlier in Section 4 as part of the definition of the symbolic execution tree (resp., GSET) the following two features: (1) a Boolean attribute `needsUpdate` that indicates whether a given symbolic state needs to be updated or not and (2) a hashmap `branchesToUpdate : S → P(T)` of states with their corresponding outgoing transitions that need to be updated.

To be able to identify the nodes and the branches that need to be updated, we need to provide the technique with a list of all the differences between the modified mode version and its previous version. Our technique only considers a subset of model changes that can be mapped to changes to transitions in the MLM (resp., CMLMs) representation. This includes (1) adding states or transitions, (2) removing states or transitions, (3) modifying entry or exit actions of states, and (4) modifying events, guards, or actions of transitions. For example, adding or removing states can be represented by an addition or a deletion of transitions connecting these states with other states in the model; also, updating the entry actions (resp., exit actions) of states can be represented by an update of their incoming transitions (resp., outgoing transitions). Therefore, we manage to represent these differences in terms of a set of updated transitions $T^{updated}$ including all transitions that have been added to, deleted from or updated in the modified version of the model. Other types of model changes such as the addition, deletion, or modification of variables (i.e., class attributes) are not supported since these changes impacts the variable valuation of every state in the model and hence requires a complete regeneration of the entire SET of the model after the change.

To find the differences between two Rhapsody models, we use the Rhapsody DiffMerge tool that we mentioned in Section 2.4. These differences are mapped to their MLM (resp., CMLMs) correspondences using our "`SC2MLM DiffMap`" (resp., "`RhapM2CMLM DiffMap`") component, which outputs the set of all transitions $T^{updated}$ that are found to have been updated in the modified MLM (resp., CMLMs) version. Each element in the set $T^{updated}$ indicates the name of a transition that has been affected by the change, the name of the source state of this transition, and the change type applied to this transition (i.e., "*Addition*", "*Deletion*", "*Modification*").

The main component of our MSE technique for individual Statecharts is the "`KLEE-based MLM2MSET Generation`" component shown in Fig. 15A. Similarly, the main component for communicating Statecharts is the "`KLEE-based CMLM2MGSET Generation`" component shown in Fig. 15B. The detailed steps of implementing each component are listed in Algorithms 7 and 8, respectively. In the sequel, we describe the details of both components.

Algorithm 7 Memoized-Based Symbolic Execution (MSE) of an Evolving MLM

Inputs:

1: A modified MLM model version: $MLM^{modVer} = (S, V, E, EA, T, s_0, V_0)$
2: A SET of the previous MLM model version $MLM^{baseVer}$: $SET^{baseVer} = (SS, L, ST, ss_0)$
3: A set of all changed transitions between MLM^{modVer} and $MLM^{baseVer}$: $T^{updated}$

Output:

1: A Memoized-based SET of MLM^{modVer}: $MSET^{modVer}$

Steps:

1: //Task 1: Explore $SET^{baseVer}$ in breadth-first-search fashion to remove the parts (i.e., the subtrees) that need to be updated and to mark the nodes in $SET^{baseVer}$ to be reexplored when we rerun the standard SE on MLM^{modVer}
2: $ss_0 \leftarrow (s_0, V_0, true)$
3: Create a queue q
4: $q \leftarrow [ss_0]$
5: **while** q is not empty **do**
6:     Remove the first element $ss = (s, val, pc)$ from q
7:     **if** there is t in $T^{updated}$ such that $s = sourceState(t)$ and t is marked as "Added" **then**
8:         //Mark ss as "needsUpdate"
9:         $ss.needsUpdate \leftarrow true$
10:         //Add state s and its outgoing transition t to the hashmap $branchesToUpdate$ of ss
11:         $ss.branchesToUpdate \leftarrow ss.branchesToUpdate \cup \{s \rightarrow \{\{t\}\}\}$
12:     **end if**
13:     **for all** outgoing symbolic transitions $st = (ss, l, ss')$ of ss **do**
14:         **if** there is $t \in T^{updated}$ such that $sourceState(t) = ss.s$ and $targetState(t) = ss'.s$ **then**
15:             Assert t is marked as "Modified" or "Deleted"
16:             Remove l and ss' with all its children from $SET^{baseVer}$
17:             **if** t is marked as "Modified" **then**
18:                 //Mark ss as "needsUpdate"
19:                 $ss.needsUpdate \leftarrow true$
20:                 //Add state s and its outgoing transition t to the hashmap $branchesToUpdate$ of ss
21:                 $ss.branchesToUpdate \leftarrow ss.branchesToUpdate \cup \{s \rightarrow \{\{t\}\}\}$
22:             **end if**
23:         **else**
24:             $q \leftarrow enqueue(q, ss')$
25:         **end if**
26:     **end for**
27: **end while**

28: //Task 2: Iterate over $SET^{baseVer}$ and set the attributes *needsUpdate* and *branchesToUpdate* of all leaf symbolic states that are no longer subsumed by any of previsited nodes to *"true"* and *"All"*, respectively

29: $SET^{baseVer} = Task2\_helperMethod(SET^{baseVer})$

30: //Task 3: Reexplore $SET^{baseVer}$ to merge the results from the standard symbolic execution of MLM^{modVer} for the parts that need to be updated

31: $q \leftarrow [ss_0]$

32: //Create a set of previsited symbolic states for the purpose of subsumption checking

33: $SS^{preVisited} \leftarrow \{\}$

34: **while** q is not empty **do**

35:     Remove the first element $ss = (s, val, pc)$ from q

36:     **if** ss has one or more outgoing symbolic transitions **then**

37:         //Record ss in the set of previsited symbolic states

38:         $SS^{preVisited} \leftarrow SS^{preVisited} \cup \{ss\}$

39:         //Add all the target symbolic states of the outgoing symbolic transitions of ss to the queue

40:         **for all** outgoing symbolic transitions $st = (ss, l, ss')$ of ss **do**

41:             $q \leftarrow enqueue(q, ss')$

42:         **end for**

43:     **end if**

44:     **if** $ss.needsUpdate$ **then**

45:         //Rerun the SE on the model version MLM^{modVer} taking into consideration only transitions in $ss.branchesToUpdate$

46:         **for all** outgoing transitions $t = (s, e, eA, G, A, s')$ of $s \wedge t \in ss.branchesToUpdate$ **do**

47:             Create a unique set V^s of new variables in one-to-one relation with the transition's event arguments eA

48:             Create a map σ between eA and V^s

49:             Substitute every occurrence of a variable $x \in eA$ in the guard G and the action code A by $\sigma(x)$ to obtain $G^s \leftarrow \sigma(G)$ and $A^s \leftarrow \sigma(A)$

50:             **if** G^s is satisfiable given val and pc **then**

51:                 $feasiblePaths \leftarrow SymbolicExecutionOfCode(A^s, val, G^s, pc)$

52:                 **for all** $feasiblePath = (val', pc', out') \in feasiblePaths$ **do**

53:                     Create a new symbolic state, $ss' = (s', val', pc')$

54:                     **if** ss' is not subsumed by any previously generated symbolic state in $SS^{preVisited}$ **then**

55:                         $ss'.needsUpdate \leftarrow true$

56:                         $ss'.branchesToUpdate \leftarrow "All"$

57:                         $SS \leftarrow SS \cup \{ss'\}$

58:                         Create a new label, $l' = (e, V^s, \sigma, out')$

59:                         $L \leftarrow L \cup \{l'\}$

60:                         Create a new symbolic transition, $st' = (ss, l', ss')$

61: $\quad\quad\quad\quad\quad ST \leftarrow ST \cup \{st'\}$
62: $\quad\quad\quad\quad\quad q \leftarrow enqueue(q, ss')$
63: $\quad\quad\quad\quad$ **end if**
64: $\quad\quad\quad$ **end for**
65: $\quad\quad$ **end if**
66: $\quad\quad$ **end for**
67: $\quad$ **end if**
68: **end while**
69: //Task 4 (Optional): Iterate over $SET^{baseVer}$ and remove any node that can be subsumed by any other nodes (either at the same hierarchical level or higher)
70: $SET^{modVer} = Task4\_helperMethod(SET^{baseVer})$
71: $MSET^{modVer} \leftarrow SET^{baseVer}$

Algorithm 8 Memoization-based Symbolic Execution (MSE) of an Evolving CMLM

Inputs:

1: A modified CMLM model version of n communicating MLMs: $M2 = (MLM^{M2}, Q^{M2}, s_0^{M2}, V^{M2}, V_0^{M2})$
2: A GSET of the previous CMLM model version $M1$: $GSET^{M1} = (SS^{M1}, L, ST, ss_0^{M1})$
3: A set of all changed transitions between $M2$ and $M1$: $T^{updated}$

Output:

1: A memoization-based GSET of $M2$: $MGSET^{M2}$

Steps:

1: //Task 1: Explore $GSET^{M1}$ in breadth-first-search fashion to remove the parts (i.e., the subtrees) that need to be updated and to mark the nodes in $GSET^{M1}$ to be reexplored when we rerun the standard SE on $M2$
2: $ss_0^{M1} \leftarrow \left(s_0^{M1}, V_0^{M1}, true, q_0^{M1}\right)$
3: Create a queue q
4: $q \leftarrow \left[ss_0^{M1}\right]$
5: **while** q is not empty **do**
6: $\quad$ Remove the first element $ss^{M1} = (s^{M1}, val, pc, q^{M1})$ from q
7: $\quad$ **if** there is $s1$ in s^{M1} and there t in $T^{updated}$ such that $s1 = sourceState(t)$ and t is marked as "Added" **then**
8: $\quad\quad$ //Mark ss^{M1} as "needsUpdate"
9: $\quad\quad$ $ss^{M1}.needsUpdate \leftarrow true$
10: $\quad\quad$ //Add state $s1$ and its outgoing transition t to the hashmap $branchesToUpdate$ of ss^{M1}
11: $\quad\quad$ $ss^{M1}.branchesToUpdate \leftarrow ss^{M1}.branchesToUpdate \cup \{s \rightarrow \{t\}\}$
12: $\quad$ **end if**

13:     **for all** outgoing symbolic transitions $st = (ss^{M1}, l, ss'^{M1})$ of ss^{M1} **do**

14:       **if** there is $t \in T^{updated}$ such that $sourceState(t) \in ss^{M1}.s^{M1}$ and $targetState(t) \in ss'^{M1}.s^{M1}$ **then**

15:         Assert t is marked as "Modified" or "Deleted"

16:         Remove l and ss'^{M1} with all its children from SET^{M1}

17:         **if** t is marked as "Modified" **then**

18:           //Mark ss^{M1} as "needsUpdate"

19:           $ss^{M1}.needsUpdate \leftarrow true$

20:           //Add state $s1$ and its outgoing transition t to the hashmap $branchesToUpdate$ of ss^{M1}

21:           $ss^{M1}.branchesToUpdate \leftarrow ss^{M1}.branchesToUpdate \cup \{s \rightarrow \{t\}\}$

22:         **end if**

23:       **else**

24:         $q \leftarrow enqueue(q, ss'^{M1})$

25:       **end if**

26:     **end for**

27: **end while**

28: //Task 2: Iterate over $GSET^{M1}$ and set the attributes $needsUpdate$ and $branchesToUpdate$ of all leaf symbolic states that are no longer subsumed by any of previsited nodes to "$true$" and "All", respectively

29: $GSET^{M1} = Task2\_helperMethod(GSET^{M1})$

30: //Task 3: Reexplore $GSET^{M1}$ to merge the results from the symbolic execution of $M2$ for the parts that need to be updated

31: $q \leftarrow [ss_0^{M1}]$

32: //Create a set of previsited symbolic states for the purpose of subsumption checking

33: $SS^{preVisited} \leftarrow \{\}$

34: **while** q is not empty **do**

35:     Remove the first element $ss^{M1} = (s^{M1}, val, pc, q^{M1})$ from q

36:     **if** ss^{M1} has one or more outgoing symbolic transitions **then**

37:       //Record ss^{M1} in the set of previsited symbolic model states

38:       $SS^{preVisited} \leftarrow SS^{preVisited} \cup \{ss^{M1}\}$

39:       //Add all the target symbolic states of the outgoing symbolic transitions of ss^{M1} to the queue

40:       **for all** outgoing symbolic transitions $st = (ss^{M1}, l, ss'^{M1})$ of ss^{M1} **do**

41:         $q \leftarrow enqueue(q, ss'^{M1})$

42:       **end for**

43:     **end if**

44:     **if** $ss^{M1}.needsUpdate$ **then**

45:       //Rerun the SE on the model version $M2$ taking into consideration only the states and the transitions in the hashmap $branchesToUpdate$ of ss^{M1}

46:    **for all** MLM_i in MLM^{M2} **do**

47:       $s_i \leftarrow s^{M2}(i)$ where $s_i \in S_i(MLM_i)$

48:       **if** s_i is found in the *branchesToUpdate* map of ss^{M1} **then**

49:           **for all** outgoing transitions $t_i = (s_i, e, eA, G, A, s_i')$ of s_i **do**

50:               **if** t_i is found in the *branchesToUpdate* map of ss^{M1} **then**

51:                  **if** $((q^M = \phi) \wedge (e \in E_i^{inp})) \vee ((head(q_i^M) = e) \wedge (e \in E_i^{int}))$ **then**

52:                     Create a unique set V^s of new variables in one-to-one relation with the transition's event arguments eA

53:                     Create a map σ between eA and V^s

54:                     Substitute every occurrence of a variable $x \in eA$ in the guard G and the action code A by $\sigma(x)$ to obtain $G^s \leftarrow \sigma(G)$ and $A^s \leftarrow \sigma(A)$

55:                  **if** G^s is satisfiable given val and pc **then**

56:                     *feasiblePaths* $\leftarrow$ *SymbolicExecutionOfCode*(A^s, *val*, G^s, *pc*)

57:                     **for all** *feasiblePath* = (*val'*, *pc'*, *out'*) $\in$ *feasiblePaths* **do**

58:                        $q'^{M2} \leftarrow out'$

59:                        Create a new symbolic model state $ss'^{M2} = (s'^{M2}, val', pc', q'^{M2})$ where $s'^{M2} = s^{M1}.replace(s_i, s_i')$

60:                        **if** ss'^{M2} is not subsumed by any previously generated symbolic model state in $SS^{preVisited}$ **then**

61:                           Mark ss'^{M2} as "needsUpdate" with respect to "*All*" states in s'^{M2} and "*All*" their outgoing transitions

62:                           $SS^{M1} \leftarrow SS^{M1} \cup \{ss'^{M2}\}$

63:                           Create a new label $l' = (e, V^s, \sigma, out')$

64:                           $L \leftarrow L \cup \{l'\}$

65:                           Create a new symbolic transition $st' = (ss^{M1}, l', ss'^{M2})$

66:                           $ST \leftarrow ST \cup \{st'\}$

67:                           //Recored ss'^{M2} in the set of previsited symbolic model states

68:                           $SS^{preVisited} \leftarrow SS^{preVisited} \cup \{ss'^{M2}\}$

69:                           $q \leftarrow enqueue(q, ss'^{M2})$

70:                    **end if**

71:                 **end for**

72:              **end if**

73:           **end if**

74:        **end if**

75:       **end for**

76:     **end if**

77:    **end for**

78: **end if**

79: **end while**

80: //Task 4 (Optional): Iterate over $GSET^{M1}$ and remove any node that can be subsumed by any other nodes (either at the same hierarchical level or higher)
81: $GSET^{M1} = Task4\_helperMethod(GSET^{M1})$
82: $MGSET^{M2} \leftarrow GSET^{M1}$

The three inputs to both components are (1) the MLM (resp., CMLMs) representation of the modified model version MLM^{modVer} (resp., $CMLM^{modVer}$), (2) the set of transitions that have been updated in the modified version of the model $T^{updated}$, and (3) the $SET^{baseVer}$ (resp., $GSET^{baseVer}$) to be reused.

Four successive tasks are performed by both components.

Task 1: The first task is to load and explore the input $SET^{baseVer}$ (resp., $GSE-T^{baseVer}$) in order to: (1) remove all the edges representing any transition belonging to the set of updated transitions $T^{updated}$ (note that removing an edge leads to removing the entire subtree rooted at the target node of the edge) and (2) mark all symbolic states (resp., symbolic model states) representing states with outgoing transitions belonging to $T^{updated}$ as "*needing an update*" with respect to these outgoing transitions. This is done by updating the attributes `needsUpdate` and `branchesToUpdate` of each such symbolic state (resp., symbolic model state).

Task 2: As a result of the previous task, we may find some leaves in the resulting $SET^{baseVer}$ (resp., $GSET^{baseVer}$) that lost their subsuming nodes. In this case, we need also to mark these leaves as "needing an update" with respect to all outgoing transitions of their corresponding model states. Therefore, the second task of both components is to iterate over the nodes in the $SET^{baseVer}$ (resp., $GSET^{baseVer}$) resulting from the previous task and to set the attributes `needsUpdate` and `branchesToUpdate` of all leaf symbolic states (resp., symbolic model states) that are no longer subsumed to "*true*" and "*All*", respectively. A detailed description of this task is omitted for brevity.

Task 3: The third task of both components is to reexplore the $SET^{baseVer}$ (resp., $GSET^{baseVer}$) resulting from the previous task to find all symbolic states (resp., symbolic model states) that need an update. For each such symbolic state (resp., symbolic model state), we rerun the SE technique on the modified model version MLM^{modVer} (resp., $CMLMs^{modVer}$) targeting only the parts of the model identified by the hashmap `branchesToUpdate` of the symbolic state (resp., symbolic model state) and taking into consideration the set of symbolic states (resp., symbolic model states) that have been previously explored to be used for the subsumption checking. Our strategy to implement this task differs from the one that we initially used to implement the "KLEE-based MLM2MSET Generation" component in an earlier version of this work [94]. This modification is done based on the intuition that we got from applying the MSE technique for individual Statecharts where we found that our *Initial Nodes Update Strategy* may generate, in some cases, extra symbolic states. The process of detecting and removing these extra symbolic states may reduce the efficiency of the technique especially for large SETs. Additionally, when we followed the same strategy in the implementation of the "KLEE-based CMLM2MGSET Generation" component, we found, based on an initial evaluation of the technique, that it generates larger MGSETs with many extra symbolic model states. Therefore, we have adapted this initial nodes update strategy in the implementation of the two components: "KLEE-based MLM2MSET Generation" and "KLEE-based CMLM2MGSET Generation" presented here such that we overcome, to some extent, this problem. The difference between both strategies lies in the way we generate the subtrees required to update the symbolic states (resp., symbolic model states) that need an update. In the following, we explain this in details.

(1) *Initial Nodes Update Strategy (INUS)*: Our initial nodes update strategy depends on generating the subtrees required to update symbolic states (resp., symbolic model states) that are marked as "needing an update" independently (i.e., one at a time and in a breadth-first manner) by running an adapted version of the SE on the modified model version based on some additional parameters. These additional parameters determine where the exploration starts (i.e., the symbolic state to begin the exploration at), which updated transitions to consider when exploring the state representing the start symbolic state for the first time, and the set of symbolic states that have been previously explored to be used for the subsumption checking. The resulting sub-SETs for running this adapted version of the SE are to be merged with their corresponding symbolic states in the $SET^{baseVer}$. One limitation of this strategy is that it prevents the detection of existing subsumption relationships between symbolic states in the sub-SETs that were generated first and symbolic states in the sub-SETs that are generated later.

(2) *Adapted Nodes Update Strategy (ANUS)*: One possible solution to overcome the limitation presented in the initial nodes update strategy is to run the SE on the modified model version for symbolic states (resp., symbolic model states) that need an update on-the-fly as they are

identified. In this case, the construction of the sub-SETs for the nodes that need an update is carried out jointly (concurrently) at the same time rather than successively as in the initial strategy. This procedure is implemented in Algorithm 7, Lines 30–68 and in Algorithm 8, Lines 30–79. This strategy can overcome part (but not all) of the limitation raised by the initial strategy (i.e., we may also find some extra symbolic model states in the generated SETs following this strategy as well).

A step-by-step illustration of both update strategies is shown in Figs. 16 and 17, respectively, for updating nodes 2 and 5 in the example SET shown in Fig. 18. Please note that: (1) nodes are labeled based on their creation order and (2) leading zeros in a node label indicate the number of reruns of the SE technique that created the node. A comparison between the SETs generated after applying each update strategy is shown in Fig. 19. A demonstration of a scenario of an undetected subsumption relation between existing and newly created nodes following any of the two presented nodes update strategies is shown in Fig. 20. In this scenario, the root node "0" is marked as "needing an update" with respect to a newly added branch which is colored red. Following either update strategies will generate the subtree consisting of nodes "001", "002", and "003". Assume the case where a subsumption relation now exists between the already existing node "5" and the newly created node "001". In this case, nodes "02", "03", "06", and "07" are considered extra nodes. This is because the newly detected subsumption relation between nodes "5" and "001" does not exist before the creation of node "001" and also when the generation of nodes "02", "03", "06", and "07" took place.

Task 4: Since the resulting SETs (resp., GSETs) of the previous task may contain some extra symbolic states (resp., symbolic model states), the last task to be performed in both components is to detect and remove such extra symbolic states (resp., symbolic model states). A detailed description of this task is straightforward and omitted for brevity.

In Table 1, we highlight the steps of the algorithms implementing each one of the aforementioned four tasks of both components.

The output from both components is a memoization-based SET (resp., GSET) of the modified model version, $MSET^{modVer}$ (resp., $MGSET^{modVer}$), that shares all what can be reused from the old tree $SET^{baseVer}$ (resp., $GSET^{baseVer}$) but also contains the modifications resulting from the SE of the parts of the new version of the model that are found changed. We use a tree-based data structure for representing, storing, retrieving, and manipulating these SETs.

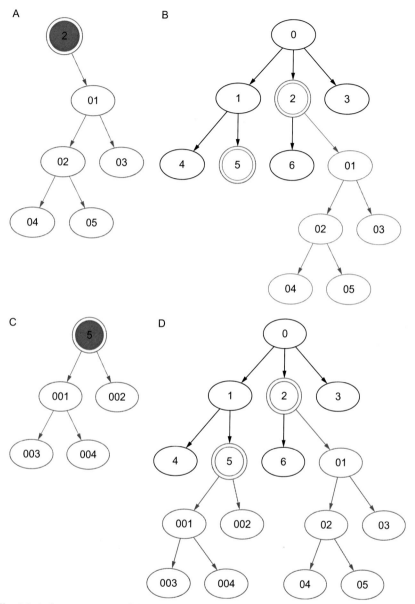

Fig. 16 A demonstration of the *Initial Nodes Update Strategy* used in the implementation of an earlier version of MSE [94, 95]. Nodes are labeled based on their creation order and leading zeros in a node label indicate the number of reruns of the SE technique that created the node. (A) Step 1: Reexploring node 2 given previsited nodes = {0, 1, 2, 3, 4, 5, 6}. (B) Step 2: Merging the subtree updating node 2 with the original SET. (C) Step 3: Reexploring node 5 given previsited nodes = {0, 1, 2, 3, 4, 5, 6, 01, 02, 03, 04, 05}. (D) Step 4: Merging the subtree updating node 5 with the original SET.

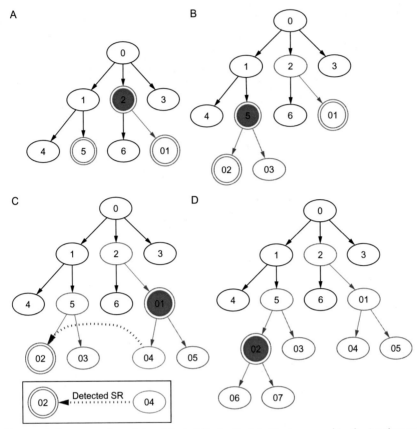

Fig. 17 A demonstration of the *Adapted Nodes Update Strategy* used in the implementation of the current version of MSE as presented in Algorithms 7 and 8—one-level depth exploration with marking generated nodes (if not subsumed by a previsited node) as "needing an update". Nodes are labeled based on their creation order and leading zeros in a node label indicate the number of reruns of the SE technique that created the node. (A) Step 1: One-level depth exploration of node 2 given previsited nodes = {0, 1, 2, 3, 4, 5, 6}. The resulting node 01 is marked (with *double circles*) as "needing an update". (B) Step 2: One-level depth exploration of node 5 given previsited nodes = {0, 1, 2, 3, 4, 5, 6, 01}. The resulting node 02 is marked (with *double circles*) as "needing an update", while the resulting node 03 is found to be subsumed by one of the given previsited nodes. (C) Step 3: One-level depth exploration of node 01 given previsited nodes = {0, 1, 2, 3, 4, 5, 6, 01, 02, 03}. The resulting nodes 04 and 05 are both found to be subsumed by one of the given previsited nodes. Note that subsumption relation between nodes 04 and 02 is detected. (D) Step 4: One-level depth exploration of node 02 given previsited nodes = {0, 1, 2, 3, 4, 5, 6, 01, 02, 03, 04, 05}. The resulting nodes 06 and 07 are both found to be subsumed by one of the given previsited nodes.

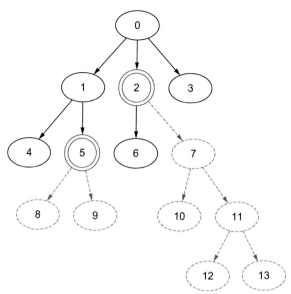

Fig. 18 An example SET with nodes 2 and 5 are marked as *"needing an update"* with respect to some of their outgoing branches. Nodes are labeled based on their creation order.

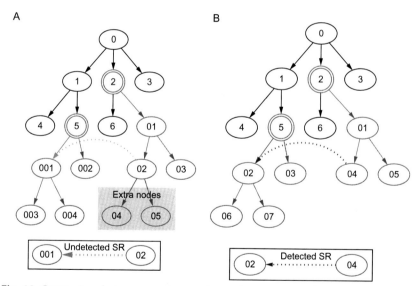

Fig. 19 Comparison between our two nodes update strategies showing (in A) a demonstration of undetected subsumption relation (SR) between nodes 02 and 001 resulting in having extra nodes 04 and 05. (A) Less optimized MSET resulting from our *Initial Nodes Update Strategy*. (B) A more optimized MSET resulting from our *Adapted Nodes Update Strategy*.

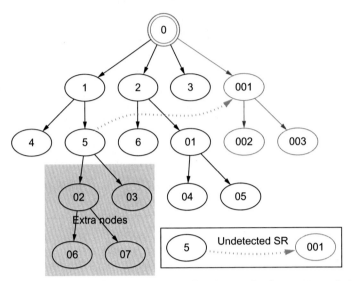

Fig. 20 A demonstration of a possible case of undetected subsumption relation (SR) between nodes 5 and 001 resulting in having extra nodes 02, 03, 06, and 07.

Table 1 Components' Tasks and Their Corresponding Algorithms' Steps

	Component	
Task	KLEE-Based CMLM2MSET Generation	KLEE-Based CMLM2MGSET Generation
Task 1	Algorithm 7, Lines 1–27	Algorithm 8, Lines 1–27
Task 2	Algorithm 7, Lines 28–29	Algorithm 8, Lines 28–29
Task 3	Algorithm 7, Lines 30–68	Algorithm 8, Lines 30–79
Task 4 (Optional)	Algorithm 7, Lines 69–70	Algorithm 8, Lines 80–81

Example 5. Applying MSE on the modified model of individual Statechart V1 shown in Fig. 9B and on the modified model of communicating Statecharts AB^{V1} shown in Fig. 12 generates the memoization-based SET (resp., GSET) that are shown in Figs. 11A and 13, respectively. The amount of savings gained from applying MSE on the modified model V1 (or AB^{V1}) compared to applying standard SE on the same models is measured by the percentage of reused nodes (which are colored black) to the total number of nodes in the resulting MSET (or MGSET). We can see that the savings obtained in the MSET(V1) in Fig. 11A is equal to 6/14 (i.e., approximately 43%), while

the savings gained in the MGSET(AB^{V1}) in Fig. 13 is equal to 11/21 (i.e., approximately 52%).

It is quite clear that in some cases the savings gained from MSE can be very minimal depending on how many tree nodes are reused and the location of tree nodes that need to be updated. In this case, it is the analyst decision to choose between using MSE or standard SE.

5.2.1.2 Discussion

We consider MSE an example of an artifact-oriented optimization technique because it can be used to optimize the generation of one type of intermediate artifact (in our case they are in the form of SETs) as the system evolves. At any point in time, these intermediate artifacts are maintained (i.e., up-to-date) and complete (i.e., they represent all execution paths of their corresponding models) and thus they can be used to run any SET-based analyses.

Nevertheless, by marking the new parts of the generated MSETs (resp., MGSETs) resulting from MSE from the reused parts, we can use the new parts to run regression-based analyses. In that sense, MSE can also be considered as an example of an analysis-oriented optimization technique meaning that if a given analysis has been performed on a given model before the change, then using only the new parts of the generated MSETs (resp., MGSETs) of the model after the change will allow us to rerun the same analysis and to conclude the overall correctness of the changed model.

5.2.2 Dependency-Based Symbolic Execution

In this section we present our DSE technique [e] for both individual and communicating Statecharts. The architectures of the implementation of the proposed approach are shown in Fig. 21A and B, respectively.

5.2.2.1 Technique Description

The main idea of this technique is to generate a subset (or a partial view) of the complete SET of a modified model version that characterizes all model paths that are impacted by the change which is sufficient to perform regression-based analyses. To achieve this, we integrate a dependency-based change impact analysis with our standard SE technique such that we can direct the successive runs of the SE technique to fully explore model transitions that are impacted by the change and to partially explore model

---

[e] Note that we can also call it regression or partial symbolic execution.

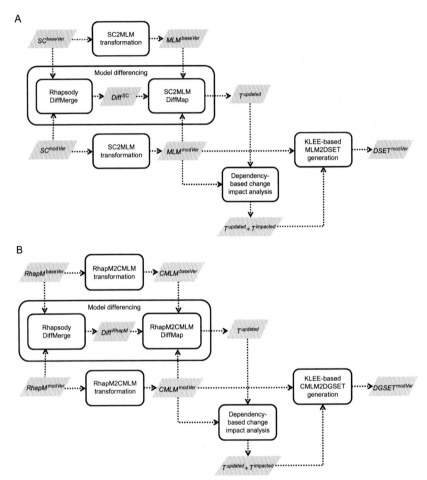

Fig. 21 Dependency-based symbolic execution (DSE) proposed architecture. (A) DSE architecture for evolving individual Statecharts. (B) DSE architecture for evolving communicating Statecharts.

transitions that are not impacted by the change. A full exploration of a model transition means to consider all the symbolic execution paths resulting from symbolically executing the transition guard and action code, while a partial exploration means to consider only one of these symbolic paths, namely the first one. Note that only models with multipath transitions (i.e., transitions with disjoint guard conditions and/or conditional statements in their action code) can benefit from this technique. We found that this model feature is extensively used in many of existing Simulink/Stateflow models. In our model examples in Section 5.1, transitions T1 and T2 in Fig. 9 (resp., transition itsA-T1 in Fig. 12A) are all examples of multipath transitions. The SE of each one of these transitions results in two symbolic paths originating at the same symbolic state (resp., symbolic model state) representing the source state of the transition and ending at two different symbolic states (resp., symbolic model states) representing the target state of the transition, each of which is characterized by the valuations of variables, the path constraints and the updates of queues resulting from the symbolic evaluation of each branch in the guard condition and in the conditional and iterative statements found in the action code of the transition.

Based on this idea we can conclude that the key requirements of this technique are first to identify the differences between two model versions and second to perform a dependency analysis on the modified model version to compute the impact of these differences. These two requirements are achieved by the two components "Model Differencing" and "Dependency-based Change Impact Analysis" shown in Fig. 21A and B.

To implement the "Model Differencing" component, we follow the same procedure as in the previous section. First we use the Rhapsody DiffMerge to find the differences between two Rhapsody models, then we use our "SC2MLM DiffMap" (resp., "RhapM2CMLM DiffMap") component to map these differences to their MLMs (resp. CMLMs) correspondences. The output in this case, as we explained earlier, represents the set $T^{updated}$ of all transitions that are found to have been updated in the modified MLM (resp., CMLMs) version. Each element in the set $T^{updated}$ indicates the name of a transition that has been affected by the change, the name of the source state of this transition, and the change type applied to this transition (i.e., "*Addition*", "*Deletion*", "*Modification*").

To implement the "Dependency-based Change Impact Analysis" component, we considered the two types of dependences discussed in Section 2.3: data dependence (DD) and communication dependence (COMD). The formal definitions of both types of dependences and the algorithms to compute them are provided in Section 2.3. The two inputs to this component are (1) the MLM (resp., CMLMs) representation of the modified model version MLM^{modVer} (resp., $CMLM^{modVer}$), and (2) the set of transitions $T^{updated}$ that have updated in the modified model version. The first task to be performed by this component is to compute the data dependences and the communication dependences between the transitions of the input model version. The second task to be performed is to identify the set of transitions

$T^{impacted}$ that have a dependency relation (DD or COMD) with any of the transitions in $T^{updated}$. The output of this component is the union set of $T^{updated}$ and $T^{impacted}$; these transitions are the ones that we need to fully explore when we run our DSE technique on the modified model version MLM^{modVer} (resp., $CMLM^{modVer}$).

The last task to be performed is to run the symbolic execution of the modified version of the model guided by the list of updated/impacted transitions resulting from the "Dependency-based Change Impact Analysis" component. Two modes of exploration are defined: "*full*" and "*partial*". The "*full*" exploration mode requires a complete exploration of all execution paths (i.e., symbolic transitions) of a given transition and is applied to all transitions that are found to have been updated or impacted. However, the "*partial*" exploration mode requires the execution of only one path (i.e., symbolic transition) of an explored transition and is applied to transitions that are found neither updated nor impacted. The components performing this task are the "KLEE-based MLM2DSET Generation" component shown in Fig. 21A for individual Statecharts and the "KLEE-based CMLM2DGSET Generation" component shown in Fig. 21B for communicating Statecharts. The algorithms implementing this task for each component are listed in Algorithm 9 and Algorithm 10, respectively.

Algorithm 9 Dependency-Based Symbolic Execution (DSE) of an Evolving MLM

Input:
1: A modified MLM model version: $MLM^{modVer} = (S, V, E, EA, T, s_0, V_0)$
2: A set of all changed transitions and their imapcted ones: $T^{ofInterest} = T^{updated} \cup T^{impacted}$

Output:
1: A Dependency-based SET of MLM^{modVer}: $DSET^{modVer} = (SS, L, ST, ss_0)$

Steps:
1: $ss_0 \leftarrow (s_0, V_0, \text{true})$
2: $SS \leftarrow \{ss_0\}$
3: $L \leftarrow \{\}$
4: $ST \leftarrow \{\}$
5: Create a queue q
6: $q \leftarrow [ss_0]$
7: //Explore the MLM in Breadth-First-Search fashion
8: **while** q is not empty **do**

```
 9:    Remove the first element ss = (s, val, pc) from q
10:    for all outgoing transitions t = (s, e, eA, G, A, s') of s do
11:        Create a unique set V^s of new variables in one-to-one relation
       with the transition event's arguments eA
12:        Create a map σ between eA and V^s
13:        Substitute every occurrence of a variable x ∈ eA in the guard G
       and the action code A by σ(x) to obtain G^s ← σ(G) and A^s ← σ(A)
14:        if G^s is satisfiable given val and pc then
15:            feasiblePaths ← SymbolicExecutionOfCode(A^s, val, G^s, pc)
16:            if t ∈ T^{ofInterest} then
17:                //Full-exploration mode
18:                for all feasiblePath = (val', pc', out') ∈ feasiblePaths do
19:                    Create a new symbolic state, ss' = (s', val', pc')
20:                    if ss' is not subsumed by any previously generated
       state then
21:                        SS ← SS ∪ {ss'}
22:                        Create a new label, l' = (e, V^s, σ, out')
23:                        L ← L ∪ {l'}
24:                        Create a new symbolic transition, st' = (ss, l', ss')
25:                        ST ← ST ∪ {st'}
26:                        q ← enqueue(q, ss')
27:                    end if
28:                end for
29:            else
30:                //Partial-exploration mode
31:                Select only one feasiblePath = (val', pc', out') ∈ feasiblePaths
32:                Create a new symbolic state, ss' = (s', val', pc')
33:                if ss' is not subsumed by any previously generated state then
34:                    SS ← SS ∪ {ss'}
35:                    Create a new label, l' = (e, V^s, σ, out')
36:                    L ← L ∪ {l'}
37:                    Create a new symbolic transition, st' = (ss, l', ss')
38:                    ST ← ST ∪ {st'}
39:                    q ← enqueue(q, ss')
40:                end if
41:            end if
42:        end if
43:    end for
44: end while
```

Algorithm 10 Dependency-based Symbolic Execution (DSE) of an Evolving CMLM

Input:

1: A modified CMLM model version of n communicating MLMs: $M = (MLM^M, Q^M, s_0^M, V^M, V_0^M)$
2: A set of all changed transitions and their imapcted ones: $T^{of\,Interest} = T^{updated} \cup T^{impacted}$

Output:

1: A Dependency-based GSET of M: $DGSET^M = (SS^M, L, ST, ss_0^M)$

Steps:

1: $ss_0^M \leftarrow (s_0^M, V_0^M, \text{true}, q_0^M)$
2: $SS^M \leftarrow \{ss_0^M\}$
3: $L \leftarrow \{\}$
4: $ST \leftarrow \{\}$
5: Create a queue q
6: $q \leftarrow [ss_0^M]$
7: //Breadth-First-Search fashion exploration for each MLM in M
8: **while** q is not empty **do**
9:    Remove the first element $ss^M = (s^M, val, pc, q^M)$ from q
10:    **for all** MLM_i in MLM^M **do**
11:       $s_i \leftarrow s^M(i)$ where $s_i \in S_i(MLM_i)$
12:       **for all** outgoing transitions $t_i = (s_i, e, eA, G, A, s_i')$ of s_i **do**
13:          **if** $((q_i^M = \phi) \wedge (e \in E_i^{inp})) \vee ((head(q_i^M) = e) \wedge (e \in E_i^{int}))$ **then**
14:          Create a unique set V^s of new variables in one-to-one relation with the transition's event arguments eA
15:          Create a map σ between eA and V^s
16:          Substitute every occurrence of a variable $x \in eA$ in the guard G and the action code A by $\sigma(x)$ to obtain $G^s \leftarrow \sigma(G)$ and $A^s \leftarrow \sigma(A)$
17:          **if** G^s is satisfiable given val and pc **then**
18:             $feasiblePaths \leftarrow SymbolicExecutionOfCode(A^s, val, G^s, pc)$
19:             **if** $t \in T^{of\,Interest}$ **then**
20:                //Full-exploration mode
21:                **for all** $feasiblePath = (val', pc', out') \in feasiblePaths$ **do**
22:                   $q'^M \leftarrow out'$
23:                   Create a new symbolic state, $ss'^M = (s'^M, val', pc', q'^M)$ where $s'^M = s^M.replace(s_i, s_i')$
24:                   **if** ss'^M is not subsumed by any previously generated symbolic model state in SS^M **then**
25:                      $SS^M \leftarrow SS^M \cup \{ss'^M\}$
26:                      Create a new label, $l' = (e, V^s, \sigma, out')$
27:                      $L \leftarrow L \cup \{l'\}$
28:                      Create a new symbolic transition, $st' = (ss^M, l', ss'^M)$
29:                      $ST \leftarrow ST \cup \{st'\}$

```
30:                          q ← enqueue(q, ss'^M)
31:                   end if
32:                end for
33:             else
34:                //Partial-exploration mode
35:                Select only one feasiblePath = (val', pc', out') ∈
          feasiblePaths
36:                q'^M ← out'
37:                Create a new symbolic state, ss'^M = (s'^M, val', pc', q'^M)
          where s'^M = s^M.replace(s_i, s'_i)
38:                if ss'^M is not subsumed by any previously generated
          symbolic model state in SS^M then
39:                   SS^M ← SS^M ∪ {ss'^M}
40:                   Create a new label, l' = (e, V^s, σ, out')
41:                   L ← L ∪ {l'}
42:                   Create a new symbolic transition, st'=(ss^M, l', ss'^M)
43:                   ST ← ST ∪ {st'}
44:                   q ← enqueue(q, ss'^M)
45:                end if
46:             end if
47:          end if
48:       end if
49:    end for
50:  end for
51: end while
```

Example 6. Applying our dependency-based change impact analysis on the model version V1 will not report transition T1 among the list of transitions that are data dependent on the modified transition T2. As a result, only a partial exploration of transition T1 is performed and since it is a multipath transition, only one feasible path (i.e., one symbolic transition) is considered and the rest are pruned. Similarly, applying our dependency-based change impact analysis on the model version AB^V1 will not report transition itsA-T1 among the list of transitions that are data or communication dependent on the modified transition itsA-T2. As a result, only a partial exploration of transition itsA-T1 is performed and since it is a multipath transition, only one feasible path (i.e., one symbolic transition) is considered and the rest are pruned. The amount of savings gained from applying DSE is measured by the percentage of nodes that are pruned during the partial exploration mode. We can see that the savings gained in the DSET(V1) in Fig. 11B is approximately equal to 43% which is the percentage of the 6 pruned nodes (these which are colored gray) to the total number of nodes of the complete SET which is 14 nodes. Similarly, the savings gained in the DGSET(AB^{V1}) in Fig. 14 is approximately equal to 43% which is the percentage of the 9 pruned nodes (these which are colored gray) to the total number of nodes of the complete GSET which is 21 nodes.

5.2.2.2 Discussion

Assume that the SET-based type of analysis required here is to check that the system has no dead ends.[f] In a given SET, a tree leaf that is not subsumed by any other node in the tree may indicate (1) reaching a previsited symbolic state, (2) reaching a depth bound of some loop, or (3) reaching a dead end symbolic state. In our example, we neither have termination states nor loops in any of the action code of transitions. Therefore, in this case, a SET leaf node that is not subsumed by any other node will indicate a dead end symbolic state. Also, assume that an analyst has performed this type of analysis on the GSET (AB^{V0}) and has concluded that AB^{V0} has no dead ends. In this case, applying the same type of analysis on the DGSET(AB^{V1}), which is a partial view of the GSET (AB^{V1}), will be sufficient to check whether AB^{V1} has a dead end or not. By checking the leaves of the DGSET(AB^{V1}), we can see that the symbolic model state SS15 is not subsumed by any other node in the tree and thus we can conclude that AB^{V1} has a dead end. By inspecting (1) the valuations of the variables and the contents of the queues of this symbolic model state (i.e., SS15) and (2) the outgoing transition itsB-T3 of state bS2 that is fired by the event e5, we can see that the guard of this transition is unsatisfiable with respect to the given value of the event parameter. Since the DGSET(AB^{V1}) represents only a subset of the complete GSET(AB^{V1}), we cannot use it to perform a different type of analysis. In that sense, DSE is considered as an example of an analysis-oriented optimization approach.

5.3 MSE or DSE?

Based on our discussion of the idea behind each technique, we conclude

- For MSE, we need to have access to the SET (resp., GSET) of the original model (i.e., the model before the change) to be reused and we need to identify the differences between the new version of the model and the original one. We speculate that the effectiveness of this technique may be highly dependent on the change location. The resulting MSET (resp.,

---

[f] Dead ends are states that have no feasible outgoing transitions and they are not end states.

MGSET) represents all execution paths of the modified model (i.e., the model after the change) and can thus be used to perform all SET-based analyses. However, if reestablishing a previous analysis result as the goal, a regression analysis can also be run on just the changed parts of the SET.

- For DSE, we need to identify the differences between the new version of the model and the original one as well as to integrate a dependence analysis to identify the impact of the change. We speculate that models with complex guards and action code may benefit most from this technique. As the resulting DSET (resp., DGSET) may not be complete and omit executions that are guaranteed to not have been impacted by the change, it can be used only to run regression-types of analysis.

Having said that, we think that both optimization techniques complement each other, in the sense that they both can serve the same purpose but are most useful under different circumstances, and it is the role of the analyst to choose among them. For instance, the availability of the SET from a previous analysis is a basic requirement for applying MSE, while the complexity of the guards or action code of the transitions in the model under study is a key factor for the effectiveness of DSE.

5.4 Summary

In this section we have presented two symbolic execution optimization techniques, MSE and DSE, for both evolving individual and communicating Statecharts. We first explained with examples the idea behind each technique. Second, we introduced our proposed architectures for both techniques and provided the required description of the different components in each architecture.

6. IMPLEMENTATION

In this section we provide a high-level description of the implementation decisions that we have made to build the architectures presented in Section 4 for Standard SE and in Section 5 for MSE and DSE. Fig. 22 shows the operating system required and the technology used to implement each of the components presented in these architectures. We first highlight the implementation decisions that we made based on the initial requirements of this research work. Second, we provide the details of three core tasks. These are (1) the interaction with the symbolic execution engine KLEE, (2) building the model transformation and text generation in the context

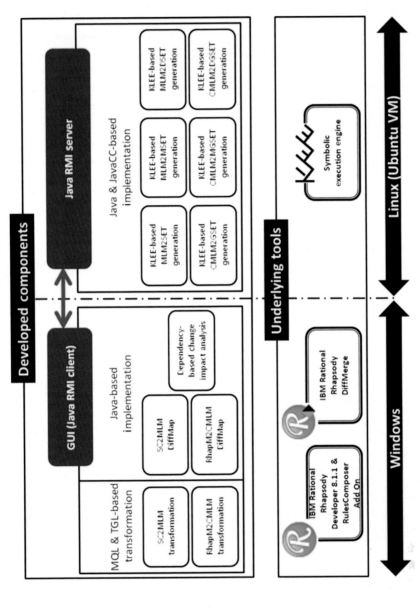

Fig. 22 Implementation illustration of our developed components.

of IBM Rational Rhapsody RulesComposer, and (3) the interaction with the IBM Rational Rhapsody DiffMerge tool.

6.1 Implementation Decisions

Most of the implementation decisions that we made are based on the following two main requirements.

(1) The context and the type of state-based models that we are going to use to develop our research. The decision of this aspect was made based on an interest from our industrial partner to provide such analysis capability for the state-based models developed in the context of the IBM Rational Rhapsody MDE tool, which are known as Rhapsody Statecharts. We found that although working with proprietary and commercial tools has its advantage in terms of quality and documentation, they limit the choices one may have to develop some external functionality or process and manipulate the artifacts they create to some extent. Therefore, we decided to implement our required components using IBM related tools (e.g., IBM Rational Rhapsody Developer RulesComposer Add-On for model transformation and custom code generation, and IBM Rational Rhapsody DiffMerge for model differencing).

(2) The symbolic execution engine to be reused for the action code encountered in the state-based models to be analyzed. Our decision was to use the KLEE symbolic execution engine—an open-source symbolic execution engine for C code. We made this decision after we examined the feasibility of building an API that enables the interaction with KLEE. In Section 6.2, we provide a more detailed description of this process. To be able to use KLEE for the symbolic execution of the fragments of action code encountered in Rhapsody Statecharts, we only consider a subset of Java and C++ code that has the same syntax as C code, including the basic assignment statements, conditional statements, and iterative statements. The data types supported are the basic types including characters, Booleans, and integer numbers. We also consider the type of statements that are used in the action code for sending events to other objects.

Unfortunately, KLEE does not run under Windows. It runs only on certain types of Linux distributions including Ubuntu but not Red Hat. On the other hand, the IBM Rational Rhapsody tools can be installed on Windows and only on the Red Hat Linux distribution. Therefore, we decided to add the Ubuntu VM to our Windows machine. The IBM Rational Rhapsody tools are to be installed on the Windows machine and the KLEE symbolic execution engine is to be installed on the Ubuntu VM. The decision of which component is to be implemented on which VM was made based on the functionality of each component and the development environment it requires. Fig. 22 highlights this categorization. For example, components that require the IBM infrastructure to operate are developed and run on Windows. Examples of these components are the "SC2MLM Transformation" and "RhapM2CMLM Transformation", which are implemented in the context of the Rhapsody RulesComposer Add-On to perform the following two tasks: (1) to translate from Rhapsody Statecharts to MLM and (2) to compute the data and communication dependencies in and between the generated MLMs. Similarly, components "SC2MLM DiffMap", "RhapM2CMLM DiffMap", and "Dependency-based Change Impact Analysis" require the manipulation of the files representing the MLMs generated from the transformation components and this requires access to the MLM meta–model and its related schema definitions which are tool-dependent generated artifacts. Thus, we have implemented these components as a Java project in the context of the Rhapsody RulesComposer Add-On. All other components which are the KLEE-based ones require the interaction with the KLEE symbolic execution engine and therefore they are developed and run on the Ubuntu VM.

In order to facilitate the communication between the part of the implementation that runs on the Ubuntu VM (where KLEE works) and the part that runs on the Windows VM (where the IBM Rational Rhapsody tool works), we use the Java Remote Method Invocation (RMI) system. We developed the part of the implementation that runs on the Ubuntu VM side as an RMI server API and the part of the implementation that runs on Window VM as an RMI client API which has been integrated with the IBM Rational Rhapsody tool as an external helper.

6.2 Interaction With KLEE

We referred to this task in the algorithms provided in Sections 4 and 5 using the pseudo-code function: SymbolicExecutionOfCode(A^s, val, G^s, pc). It identifies the information necessary to symbolically execute the action code and the guard of a transition given the variables' valuation and the path constraints collected to reach a symbolic state corresponding to the source state of a transition. The output from executing this statement is a set of all feasible paths resulting from parsing the output from the KLEE symbolic execution engine which we use to create a set of symbolic transitions and their target symbolic states. In the sequel we explain in details how we implement this task using the illustrations in Fig. 23.

In Fig. 23A, we show (1) on the left, part of an example Statechart model that consists of two states: S2 and S3 which are connected by a transition T2 and (2) on the right, an example symbolic state representation SS4 of the transition's source state S2. We also highlight the necessary information about the attribute(s) defined for the given model, the event e2 of transition T2 and its arguments (if any) and the symbolic variable(s) used in the given symbolic state representation SS4. In Fig. 23B, we show the flow diagram of

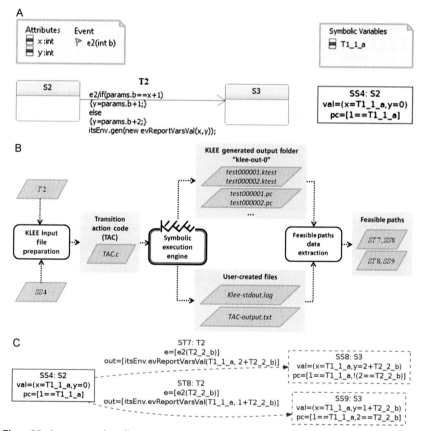

Fig. 23 An example illustrating an execution of the pseudo-code function SymbolicExecutionOfCode(A^s, val, G^s, pc). (A) On the *left* is part of a Statechart showing a transition T2 and its source and target States S2 and S3. On the *right* is a symbolic state SS4 of the source state S2. (B) Illustration of our interface components with KLEE. (C) Highlighted (in *red*) are the generated symbolic transitions ST7 and ST8 and their target symbolic states SS8 and SS9 resulting from the SE of transition T2 given a symbolic state SS4 of its source state S2.

how we interact with the KLEE symbolic execution engine to symbolically execute the action code of transition T2 given the symbolic state SS4. In this flow diagram, we identify two components: "KLEE Input File Preparation" and "Feasible Paths Data Extraction".

The "KLEE Input File Preparation" component takes as inputs a transition to be symbolically explored and a symbolic state representing the source state of the transition and generates as an output C file named "TAC.c" which has at the beginning the declarations of all required variables, including those which are symbolic, and the action code statements of the input transition as well as the path constraints definition of the input symbolic state. Following this, we add, for every declared variable, a conditional statement to check whether its last assigned value is concrete or symbolic and to record in a user-defined output file named "TAC-output.txt" the variable name and its value if it is concrete or the keyword "SYMBOLIC" if it is symbolic. In order to get the symbolic expression corresponding to a symbolic variable, we use KLEE's special function klee_print_expr which prints its output to the STDOUT stream of the running process. In order to manipulate this output, we redirect the STDOUT stream into a file named "klee-stdout.log". The output C file "TAC.c" is to be compiled and executed by the KLEE symbolic execution engine. The output from the execution is recorded in an output directory named "klee-out-0" which is created by KLEE. We also use this directory to record the aforementioned user-created files: "TAC-output.txt" and "klee-stdout.log".

Having the output files generated from KLEE ready, the next step is to parse these files to extract the required information about each feasible path reported by KLEE. This step is performed by the "Feasible Paths Data Extraction" component which consists of a series of shell script calls and a parser generated by JavaCC. Among the list of global files that KLEE generates for each execution, there is a set of per-path files to record information about the test case generated by KLEE for each path and the constraints associated with each path. The extensions for these files are .ktest and .pc, respectively. The test case files are binary files which require a specific tool called ktest-tool to read their contents, however the path constraint files are textual files written in the *KQuery* language—the input language of the *Kleaver* constraint solver used by KLEE. In order to convert these constraints into Java Boolean expressions, we have developed a JavaCC parser for a relevant subset of the KQuery language (the token definition of the lexical analyzer of our JavaCC parser is shown in Listing 1). We use this parser to parse the path constraints files as well as the symbolic expressions found in the user-created file "klee-stdout.log".

Example 7. In Listing 2 we show an excerpt of the file "TAC.c" that is generated by the component "KLEE Input File Preparation" for transition T2 and its symbolic source state SS4 in Fig. 23A. In Listings 3 and 4, we show the two path constraints files generated by KLEE as a result of executing the input file "TAC.c" shown in Listing 2. The Java Boolean expression representation corresponding to each path constraint is shown in Listings 5 and 6, respectively. Similarly, the files "TAC-output.txt" and "klee-stdout.log" resulting from executing KLEE on the input file "TAC.c" are shown in Listings 7 and 8. The Java expression representation corresponding to the symbolic variables found in the file "klee-stdout.log" in Listing 8 is shown in Listing 9. In Fig. 23C we show the symbolic transitions ST7 and ST8 and their symbolic target states SS8 and SS9 resulting from the parsing and the extraction step implemented by the component "Feasible Paths Data Extraction" for the files in Listings 3, 4, 7, and 8.

Listing 1: The Token Definition of the Lexical Analyzer of Our JavaCC Parser for a Subset of the KQuery Language

```
TOKEN :
{
    < LBR : "(" >
  | < RBR : ")" >
  | < QUERY : "query" >
  | < LSBR : "[" >
  | < RSBR : "]" >
  | < FALSE : " false" >
  | < TRUE : " true " >
  | < VarDeclStart : ("ReadLSB") >
  | < AEK : ("Add" | "Sub" | "Mul" | "UDiv" | "URem" | "SDiv" | "SRem") >
  | < CEK : ("Eq" | "Ne" | "Ult" | "Ule" | "Ugt" | "Uge" | " Slt " |
        " Sle " | "Sgt" | "Sge") >
  | < type : ["w"] (["0"-"9"])+ >
  | < identifier : ["a"-"z", "A"-"Z","_"] (["a"-"z","A"-"Z",
        "0"-"9", "_"])*>
  | < colon : " : " >
  | < number : (<decConstant> | <signedDecConstant>) >
  | < signedDecConstant : ("+") | ("-")<decConstant>>
  | < decConstant : (<digit>)+ >
  | < digit : ["0"-"9"] >
}
```

Listing 2: **Excerpt From** TAC.c

```
# include <stdio . h >
# include <klee / klee . h >
int main ()
{
    FILE * file;
    // variable declaration and initialization
    int T1_1_a;
    klee_make_symbolic(&T1_1_a, sizeof (T1_1_a), "T1_1_a");
    int T2_2_b;
    klee_make_symbolic(&T2_2_b, sizeof (T2_2_b), "T2_2_b");
    int x = T1_1_a;
    int y = 0;
    //user-defined variables to represent output events to be
    generated
    char* output1_eventName = "";
    char* output1_eventArg_1 = "";
    char* output1_eventArg_2 = "";
    //pre-defined path constraint (PC) declaraion
    klee_assume(1==T1_1_a);
    // action code symbolic representation
    if (T2_2_b==x+1) {
            y=T2_2_b+1;
    }
    else {
            y=T2_2_b+2;
    }
    output1_eventName = "itsEnv . evReportVarsVal";
    output1_eventArg_1 = x;
    output1_eventArg_2 = y;
    // logging all variables ' valuation
    file = fopen (" ./ klee-out-0/TAC-output.txt","a+");
    fprintf (file, "%s\n", "Output = ");
    printf ("%s\n", "Output = ");
    if (klee_is_symbolic (T1_1_a)) {
            fprintf (file, "%s\n", "T1_1_a = SYMBOLIC");
            klee_print_expr ("T1_1_a = ", T1_1_a);
    }
    else {
            fprintf (file, "%s%d\n", "T1_1_a = ", T1_1_a);
    }
```

```
\ ldots
if (! (output1_eventName== "")) {
        fprintf (file, "%s%s\n", "output1_eventName = " ,
            output1_eventName);
        if (klee_is_symbolic (output1_eventArg_1)) {
                fprintf (file, "%s\n", "output1_eventArg_1 =
                    SYMBOLIC");
                klee_print_expr ("output1_eventArg_1 = ",
                    output1_eventArg_1);
        }
        else if (output1_eventArg_1== "")
                fprintf (file, "%s\n", "output1_event
                    Arg_1 = ");
        else
                fprintf (file, "%s%d\n", "output1_
                    eventArg_1 = ", output1_eventArg_1);
        if (klee_is_symbolic (output1_eventArg_2)) {
                fprintf (file, "%s\n", "output1_eventArg_2 =
                    SYMBOLIC");
                klee_print_expr ("output1_eventArg_2 = ",
                    output1_eventArg_2);
        }
        else if (output1_eventArg_2== "")
                fprintf (file, "%s\n", "output1_event
                    Arg_2 = ");
        else
                fprintf (file, "%s%d\n", "output1_
                    eventArg_2 = ", output1_eventArg_2);
}
fclose (file);
}
```

Listing 3: KLEE Generated KQuery Constraint File Associated With the First Feasible Path of the Code in Listing 2: test000001.pc

```
array T1_1_a [4] : w32 -> w8 = symbolic
array T2_2_b [4] : w32 -> w8 = symbolic
(query [((Eq 1
            (ReadLSB w32 0 T1_1_a))
        (Eq false
            (Eq 2
                (ReadLSB w32 0 T2_2_b)))]
        false)
```

Listing 4: KLEE Generated KQuery Constraint File Associated With the Second Feasible Path of the Code in Listing 2: test000002.pc

```
array T1_1_a [4] : w32 -> w8 = symbolic
array T2_2_b [4] : w32 -> w8 = symbolic
(query [(Eq 1
            (ReadLSB w32 0 T1_1_a))
        (Eq 2
            (ReadLSB w32 0 T2_2_b))]
        false)
```

Listing 5: The Boolean Expressions of the KQuery Path Constraint in Listing 3

```
(1==T1_1_a) && !(2==T2_2_b)
```

Listing 6: The Boolean Expressions of the KQuery Path Constraint in Listing 4

```
(1==T1_1_a) && (2==T2_2_b)
```

Listing 7: User-Created File From Running KLEE on the LLVM Bitcode Representation of File TAC.c: TAC-output.txt

```
Output =
T1_1_a = SYMBOLIC
T2_2_b = SYMBOLIC
x = SYMBOLIC
y = SYMBOLIC
output1_eventName = itsEnv . evReportVarsVal
output1_eventArg_1 = SYMBOLIC
output1_eventArg_2 = SYMBOLIC
Output =
T1_1_a = SYMBOLIC
T2_2_b = SYMBOLIC
x = SYMBOLIC
y = SYMBOLIC
output1_eventName = itsEnv . evReportVarsVal
output1_eventArg_1 = SYMBOLIC
output1_eventArg_2 = SYMBOLIC
```

Listing 8: User-Created File of the Resulting stdout **From Running KLEE on the LLVM Bitcode Representation of File** TAC.c: klee-stdout.log

```
KLEE: output directory = " klee—out—0"

Output =
T1_1_a = : (ReadLSB w32 0 T1_1_a)
T2_2_b = : (ReadLSB w32 0 T2_2_b)
x = : (ReadLSB w32 0 T1_1_a)
y = : (Add w32 2
        (ReadLSB w32 0 T2_2_b))
output1_eventArg_1 = : (ReadLSB w32 0 T1_1_a)
output1_eventArg_2 = : (Add w32 2
        (ReadLSB w32 0 T2_2_b))
Output =
T1_1_a = : (ReadLSB w32 0 T1_1_a)
T2_2_b = : (ReadLSB w32 0 T2_2_b)
x = : (ReadLSB w32 0 T1_1_a)
y = : (Add w32 1
        (ReadLSB w32 0 T2_2_b))
output1_eventArg_1 = : (ReadLSB w32 0 T1_1_a)
output1_eventArg_2 = : (Add w32 1
        (ReadLSB w32 0 T2_2_b))

KLEE: done : total instructions = 194
KLEE: done : completed paths = 2
KLEE: done : generated tests = 2
```

Listing 9: The Boolean Expressions of the KQuery Path Constraint in Listing 8

```
T1_1_a = T1_1_a
T2_2_b = T2_2_b
x = T1_1_a
y = 2 + T2_2_b
output1_eventArg_1 = T1_1_a
output1_eventArg_2 = 2 + T2_2_b

T1_1_a = T1_1_a
T2_2_b = T2_2_b
x = T1_1_a
y = 1 + T2_2_b
output1_eventArg_1 = T1_1_a
output1_eventArg_2 = 1 + T2_2_b
```

6.3 Implementation of the MQL and TGL-Based Transformation Components

Our transformation components are implemented in the context of the IBM Rational Rhapsody Developer RulesComposer Add-On which is an Eclipse-based development environment for model transformation and text generation. In such an environment, a model transformation is defined using an imperative transformation language called MQL. The basic element of a model transformation is a ruleset which is a group of logically inter-dependent rules, each of which specifies a set of procedural expressions that query source and target model elements to create new elements, update the attributes and the references of existing elements, or launch text generation templates. A model transformation may be implemented using one or more rulesets. Text generation is defined using templates written in a language called the TGL which uses MQL expressions (i.e., placeholders) for dynamic text generation. Scripts are methods written in MQL, TGL, or Java to extend and customize a meta-type of a source or a target meta-model for a particular model transformation or text generation. They are called from MQL and TGL code on model elements in the same way as regular Java methods.

To implement our transformation components, we first defined the meta-model of the CMLMs representation defined in Section 4 as an Ecore model which we show in Appendix. We also imported this Ecore model as a meta-model project that is added to the default set of meta-models defined within the RulesComposer Add-On. Then, we created a RulesComposer project with references to the Rhapsody meta-model which is already defined and integrated within the tool and the CMLMs meta-model that we have defined and imported into the tool. Our implementation of this model transformation and text generation RulesComposer project consists of:

- 19 rules grouped into 2 rulesets. The first ruleset groups a set of 17 rules that specify how to flatten and map a Rhapsody source model into a CMLMs target model. The second ruleset groups a set of 2 rules that computes the data and the communication dependencies in a given CMLMs model.
- 48 MQL scripts distributed among the following meta-types:
  - 9 scripts for "`rhapsody_Statechart`",
  - 21 scripts for "`rhapsody_State`",
  - 1 script for "`rhapsody_Transition`",

- • 5 scripts for "`cmlms_CMLM`",
  - • 5 scripts for "`cmlms_MLM`", and
  - • 7 scripts for "`cmlms_Transition`".
- • 6 Java scripts equally distributed between the "`cmlms_CMLM`" and "`cmlms_MLM`" meta-types. We defined these scripts to be able to create specialized ArrayList, HashMap, and HashSet objects to be used within our rules.
- • 2 TGL templates. These templates are used to generate a Java representation for our CMLMs representations; one for the individual MLMs and one for the communicating MLMs.
- • 4948 lines of code (LOC) distributed as follows:
  - • 2761 LOC for the rulesets,
  - • 1516 LOC for the MQL scripts,
  - • 102 LOC for the Java scripts, and
  - • 568 LOC for the TGL templates.

6.4 Interaction With the IBM Rational Rhapsody DiffMerge and Its Related Components

As we decided to use the IBM Rational Rhapsody DiffMerge tool to find the differences between evolving versions of Rhapsody models, we needed to find a way to process and parse the output generated by the tool. Two types of reports can be exported from the tool that summarize the differences found between compared models. The first report is a Rich Text Format (RTF) report and it is designed to contain a greater level of detail but it is not suitable for further processing to extract difference data. On the other hand, the second report is a Comma Separated Values (CSV) format report which is easier to process but it is designed to contain a lower level of detail. In order to extend the information included in the CSV format reports about the difference data, we had to extend the underlying CSV report template of the tool. The default template file we found is called "`DiffReport_csv.dpl`" and it is written in a language that is embedded in the IBM Rational Rhapsody ReporterPLUS for generating CSV format reports about the difference data found by the Rhapsody DiffMerge tool.

The original version of this template file contains 156 LOC and it reports only on six specified attributes for a given difference. This includes information about the meta-class of the difference elements, their names, their locations, the difference type indicating whether the difference element is an added element, a deleted element or a modified element, the differences count (i.e., the number of attributes in the difference elements with different values) and whether a difference is a trivial (i.e., nonconflicting) difference or not. This latter feature is mainly used when the three-way comparison with the base unit option is performed. In this case if a difference is found in the two compared units with respect to their base unit, then this difference is a nontrivial (i.e., conflicting) difference which needs manual resolution; the automatic merging operation of the tool cannot resolve nontrivial differences. The merging operation is outside the scope of our work and we do not need the three-way comparison option. Therefore, all the differences reported were trivial differences.

Our adapted version of the same template file contains 205 LOC and it adds three extra attributes to be reported to the original list. These are the names of the attributes in the difference elements with different values and their values in both compared models, namely, the "left value" and the "right value". Recording the values of these attributes enabled us to uniquely identify specific Rhapsody model elements which may have the same value assigned to their "name" attribute. Transitions are an example of such elements where two or more transitions can have the same name. Users are not allowed to set the value of the "name" attribute of a transition in the IBM Rhapsody tool. Instead, the tool uses the values assigned to the "trigger", the "guard", and the "action" attributes of a transition to generate and assign a compound value to its "name" attribute. For the unique identification of transitions in a model, the tool uses an internal global unique identifier number to be assigned to each newly created transition. The attribute that is used for this purpose is a private variable which cannot be accessed by users. We also noticed that the value of the "name" attribute of a difference transition recorded by the Rhapsody DiffMerge tool does not exactly match the value recorded for the attribute "name" of the same transition in the corresponding Rhapsody Statechart model. Only the "trigger" and the "guard" parts of the name of a transition are used by the Rhapsody DiffMerge tool to report the name of a difference transition. This information alone is not enough to uniquely identify a transition in a model especially in the case of poorly designed Statecharts that may have nondeterministic transitions. Therefore, we believe that the addition we have made to the default CSV report template of the Rhapsody DiffMerge tool was mandatory and it enabled us to uniquely identify the difference elements reported including states and transitions.

Given our adapted CSV format difference reports, the next step for us was to develop a parser such that all the difference information recorded in a given CSV difference report are mapped to our user-defined data structure objects. We use this parsed information and the MLM or CMLMs representations of compared Rhapsody models as the inputs to components "SC2MLM DiffMap" and "RhapM2CMLMs" in order to map the identified difference elements in the compared Rhapsody models into their MLM or CMLMs correspondences. The generated difference report from either of the two aforementioned components is used as an input to component "Dependency-based Change Impact Analysis" to identify the parts of an input MLM or CMLMs representation that have a data or communication dependence with a difference element in the input difference report. The implementation of the CSV parser, the two DiffMap components and the "Depend-ency-based Change Impact Analysis" component are all built as a Java project in the context of the IBM Rational Rhapsody RulesComposer Add-On. The project consists of 10 Java classes and approximately 1581 lines of code (LOC).

6.5 Summary

In this section we have provided an overview of the implementation of the architectures presented in the previous sections, showing the operating systems required and the technologies used and how we realized the interaction between the different components to fulfill the required tasks.

7. EVALUATION

In this section we present the evaluation of the effectiveness of the two proposed optimization techniques: MSE and DSE, on both individual Statecharts and communicating Statecharts. We first state the research questions considered in our evaluation. Second, we list the set of artifacts used as case studies for individual and communicating Statecharts. Third, we describe the evaluation setup. Finally we discuss the results and draw conclusions.

It is worth mentioning that we performed our evaluation in two stages. The first stage considered the evaluation of individual Statecharts, while the second stage considered the evaluation of communicating Statecharts. In the first stage we consider only modification changes, while in the second stage we consider also addition and deletion changes.

7.1 Research Questions and Variables of Interest

We consider the following three research questions:

- *RQ1*. How effective are our optimizations (i.e., how much do they reduce the resource requirements of the symbolic execution of a changed state machine model) compared to standard SE of a changed Statechart model?

    For RQ1, we consider the following three correlated variables: (1) the time taken to run each technique, (2) the number of symbolic states, and (3) the number of execution paths in the resulting SETs.

- *RQ2*. Does the SET generated from MSE match the one generated from standard SE?

    For RQ2, we compare the total number of symbolic states in the SETs resulting from standard SE and MSE. We also perform manual inspection of a subset of the two SETs to ensure their equivalence.

- *RQ3*. Which aspects influence the effectiveness of each technique?

    For RQ3, we consider the following two aspects: (1) *change impact* and (2) *model characteristics*.

    To measure the impact of the change, we consider the following two criteria:

    (1) we define a change impact metric (CIM) as the percentage of the maximal paths of the MLM model involving the change. The lower the value of this metric is, the lower the impact of the change on the model and the higher the opportunity to benefit from a previous analysis results, and vice versa. This metric is used only for individual Statecharts (i.e., it is applied only on the AQS, LGS, and ACCS models) where the notion of maximal path is defined and, thus, can be computed.

    (2) we identify the type of the change applied to create each modified version of the subject models. The change types we consider are modification (M), addition (A), and deletion (D).

    For model characteristics, we identify the following two metrics that appear likely to have an impact on the effectiveness of DSE:

    (1) the number of transitions in the subject model with multipath guards or action code.

    (2) the number of transitions that have a dependency with other transitions in the model.

    We speculate that the change impact is a better predictor for the effectiveness of MSE, whereas the model characteristics are more suitable as indicators of the effectiveness of DSE.

7.2 Case Study Artifacts

To evaluate the effectiveness of our optimization techniques on both individual and communicating Statecharts, the following artifacts are selected:

(1) *For individual Statecharts:* We chose three industrial-sized models from the automotive domain. The first model, the Air Quality System (AQS), is a proprietary model that we obtained from our industrial partner that is responsible for air purification in the vehicle's cabin. The second and the third models, the Lane Guide System (LGS) and the Adaptive Cruise Control System (ACCS), are nonproprietary models designed at the University of Waterloo [96]. The LGS is an automotive feature used to avoid unintentional lane departure by providing alerts when certain events occur. The ACCS is an automotive feature used to automatically maintain the speed of a vehicle set by the driver through the automatic operation of the vehicle. The three models were developed as Simulink/Stateflow models and we manually converted them to behaviorally equivalent Rhapsody Statecharts. Table 2a summarizes the characteristics of the three models and their MLM representations, including the two model characteristics metrics used to answer RQ3. The models contain 4, 7, and 3 variables of type integer, respectively.

(2) *For communicating Statecharts:* We used an adapted version of a system model implementing a well-known communication protocol called full-duplex alternating protocol (FABP) found in [97]. The model consists of three communicating objects: Transmitter, Receiver, and Buffer. The behavior of each object is modeled as a Statechart. A summary of the number of states and transitions in each state machine is shown in Table 2b. We also listed the values of the model characteristics metrics, namely, the number of transitions with multipath action in each Statechart and the number of transition with data or communication dependencies. There are four integer variables in this model.

7.3 Evaluation Setup

To perform our study, we followed the following procedures:

(1) *For individual Statecharts:* We first prepared a set of different versions for each artifact: the base version and a number of modified versions. Each modified version introduces a single change to one simple or group transition in the base version. Each change is made in the form of an alteration to the event name or the addition of a send event statement

Table 2 Characteristics of the Artifacts Used in Our Evaluation
(a) Three individual Statecharts models

Numbers of	Air Quality System (AQS)		Lane Guide System (LGS)		Adaptive Cruise Control System (ACCS)	
	Rhapsody Statechart (Before Flattening)	MLM (After Flattening)	Rhapsody Statechart (before flattening)	MLM (after flattening)	Rhapsody Statechart (Before Flattening)	MLM (after flattening)
Concurrent regions	1	1	1	1	2	1
Hierarchical levels	3	1	3	1	3	1
States	3 CS+14 SS	14 SS	2 CS+10 SS	10 SS	2 CS+6 SS	3 SS 19 SS
Transitions	8 GT+22 ST	55 ST	6 GT+16 ST	40 ST	9 GT+5 ST	9 ST 73 ST
Transitions with multipath guards or action code	5 ST	5 ST	2 GT	10 ST	1 ST	0 1 ST
Transitions with data dependencies	3 ST	3 ST	0	0	2 GT+2 ST	4 ST 16 ST

CS, composite state; GT, group transition; SS, simple state; ST, simple transition.

(b) One Communicating Statecharts Model

Full-Duplex Alternating Bit Protocol (FABP)

Numbers of	Transmitter (Trans)		Receiver (Rcv)		Buffer (Buff)	
	Rhapsody Statechart (before flattening)	MLM (after flattening)	Rhapsody Statechart (before flattening)	MLM (after flattening)	Rhapsody Statechart (Before Flattening)	MLM (after flattening)
Concurrent regions	1	1	1	1	1	1
Hierarchical levels	1	1	1	1	1	1
States	4 SS	4 SS	4 SS	4 SS	2 SS	2 SS
Transitions	14 ST	14 ST	14 ST	14 ST	4 ST	4 ST
Transitions with multipath guards or action code	0	0	6 ST	6 ST	0	0
Transitions with data dependencies	8 ST	8 ST	0	0	0	0
Transitions with communication dependencies	10 ST	10 ST	4 ST	4 ST	4 ST	4 ST

SS, simple State; ST, simple transition.

to the action code of the subject transition. The main reason for selecting these specific types of changes is to keep the number of modified models manageable and to facilitate the manual correctness check of the results. Another important reason is that some of the tools used in the implementation (e.g., the IBM Rhapsody DiffMerge) do not have command line access which made it impossible to fully automate the collection of the performance data and evaluate the techniques more comprehensively using, e.g., a large number of automatically modified models. The total numbers of the modified versions created for each model are 26 for the AQS model, 19 of the LGS, and 19 for the ACCS model.

For each modified version, we first recorded the following information: (1) the change ID [g] and the number of transitions in the MLM representation of the subject model corresponding to the changed transition in Rhapsody, (2) the number of multipath transitions in the MLM representation of the subject model that have been found to impact or be impacted by the changed transition, and (3) the value of the CIM metric computed for the study of RQ3. Second, we ran our standard SE on all the versions of the three selected artifacts, while we ran both the MSE and the DSE only on the modified versions, given the results of the SE of their base versions in the case of MSE.

For each run, we recorded the time to generate the SET and the total number of symbolic states in the generated SET. Additionally, in case of an MSE run, we recorded the number of symbolic states that are newly created. This is to differentiate between the symbolic states generated by the MSE technique and those that have been reused from the SET of the base version. To measure the effectiveness of MSE and DSE compared to standard SE, we computed the ratios between the execution times, the number of symbolic states, and the number of the execution paths recorded for the standard SE and their correspondences in the MSE and DSE, respectively. From these ratios, we computed the average savings and the standard deviation. We also computed the correlation coefficients [h] between the savings gained from each technique and the computed CIM metric values.

---

[g] A compound value of the model name and the ID of the transition in Rhapsody where the change is made prefixed with the change type (M for a modification change, A for an addition change, or D for a deletion change).

[h] We use the sample Pearson correlation coefficient formula implemented in the Microsoft Excel function *CORREL* to measure the degree of linear correlation between two variables with a range between $+1$ and -1 inclusive. A value of $+1$ indicates a perfect positive correlation, while a value of -1 indicates a perfect negative correlation. A value of 0 indicates no correlation.

(2) *For communicating Statecharts:* We followed the same procedure described above for individual Statecharts; however, here we prepared three sets of different versions of the base version of the FABP model. Each set has 32 modified versions with an added transition, a deleted transition, or a modified transition. All changes in the modified versions are atomic. However, they cover all the transitions in the FABP model. For each model version, we recorded the same information listed earlier except the information about the CIM metric which is not applicable for communicating Statecharts. We ran our standard SE and the MSE on all versions, while we ran the DSE only on the versions where the change is found to not impact all of the multipath transitions in the model. Performing the DSE on a model version with a change that has an impact on all multipath transitions will result in no savings compared to performing the standard SE on the same model. For each run, we recorded the same information listed earlier about the characteristics of the generated GSETs.

The characteristics of the SETs generated from running the standard SE on the base versions of the AQS, LGS, ACCS, and FABP models are shown in Table 3. They include the time taken to run the standard SE technique and the number of symbolic states and execution paths in the resulting SETs.

Tables 4, 5, and 6 summarize the results of running the three symbolic execution techniques on the modified versions of the AQS, the LGS, and the ACCS models, respectively. Additionally, a summary of the results for the versions of each change type of the FABP model is depicted in Tables 7, 8, and 9, respectively.

In columns 5–6, 8 (resp., 10–12) of Tables 4–6 and columns 4–5, 7 (resp., 9–11) of Tables 7–9, we show the savings ratios in time, in the number of

Table 3 Performance of Standard SE on the Base Version of the Four Models: AQS, LGS, ACCS, and FABP

SET/GSET Characteristics	Air Quality System (AQS)	Lane Guide System (LGS)	Adaptive Cruise Control System (ACCS)	Full-Duplex Alternating Bit Protocol (FABP)
Time (min)	≈ 10	≈ 5	≈ 9	≈ 14
No. symbolic states	6039	1769	3492	2805
No. execution paths	5019	1325	2855	1342

Table 4 Results of MSE and DSE on the AQS Example

		MLM			SE:MSE					SE:DSE		
Ver.	Change ID/No. of Changed Transitions	No. of Multipath Impacted Transitions	CIM [0, 100]	Time	No. of New Symbolic States	No. of Total Symbolic States	No. of New Execution Paths	No. of Total Execution Paths	Time	No. of Total Symbolic States	No. of Total Execution Paths
V1	M[AQS-06]/7	0	30	1:0.3	1:0.08	1:1	1:0.08	1:1	1:0.02	1:0.03	1:0.02
V2	M[AQS-27]/1	0	4	1:0.3	1:0.36	1:1	1:0.36	1:1	1:0.01	1:0.03	1:0.02
V3	M[AQS-07]/1	0	100	1:0.51	1:1	1:1	1:1	1:1	1:0.02	1:0.03	1:0.02
V4	M[AQS-09]/1	0	20	1:0.26	1:0.23	1:1	1:0.23	1:1	1:0.02	1:0.04	1:0.03
V5	M[AQS-10]/1	1	30	1:0.21	1:0.01	1:1	1:0.01	1:1	1:0.02	1:0.14	1:0.13
V6	M[AQS-08]/1	1	90	1:0.44	1:0.64	1:1	1:0.66	1:1	1:0.02	1:0.14	1:0.13
V7	M[AQS-28]/1	1	0	1:0.23	1:0.06	1:1	1:0.06	1:1	1:0.02	1:0.14	1:0.13
V8	M[AQS-29]/1	0	30	1:0.27	1:0.23	1:1	1:0.23	1:1	1:0.02	1:0.03	1:0.02
V9	M[AQS-11]/1	0	0	1:0.19	1:0	1:1	1:0	1:1	1:0.02	1:0.03	1:0.02
V10	M[AQS-12]/1	0	40	1:0.24	1:0.23	1:1	1:0.23	1:1	1:0.02	1:0.04	1:0.03
V11	M[AQS-13]/1	0	30	1:0.2	1:0.06	1:1	1:0.06	1:1	1:0.02	1:0.03	1:0.02
V12	M[AQS-14]/1	0	1	1:0.22	1:0.02	1:1	1:0.02	1:1	1:0.02	1:0.03	1:0.02
V13	M[AQS-04]/5	0	97	1:0.64	1:0.87	1:1	1:0.91	1:1	1:0.02	1:0.03	1:0.02
V14	M[AQS-05]/7	0	30	1:0.38	1:0.39	1:1	1:0.43	1:1	1:0.02	1:0.03	1:0.02

Change Information

V15 M[AQS-15]/1	0	94	1:0.41	1:0.61	1:1	1:0.64	1:1	1:0.02	1:0.17	1:0.16
V16 M[AQS-16]/1	0	14	1:0.22	1:0.01	1:1	1:0.02	1:1	1:0.02	1:0.03	1:0.02
V17 M[AQS-17]/5	0	14	1:0.29	1:0.15	1:1	1:0.17	1:1	1:0.02	1:0.03	1:0.02
V18 M[AQS-18]/5	0	42	1:0.22	1:0.06	1:1	1:0.06	1:1	1:0.02	1:0.03	1:0.02
V19 M[AQS-20]/1	0	35	1:0.21	1:0.01	1:1	1:0.01	1:1	1:0.02	1:0.03	1:0.02
V20 M[AQS-21]/1	0	48	1:0.32	1:0.41	1:1	1:0.42	1:1	1:0.01	1:0.03	1:0.02
V21 M[AQS-22]/1	0	27	1:0.22	1:0.01	1:1	1:0.01	1:1	1:0.01	1:0.03	1:0.02
V22 M[AQS-23]/1	0	2	1:0.21	1:0	1:1	1:0	1:1	1:0.01	1:0.03	1:0.02
V23 M[AQS-24]/1	0	35	1:0.29	1:0.41	1:1	1:0.41	1:1	1:0.01	1:0.03	1:0.02
V24 M[AQS-25]/1	0	14	1:0.2	1:0.03	1:1	1:0.03	1:1	1:0.01	1:0.03	1:0.02
V25 M[AQS-26]/1	0	25	1:0.3	1:0.4	1:1	1:0.4	1:1	1:0.01	1:0.03	1:0.02
V26 M[AQS-03]/5	0	2	1:0.28	1:0.09	1:1	1:0.08	1:1	1:0.02	1:0.03	1:0.02
Average savings			71.0%	75.5%		74.9%		98.4%	95.6%	96.1%
Standard deviation			10.9%	28.0%		28.8%		0.2%	4.4%	4.2%
CORREL(CIM)			−0.81	−0.83		−0.84		−0.21	−0.33	−0.33

Table 5 Results of MSE and DSE on the LGS Example

		MLM			SE:MSE					SE:DSE		
Ver.	Change ID/No. of Changed Transitions	No. of Multipath Impacted Transitions	CIM [0, 100]	Time	No. of New Symbolic States	No. of Total Symbolic States	No. of New Execution Paths	No. of Total Execution Paths	Time	No. of Total Symbolic States	No. of Total Execution Paths	
V1	M[LGS–01]/1	0	100	1:0.81	1:1	1:1	1:1	1:1	1:0.28	1:0.3	1:0.3	
V2	M[LGS–02]/7	0	28	1:0.19	1:0.25	1:1	1:0.33	1:1	1:0.25	1:0.3	1:0.3	
V3	M[LGS–05]/7	0	28	1:0.19	1:0.25	1:1	1:0.33	1:1	1:0.3	1:0.3	1:0.3	
V4	M[LGS–09]/1	0	97	1:0.74	1:1	1:1	1:1	1:1	1:0.23	1:0.3	1:0.3	
V5	M[LGS–08]/4	0	13	1:0.38	1:0.5	1:1	1:0.5	1:1	1:0.22	1:0.47	1:0.47	
V6	M[LGS–06]/1	0	13	1:0.47	1:0.58	1:1	1:0.59	1:1	1:0.22	1:0.3	1:0.3	
V7	M[LGS–07]/5	0	40	1:0.55	1:0.74	1:1	1:0.74	1:1	1:0.22	1:0.73	1:0.73	
V8	M[LGS–11]/1	0	22	1:0.24	1:0.27	1:1	1:0.27	1:1	1:0.22	1:0.3	1:0.3	
V9	M[LGS–13]/1	0	22	1:0.25	1:0.27	1:1	1:0.27	1:1	1:0.22	1:0.3	1:0.3	
V10	M[LGS–19]/1	0	22	1:0.3	1:0.27	1:1	1:0.27	1:1	1:0.22	1:0.3	1:0.3	
V11	M[LGS–20]/1	0	22	1:0.28	1:0.27	1:1	1:0.27	1:1	1:0.22	1:0.3	1:0.3	
V12	M[LGS–12]/1	0	3	1:0.09	1:0.04	1:1	1:0.04	1:1	1:0.22	1:0.3	1:0.3	

V13	M[LGS-15]/1	0	13	1:0.14	1:0.12	1:1	1:0.12	1:1	1:0.22	1:0.3
V14	M[LGS-10]/1	0	3	1:0.09	1:0.04	1:1	1:0.04	1:1	1:0.22	1:0.3
V15	M[LGS-14]/1	0	13	1:0.15	1:0.12	1:1	1:0.12	1:1	1:0.22	1:0.3
V16	M[LGS-16]/1	0	13	1:0.18	1:0.12	1:1	1:0.12	1:1	1:0.22	1:0.3
V17	M[LGS-21]/1	0	3	1:0.09	1:0.04	1:1	1:0.04	1:1	1:0.22	1:0.3
V18	M[LGS-17]/1	0	13	1:0.16	1:0.12	1:1	1:0.12	1:1	1:0.22	1:0.3
V19	M[LGS-18]/1	0	3	1:0.09	1:0.04	1:1	1:0.04	1:1	1:0.22	1:0.3
Average savings				71.7%	68.2%		67.2%	76.7%	67.0%	66.6%
Standard deviation				21.6%	30.6%		30.5%	2.2%	10.5%	10.4%
CORREL(M)				−0.89	−0.88		−0.89	−0.48	−0.09	−0.08

Table 6 Results of MSE and DSE on the ACCS Example Change Information

		MLM			SE:MSE				SE:DSE		
Ver.	Change ID/No. of Changed Transitions	No. of Multipath Impacted Transitions	CIM [0, 100]	Time	No. of New Symbolic States	No. of Total Symbolic States	No. of New Execution Paths	No. of Total Execution Paths	Time	No. of Total Symbolic States	No. of Total Execution Paths
V1	M[ACCS-04]/1	0	100	1:0.82	1:1	1:1	1:1	1:1	1:1.05	1:0.92	1:0.9
V2	M[ACCS-05]/12	0	30	1:0.21	1:0.17	1:1	1:0.17	1:1	1:0.79	1:0.92	1:0.9
V3	M[ACCS-06]/12	0	30	1:0.19	1:0.15	1:1	1:0.17	1:1	1:0.78	1:0.92	1:0.9
V4	M[ACCS-19]/2	0	2	1:0.61	1:0.79	1:1	1:0.8	1:1	1:1.02	1:0.92	1:0.9
V5	M[ACCS-20]/2	0	2	1:0.43	1:0.44	1:1	1:0.46	1:1	1:1	1:0.92	1:0.9
V6	M[ACCS-13]/2	0	60	1:0.73	1:0.87	1:1	1:0.87	1:1	1:1.02	1:0.92	1:0.9
V7	M[ACCS-14]/2	0	60	1:0.45	1:0.55	1:1	1:0.55	1:1	1:1.02	1:0.92	1:0.9
V8	M[ACCS-15]/2	0	20	1:0.56	1:0.76	1:1	1:0.77	1:1	1:1.03	1:0.92	1:0.9
V9	M[ACCS-16]/2	0	20	1:0.24	1:0.18	1:1	1:0.2	1:1	1:1.03	1:0.92	1:0.9
V10	M[ACCS-17]/4	0	36	1:0.14	1:0.05	1:1	1:0.04	1:1	1:1.04	1:0.92	1:0.9
V11	M[ACCS-18]/4	0	36	1:0.26	1:0.2	1:1	1:0.18	1:1	1:1.04	1:0.92	1:0.9
V12	M[ACCS-07]/1	1	99	1:0.81	1:1	1:1	1:1	1:1	1:1.02	1:1	1:1

V13 M[ACCS-08]/6	0	14	1:0.27	1:0.22	1:1	1:0.19	1:1	1:1.02	1:0.92	1:0.9
V14 M[ACCS-09]/1	1	11	1:0.59	1:0.67	1:1	1:0.68	1:1	1:0.87	1:1	1:1
V15 M[ACCS-21]/1	0	11	1:0.11	1:0.01	1:1	1:0.01	1:1	1:0.99	1:0.92	1:0.9
V16 M[ACCS-11]/3	0	83	1:0.73	1:0.87	1:1	1:0.88	1:1	1:1	1:0.92	1:0.9
V17 M[ACCS-12]/3	1	30	1:0.23	1:0.17	1:1	1:0.2	1:1	1:0.99	1:1	1:1
V18 M[ACCS-10]/6	0	31	1:0.63	1:0.72	1:1	1:0.73	1:1	1:1.02	1:0.92	1:0.9
V19 M[ACCS-22]/6	0	31	1:0.17	1:0.09	1:1	1:0.12	1:1	1:1.02	1:0.92	1:0.9
Average savings			56.9%	53.2%		52.6%		1.4%	7.0%	8.6%
Standard deviation			24.6%	35.0%		35.1%		8.0%	3.1%	3.8%
CORREL(CIM)			−0.59	−0.56		−0.56		−0.22	−0.14	−0.14

Table 7 Results of MSE and DSE on the Versions of the FABP Example With *Modification Changes* Change Information

		CMLM		SE:MSE					SE:DSE		
Ver.	Change ID/No. of Changed Transitions	No. of Multipath Impacted Transitions	Time	No. of New Symbolic States	No. of Total Symbolic States	No. of New Execution Paths	No. of Total Execution Paths	Time	No. of Total Symbolic States	No. of Total Execution Paths
V1	M[Buff-01]/1	0	1:0.71	1:0.77	1:1	1:0.78	1:1	1:0.03	1:0.03	1:0.04
V2	M[Buff-02]/1	6	1:0.69	1:0.76	1:0.98	1:0.77	1:0.99	1:1	1:1	1:1
V3	M[Buff-03]/1	0	1:0.13	1:0.05	1:1	1:0.05	1:1	1:0.03	1:0.03	1:0.04
V4	M[Buff-04]/1	6	1:0.57	1:0.56	1:1	1:0.62	1:1	1:1	1:1	1:1
V5	M[Rcv-01]/1	6	1:0.19	1:0.17	1:1	1:0.18	1:1	1:1	1:1	1:1
V6	M[Rcv-02]/1	6	1:0.44	1:0.5	1:1	1:0.49	1:1	1:1	1:1	1:1
V7	M[Rcv-03]/1	6	1:0.37	1:0.38	1:1	1:0.39	1:1	1:1	1:1	1:1
V8	M[Rcv-04]/1	6	1:0.27	1:0.3	1:1	1:0.31	1:1	1:1	1:1	1:1
V9	M[Rcv-05]/1	6	1:0.2	1:0.17	1:1	1:0.18	1:1	1:1	1:1	1:1
V10	M[Rcv-06]/1	6	1:0.41	1:0.38	1:1	1:0.38	1:1	1:1	1:1	1:1
V11	M[Rcv-07]/1	6	1:0.42	1:0.45	1:1	1:0.44	1:1	1:1	1:1	1:1
V12	M[Rcv-08]/1	6	1:0.68	1:0.73	1:1	1:0.72	1:1	1:1	1:1	1:1
V13	M[Rcv-09]/1	6	1:0.64	1:0.76	1:1	1:0.75	1:1	1:1	1:1	1:1
V14	M[Rcv-10]/1	0	1:0.16	1:0.08	1:1	1:0.11	1:1	1:0.03	1:0.03	1:0.04
V15	M[Rcv-11]/1	0	1:0.34	1:0.28	1:0.98	1:0.27	1:0.98	1:0.03	1:0.03	1:0.04

V16 M[Rcv-12]/1	0	1:0.29	1:0.23	1:0.99	1:0.23	1:0.99	1:0.03	1:0.03	1:0.04
V17 M[Rcv-13]/1	0	1:0.47	1:0.5	1:0.7	1:0.51	1:0.71	1:0.03	1:0.03	1:0.04
V18 M[Rcv-14]/1	6	1:0.63	1:0.7	1:1	1:0.7	1:1	1:1	1:1	1:1
V19 M[Trans-01]/1	6	1:0.55	1:0.7	1:0.7	1:0.71	1:0.71	1:1	1:1	1:1
V20 M[Trans-02]/1	6	1:0.74	1:0.88	1:0.98	1:0.89	1:0.99	1:1	1:1	1:1
V21 M[Trans-03]/1	6	1:0.12	1:0.01	1:1	1:0.02	1:1	1:1	1:1	1:1
V22 M[Trans-04]/1	6	1:0.43	1:0.44	1:1	1:0.41	1:1	1:1	1:1	1:1
V23 M[Trans-05]/1	6	1:0.58	1:0.64	1:1	1:0.63	1:1	1:1	1:1	1:1
V24 M[Trans-06]/1	6	1:0.12	1:0.01	1:1	1:0.01	1:1	1:1	1:1	1:1
V25 M[Trans-07]/1	6	1:0.88	1:0.99	1:1	1:0.99	1:1	1:1	1:1	1:1
V26 M[Trans-08]/1	6	1:0.77	1:0.83	1:1	1:0.85	1:1	1:1	1:1	1:1
V27 M[Trans-09]/1	6	1:0.29	1:0.27	1:1	1:0.27	1:1	1:1	1:1	1:1
V28 M[Trans-10]/1	6	1:0.34	1:0.32	1:1	1:0.33	1:1	1:1	1:1	1:1
V29 M[Trans-11]/1	6	1:0.14	1:0.04	1:1	1:0.04	1:1	1:1	1:1	1:1
V30 M[Trans-12]/1	6	1:0.25	1:0.19	1:1	1:0.16	1:1	1:1	1:1	1:1
V31 M[Trans-13]/1	6	1:0.14	1:0.01	1:1	1:0.01	1:1	1:1	1:1	1:1
V32 M[Trans-14]/1	6	1:0.13	1:0.02	1:1	1:0.04	1:1	1:1	1:1	1:1
Average savings		59.2%	58.9%		58.6%		18.2%	18.1%	18.0%
Standard deviation		22.7%	29.8%		29.8%		38.5%	38.3%	38.2%

Table 8 Results of MSE and DSE on the Versions of the FABP Example With *Addition Changes*

	CMLM			SE:MSE					SE:DSE		
Ver.	Change ID/No. of Changed Transitions	No. of Multipath Impacted Transitions	Time	No. of New Symbolic States	No. of Total Symbolic States	No. of New Execution Paths	No. of Total Execution Paths	Time	No. of Total Symbolic States	No. of Total Execution Paths
V1	A[Buff-01]/1	0	1:0.72	1:0.78	1:1	1:0.78	1:1	1:0.03	1:0.03	1:0.04
V2	A[Buff-02]/1	6	1:0.72	1:0.8	1:0.98	1:0.86	1:0.99	1:1	1:1	1:1
V3	A[Buff-03]/1	0	1:0.13	1:0.05	1:1	1:0.05	1:1	1:0.03	1:0.03	1:0.04
V4	A[Buff-04]/1	6	1:0.53	1:0.57	1:1	1:0.62	1:1	1:1	1:1	1:1
V5	A[Rcv-01]/1	6	1:0.21	1:0.17	1:1	1:0.18	1:1	1:1	1:1	1:1
V6	A[Rcv-02]/1	6	1:0.5	1:0.5	1:1	1:0.49	1:1	1:1	1:1	1:1
V7	A[Rcv-03]/1	6	1:0.41	1:0.38	1:1	1:0.39	1:1	1:1	1:1	1:1
V8	A[Rcv-04]/1	6	1:0.35	1:0.3	1:1	1:0.31	1:1	1:1	1:1	1:1
V9	A[Rcv-05]/1	6	1:0.23	1:0.17	1:1	1:0.18	1:1	1:1	1:1	1:1
V10	A[Rcv-06]/1	6	1:0.4	1:0.38	1:1	1:0.38	1:1	1:1	1:1	1:1
V11	A[Rcv-07]/1	6	1:0.44	1:0.45	1:1	1:0.45	1:1	1:1	1:1	1:1
V12	A[Rcv-08]/1	6	1:0.72	1:0.74	1:1	1:0.73	1:1	1:1	1:1	1:1
V13	A[Rcv-09]/1	6	1:0.71	1:0.76	1:1	1:0.75	1:1	1:1	1:1	1:1
V14	A[Rcv-10]/1	0	1:0.17	1:0.09	1:1	1:0.11	1:1	1:0.03	1:0.03	1:0.04
V15	A[Rcv-11]/1	0	1:0.4	1:0.28	1:0.98	1:0.27	1:0.98	1:0.03	1:0.03	1:0.04

V16 A[Rcv-12]/1	0	1:0.26	1:0.18	1:0.99	1:0.21	1:0.99	1:0.03	1:0.03	1:0.04
V17 A[Rcv-13]/1	0	1:0.48	1:0.5	1:0.7	1:0.51	1:0.71	1:0.03	1:0.03	1:0.04
V18 A[Rcv-14]/1	6	1:0.71	1:0.7	1:1	1:0.7	1:1	1:1	1:1	1:1
V19 A[Trans-01]/1	6	1:0.58	1:0.7	1:0.7	1:0.71	1:0.71	1:1	1:1	1:1
V20 A[Trans-02]/1	6	1:0.83	1:0.88	1:0.98	1:0.89	1:0.99	1:1	1:1	1:1
V21 A[Trans-03]/1	6	1:0.13	1:0.01	1:1	1:0.02	1:1	1:1	1:1	1:1
V22 A[Trans-04]/1	6	1:0.47	1:0.44	1:1	1:0.41	1:1	1:1	1:1	1:1
V23 A[Trans-05]/1	6	1:0.62	1:0.64	1:1	1:0.64	1:1	1:1	1:1	1:1
V24 A[Trans-06]/1	6	1:0.12	1:0.01	1:1	1:0.01	1:1	1:1	1:1	1:1
V25 A[Trans-07]/1	6	1:0.9	1:0.99	1:1	1:0.99	1:1	1:1	1:1	1:1
V26 A[Trans-08]/1	6	1:0.78	1:0.83	1:1	1:0.85	1:1	1:1	1:1	1:1
V27 A[Trans-09]/1	6	1:0.32	1:0.27	1:1	1:0.27	1:1	1:1	1:1	1:1
V28 A[Trans-10]/1	6	1:0.33	1:0.33	1:1	1:0.33	1:1	1:1	1:1	1:1
V29 A[Trans-11]/1	6	1:0.15	1:0.04	1:1	1:0.04	1:1	1:1	1:1	1:1
V30 A[Trans-12]/1	6	1:0.3	1:0.19	1:1	1:0.16	1:1	1:1	1:1	1:1
V31 A[Trans-13]/1	6	1:0.14	1:0.01	1:1	1:0.01	1:1	1:1	1:1	1:1
V32 A[Trans-14]/1	6	1:0.13	1:0.02	1:1	1:0.04	1:1	1:1	1:1	1:1
Average Savings		56.6%	59.0%		60.1%		18.2%	18.1%	18.0%
Standard Deviation		23.9%	30.1%		29.1%		38.5%	38.3%	38.2%

Table 9 Results of MSE and DSE on the Versions of the FABP Example With *Deletion Changes*

		CMLM	SE:MSE					SE:DSE		
Ver.	Change ID/No. of Changed Transitions	No. of Multipath Impacted Transitions	Time	No. of New Symbolic States	No. of Total Symbolic States	No. of New Execution Paths	No. of Total Execution Paths	Time	No. of Total Symbolic States	No. of Total Execution Paths
V1	D[Buff-01]/1	0	1:0.19	1:0.07	1:1	1:0.06	1:1	1:0.08	1:0.07	1:0.08
V2	D[Buff-02]/1	6	1:0.64	1:0.61	1:1	1:0.6	1:1	1:1	1:1	1:1
V3	D[Buff-03]/1	0	1:0.09	1:0	1:1	1:0	1:1	1:0.03	1:0.04	1:0.04
V4	D[Buff-04]/1	6	1:0.18	1:0.06	1:1	1:0.06	1:1	1:1	1:1	1:1
V5	D[Rcv-01]/1	6	1:0.06	1:0	1:1	1:0	1:1	1:1	1:1	1:1
V6	D[Rcv-02]/1	6	1:0.05	1:0	1:1	1:0	1:1	1:1	1:1	1:1
V7	D[Rcv-03]/1	6	1:0.06	1:0	1:1	1:0	1:1	1:1	1:1	1:1
V8	D[Rcv-04]/1	6	1:0.06	1:0	1:1	1:0	1:1	1:1	1:1	1:1
V9	D[Rcv-05]/1	6	1:0.07	1:0	1:1	1:0	1:1	1:1	1:1	1:1
V10	D[Rcv-06]/1	6	1:0.05	1:0	1:1	1:0	1:1	1:1	1:1	1:1
V11	D[Rcv-07]/1	6	1:0.2	1:0.14	1:1	1:0.14	1:1	1:1	1:1	1:1
V12	D[Rcv-08]/1	6	1:0.24	1:0.19	1:1	1:0.19	1:1	1:1	1:1	1:1
V13	D[Rcv-09]/1	6	1:0.57	1:0.65	1:1	1:0.67	1:1	1:1	1:1	1:1
V14	D[Rcv-10]/1	0	1:0.12	1:0.12	1:1	1:0.13	1:1	1:0.03	1:0.04	1:0.04
V15	D[Rcv-11]/1	0	1:0.08	1:0.02	1:1	1:0.02	1:1	1:0.04	1:0.05	1:0.05

V16 D[Rcv-12]/1	0	1:0.2	1:0.2	1:1	1:0.18	1:1	1:0.03	1:0.04	1:0.04
V17 D[Rcv-13]/1	0	1:0.14	1:0.1	1:1	1:0.1	1:1	1:0.05	1:0.05	1:0.06
V18 D[Rcv-14]/1	6	1:0.37	1:0.32	1:1	1:0.32	1:1	1:1	1:1	1:1
V19 D[Trans-01]/1	6	1:0.33	1:0	1:1	1:0	1:1	1:1	1:1	1:1
V20 D[Trans-02]/1	6	1:0.14	1:0	1:1	1:0	1:1	1:1	1:1	1:1
V21 D[Trans-03]/1	6	1:0.11	1:0	1:1	1:0	1:1	1:1	1:1	1:1
V22 D[Trans-04]/1	6	1:0.18	1:0	1:1	1:0	1:1	1:1	1:1	1:1
V23 D[Trans-05]/1	6	1:0.16	1:0.04	1:1	1:0.03	1:1	1:1	1:1	1:1
V24 D[Trans-06]/1	6	1:0.09	1:0	1:1	1:0	1:1	1:1	1:1	1:1
V25 D[Trans-07]/1	6	1:0.29	1:0	1:1	1:0	1:1	1:1	1:1	1:1
V26 D[Trans-08]/1	6	1:0.17	1:0	1:1	1:0	1:1	1:1	1:1	1:1
V27 D[Trans-09]/1	6	1:0.14	1:0.07	1:1	1:0.09	1:1	1:1	1:1	1:1
V28 D[Trans-10]/1	6	1:0.26	1:0.24	1:1	1:0.24	1:1	1:1	1:1	1:1
V29 D[Trans-11]/1	6	1:0.1	1:0	1:1	1:0	1:1	1:1	1:1	1:1
V30 D[Trans-12]/1	6	1:0.13	1:03	1:1	1:02	1:1	1:1	1:1	1:1
V31 D[Trans-13]/1	6	1:0.09	1:0	1:1	1:0	1:1	1:1	1:1	1:1
V32 D[Trans-14]/1	6	1:0.09	1:0	1:1	1:0	1:1	1:1	1:1	1:1
Average savings		82.3%	91.0%		91.1%		18.0%	17.9%	17.8%
Standard deviation		13.9%	16.5%		16.6%		38.0%	37.9%	37.7%

symbolic states, and in the number of execution paths gained from applying MSE (resp., DSE) on the modified versions of the four given models compared to standard SE. We also show, for each such column, the overall average and the standard deviation. Additionally, we show the correlation coefficient with the CIM metric data in column 4 of Tables 4–6 for the individual Statecharts models.

The ratios between the total number of symbolic states (resp., execution paths) found in the resulting SETs generated from standard SE and MSE for the four given models are shown in column 7 (resp., 9) of Tables 4–6 and column 6 (resp., 8) of Tables 7–9. The values recorded in these four columns are computed with "Task 4" of the MSE Algorithms 7 and 8 being enabled and they are used for answering RQ2 with respect to MSE.

Our evaluation is performed on a standard PC with Intel Core i7 CPU 3.4 GHz and 8 GB of RAM and running Windows 7 as a host and Ubuntu 12.04 as VM.

7.4 Results and Analysis

In this section, we discuss the results presented in Tables 4–9 according to our three research questions.

RQ1. How effective are our optimizations compared to standard SE?

In Table 10, we show a summary of the average and the standard deviation values of the data shown in columns 5–6, 8 (resp., 10–12) of Tables 4–6 and columns 4–5, 7 (resp., 9–11) of Tables 7–9 corresponding to the amount of savings in the execution times and in the number of symbolic states and execution paths gained from applying our current implementation of MSE (resp., DSE) on all the versions of the four given models. Based on the values presented in this table, we notice the following:

(1) MSE achieved an average of savings in execution time that ranges from 56.9% (of 9 min) for the ACCS model to 71.7% (of 5 min) for the LGS model, while DSE achieved an average of savings in execution time that approximately ranges from 1.4% (of 9 min) for the ACCS model to 98.4% (of 10 min) for the AQS model. Similarly, MSE achieved an average of savings in the number of symbolic states (resp., execution paths) that ranges from 53.2% (resp., 52.6%) for the ACCS model to 75.5% (resp., 74.9%) for the AQS model, while DSE achieved an average of savings in the number of symbolic states (resp., execution paths) that ranges from 7.0% (resp., 8.6%) for the ACCS model to 95.6% (resp., 96.1%) for the AQS model.

Table 10 Summary of Statistical Measures of the Effectiveness of MSE and DSE for the Four Models: AQS, LGS, ACCS, and FABP With Respect to the Savings in Execution Time and in the Number of Symbolic States and Execution Paths
(a) Average and Standard Deviation of Savings in Time

	AVG				STDEV			
	AQS	**LGS**	**ACCS**	**FABP**	**AQS**	**LGS**	**ACCS**	**FABP**
MSE	71.0%	71.7%	56.9%	66.0%	10.9%	21.6%	24.6%	23.5%
DSE	98.4%	76.7%	1.4%	18.1%	0.2%	2.2%	8.0%	37.9%

(b) Average and Standard Deviation of Savings in Number of Symbolic States

	AVG				STDEV			
	AQS	**LGS**	**ACCS**	**FABP**	**AQS**	**LGS**	**ACCS**	**FABP**
MSE	75.5%	68.2%	53.2%	69.6%	28.0%	30.6%	35.0%	30.1%
DSE	95.6%	67.0%	7.0%	18.0%	4.4%	10.5%	3.1%	37.7%

(c) Average and Standard Deviation of Savings in Number of Execution Paths

	AVG				STDEV			
	AQS	**LGS**	**ACCS**	**FABP**	**AQS**	**LGS**	**ACCS**	**FABP**
MSE	74.9%	67.2%	52.6%	69.3%	28.8%	30.5%	35.1%	30.3%
DSE	96.1%	66.6%	8.6%	18.0%	4.2%	10.4%	3.8%	37.6%

(2) The range of the average values of the achieved savings ratios for DSE is larger than it is for MSE which means that DSE can achieve much higher savings for some models than the others. This shows a higher sensitivity of DSE to the subject models than it is for MSE.

(3) The standard deviation values of the achieved savings ratios for the AQS, LGS, and ACCS models are much higher for MSE than they are for DSE, which means that the effectiveness of MSE for these three models is more influenced by the changes made in each modified model version than it is for the DSE. Interestingly, this observation is reversed for the FABP model.

RQ2. Does the SET generated from MSE match the one generated from standard SE?

The main purpose of this research question is twofold: (1) to ensure the correctness of our MSE algorithms and their implementation and (2) to draw a precise conclusion about the relationship between the SET generated from MSE and the one generated from standard SE. Our initial expectation was that both SETs should be identical; however, the results show that, in some cases, this assumption does not hold.

(1) *For individual Statecharts:* By looking at the ratios recorded in columns 7 and 9 of Table 4–6 and our manual inspection, we notice that MSE generated the exact same SETs as standard SE for all versions of the three models: AQS, LGS, and ACCS.

(2) *For communicating Statecharts:* By looking at the ratios recorded in columns 6 and 8 of Tables 7–9 and our manual inspection, we notice that MSE generated the exact same GSETs as standard SE for 26 of the 32 FABP model versions with "modification" and "addition" changes and all the FABP model versions with "deletion" changes. However, for all other versions, MSE generated slightly different GSETs. Possible reasons for this are that: (1) the subsumption relation that may exist between the nodes of a SET (resp., GSET) is nonsymmetric and (2) modifying an existing SET (resp., GSET) may change the order of the nodes of the tree as well as the subsumption relation between the nodes. As a result, we may have two semantically equivalent SETs with different numbers of nodes for the same model if the order of generating the nodes of the tree is changed. In Fig. 24 we show an example of two differently sized SETs of the same model resulting from changing the order in which the different branches of the tree have been created. It presents a typical scenario that occurs when running the MSE algorithms where manipulating an already existing SET (resp., GSET) takes place to modify some parts of the tree or to add new ones.

Based on the results of this research question, we conclude that it is possible in some cases for MSE to generate a SET (resp., GSET) that does not exactly match the one generated from standard SE, yet they are both semantically equivalent.

RQ3. Which aspects influence the effectiveness of each technique?

We summarize the key findings related to the two aforementioned aspects as follows.

(1) For the change impact metric:
   – Change impact measure criterion (applicable only for individual Statecharts models AQS, LGS and ACCS): According to the correlation coefficients computed between the values of columns 5–6, 8, 10–12 and the values of column 4 of Tables 4–6, we notice that there is high negative correlation between the savings gained from MSE and the CIM metric for the AQS and the LGS models, meaning that

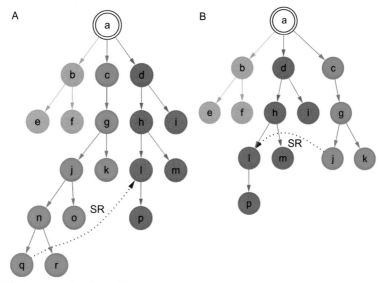

Fig. 24 An example of two differently sized SETs of the same model, each of which has a different branch order. Note the reduction in the number of nodes in the SET in (B) compared to the one in (A) as a result of the subsumption relation (SR) that is detected in the SET in (B) between node "j" and the previously visited node "l" which cannot exist if the creation order of both nodes is reversed as it is the case in the SET in (A). (A) An example SET where the branch leading to node "c" is created before the branch leading to node "d". (B) The SET resulting from modifying the SET in (A) such that the branch leading to node "c" is now created after the branch leading to node "d".

in these cases the percentage of maximal paths affected by the change is a good predictor for the effectiveness of MSE. However, it is much lower for the ACCS model and this is due to the discrepancy between the notions of maximal paths and execution paths. A maximal path in a model may correspond to zero, one, or multiple feasible execution paths in the SET of the model depending on the guard condition and the action code of the transitions in the path. This discrepancy grows with the numbers of parallel transitions in the model (i.e., transitions with the same source and target states and guards but different triggers). The ACCS model contains more of these parallel transitions than the AQS and LGS models. We also notice that there is almost no significant correlation between the savings gained from DSE and the CIM metric.

Table 11 Summary of Statistical Measures of the Effectiveness of MSE and DSE for the FABP Model Per Change Type With Respect to the Savings in Time, in the Numbers of Symbolic States and in the Numbers of Execution Paths

(a) Average and Standard Deviation of Savings in Time

	AVG			STDEV		
	FABP			FABP		
	Modification	Addition	Deletion	Modification	Addition	Deletion
MSE	59.2%	56.6%	82.3%	22.7%	23.9%	13.9%
DSE	18.2%	18.2%	18.0%	38.5%	38.4%	38.0%

(b) Average and Standard Deviation of Savings in Numbers of Symbolic States

	AVG			STDEV		
	FABP			FABP		
	Modification	Addition	Deletion	Modification	Addition	Deletion
MSE	58.9%	59.0%	91.0%	29.8%	30.1%	16.5%
DSE	18.1%	18.1%	17.9%	38.3%	38.3%	37.9%

(c) Average and Standard Deviation of Savings in Numbers of Execution Paths

	AVG			STDEV		
	FABP			FABP		
	Modification	Addition	Deletion	Modification	Addition	Deletion
MSE	58.6%	58.3%	91.1%	29.8%	30.2%	16.6%
DSE	18.0%	18.0%	17.8%	38.1%	38.1%	37.7%

- Type of change criterion (applicable only for communicating Statecharts model FABP): In Table 11, we show a summary of the average and the standard deviation values of the data shown in columns 4–5, 7 (resp., 9–11) of Tables 7–9 corresponding to the amount of savings in the execution times and in the numbers of symbolic states and execution paths gained from applying MSE (resp., DSE) on the different versions of the FABP model per change type. Based on the values presented in this table, we notice that MSE achieved higher average savings for the versions of the FABP model with deletion changes than for the versions the FABP model with modification and addition changes, while DSE achieved similar average savings for the versions of each change type. This indicates that DSE is less influenced by the type of the changes made to the subject model.

(2) For the model characteristics metrics: By looking at the characteristics of the four models in Table 2 and the summary of the average and the standard deviation values of the amount of savings in the numbers of symbolic states gained for the four models in Table 10, we notice the following:

- The differences between the average of savings gained among the four used models are not as significant for MSE as they are for DSE. As a result we could not find a clear relationship between the two defined metrics and the savings gained from MSE.

- DSE achieved the least average savings for the ACCS model. This is because ACCS has the least number of multipath transitions which decreases the opportunity for savings from the partial exploration implemented in DSE.

- DSE achieved much higher average savings for both the AQS and LGS models compared with the FABP model. This is because both AQS and LGS have less dependency between their multipath transitions compared with the FABP model which has the highest number of multipath dependent transitions. As we mentioned earlier, a high dependency between multipath transitions in a model decreases the opportunity for savings from the partial exploration implemented in DSE.

- The standard deviation of the savings gained from DSE for the FABP model is the highest among the other models and it is even higher than the average value of the savings. This is because: (1) 78 of the 96 of the FABP versions have changes in transitions that have dependencies with all the multipath transitions in the model and therefore result in 0% savings with DSE and (2) only 18 of the 96 of the FABP versions have changes in transitions that have no dependencies with any of the multipath transitions in the model and therefore result in 96% savings with DSE.

Based on the key findings of this research question, we conclude that our optimization techniques are complementary in the sense that the effectiveness of MSE depends mostly on the impact and the type of the change, while the effectiveness of DSE depends more on the numbers of transitions with multipath guards or action code as well as the numbers of transitions with dependencies with each other.

7.5 Threats to Validity

Threats to external validity for our evaluation may include (1) the use of KLEE in the implementation of our proposed techniques, (2) the selection of artifacts used in the evaluation, and (3) the types of changes applied to create the modified versions of the used models. Replacing KLEE with any other symbolic execution engine could produce slightly different SETs, depending on the searching strategy employed to explore the different paths of the code been executed, but this should have no impact on the effectiveness of our proposed techniques and rerunning the same evaluation steps with the other symbolic execution engines would mitigate this threat. The artifacts selected for our study are typical behavioral models that can be systematically explored using symbolic execution. The types of changes we used to create the modified versions were selective and may not reflect actual changes; however, rerunning the evaluation on actual model versions would address this threat.

Threats to internal validity may include (1) the originality of the proposed techniques and (2) the correctness of the implementation of our transformation components and proposed algorithms. Although the work presented here is inspired by an already existing work for evolving programs, we are the first to apply it in the context of evolving state-based models. Regarding the correctness of our implementation, we have already tried to address this threat through extensive testing and inspection; more experimentation might allow us to reduce the risk of bugs further.

7.6 Summary

In this section we have provided the evaluation of the proposed optimization techniques. We have identified the research questions and the criteria that the evaluation should investigate and the artifacts that we use to perform the evaluation. Three models of hierarchical individual Statecharts as well as one model of three communicating Statecharts are used to evaluate the optimization techniques for each type. An analysis of the results shows a considerable amount of savings resulting from both optimization techniques. In certain versions the savings have reached up to 90+%. The results of MSE have shown its sensitivity to the location and the type of the change made in the models versions, while the results of DSE have shown its sensitivity to the characteristics of the subject models that are related to the existence of multipath transitions in these models that have few dependencies with other transitions.

8. CONCLUSION: SUMMARY, FUTURE WORK, AND PERSPECTIVE

8.1 Summary and Future Work

In this chapter, we have presented two different techniques for optimizing the symbolic execution of evolving Rhapsody Statecharts. The first technique reuses the SET of a previous version of a model to improve the symbolic execution of the current version such that it avoids redundant exploration of common execution paths between the two versions, whereas the second technique uses a change impact analysis to reduce the scope of the exploration to mainly exercise the parts impacted by the change. A key contribution in this work is a proof-of-concept of the feasibility and the effectiveness of the two proposed optimization techniques such that they can be used to speed up the analysis process of such behavioral models as they evolve. The two techniques proposed in this work are applied on both individual and communicating Statecharts and are implemented and integrated in the IBM Rhapsody tool as a set of three external helpers. The first helper represents the transformation components which are built in the context of IBM Rhapsody RulesComposer Add-On and are responsible for (1) flattening a Rhapsody model into our Communicating Mealy-like machines (CMLM) internal representation, (2) compute the data and the communication dependencies in a CMLM and its MLMs, and (3) generate a Java representation of the CMLM representation. The second helper represents the component that is responsible for: (1) mapping a difference report that is generated from the Rhapsody DiffMerge tool to compare between two successive versions of a Rhapsody model into a difference report between the CMLM representation of each version and (2) identifying the impacted elements of each change found in the difference report. The last helper represents our KLEE-based symbolic execution components including standard SE and the two proposed optimizations: MSE and DSE. We performed an extensive evaluation of the two proposed techniques on a set of three individual hierarchical Statecharts and one model of three communicating Statecharts with a set of 160 different model versions and different types of atomic changes including modification, addition and deletion of the transitions in these models. The results from our experiments show a significant amount of savings up to 90+% in certain scenarios with respect to the size of the SETs generated from applying either technique and the time taken to generate them. The results also highlight the key factors that mostly influence the effectiveness of each technique. These are the change type and location for the MSE technique and the model characteristics concerning the number of multipath transitions in a model and the dependencies between its transitions for the DSE technique.

Some directions for future work include but not limited to the following:

(1) The first direction is to extend the support for more advanced features of action code including more complex data types and user-defined method calls. Both features can be easily added to the current implementation by extending our interface component with the KLEE symbolic execution engine.

(2) Another important direction is to extend the support for some user-defined or domain-specific semantics of specific Statechart features (e.g., priority schema for conflicting transitions and execution order of concurrent states) that is used in practice to override the default one of the tool. To the best of our knowledge, only practitioners who build these models can determine such refinements. In such cases, a modification to our current transformation components should be made.

(3) The third direction is to extend the SET representation to include the test case values of the symbolic variables used to reach a feasible symbolic state such that a complete set of traces can be extracted for a generated SET to be used as a driver for model-based testing and run-time monitoring. This extension is feasible and is supported by the KLEE's "`ktest-tool`" utility that enables the reading of the contents of the "`.ktest`" binary files of the test cases generated by the tool.

(4) The fourth direction is to extend the evaluation of the proposed work to consider more models with actual versions. This extension is limited by the accessibility to existing model repositories.

(5) The fifth direction is to build a SET-based query engine to facilitate user queries and visualization of subparts of the tree. This feature is very important to provide us with a more precise and automated way to inspect some features in a given SET or to compare two SETs.

(6) Finally, the proposed work in this research can be adapted to consider other state-based models used in other MDE tools, especially the open-source ones where the opportunity for extensibility is higher and is not limited by the use of some proprietary tools.

8.2 Perspective: The Road Ahead for MDE

The previous section contains suggestions how the work presented in this paper could be extended. But, the context of our work is MDE, and some more general words about MDE and its challenges and opportunities also seem appropriate.

According to Bran Selic, a key characteristic of traditional engineering disciplines is reliability, encompassing both, the "reliability of designs, that is, confidence that a proposed solution will fulfill its requirements and do so more or less within agreed-on cost, resource, and time constraints" and the "reliable transmission of design intent from specification to the implementation" [98]. The goal of MDE is to bring the reliability, predictability and productivity of traditional engineering to software development. To this end, MDE aims to facilitate the use of abstraction, automation, and analysis for software and system development. More precisely, MDE offers notations, techniques, and tools to capture relevant design information on the most appropriate level of abstraction and achieve productivity and reliability through, e.g., the automatic generation of required related artifacts and automatic analysis capabilities.

MDE thus goes beyond particular forms of notation (graphical versus textual), languages (such as UML), standards (such as OMG's MDA), processes, or tools. Instead, MDE shares with large parts of the computer science community an interest in abstraction [99]. Indeed, key advances in computer science such as virtual memory [100], high-level programming languages, and information hiding [101] rest on abstraction, automation, and analysis, and it is difficult to see how the challenges of future software systems in, e.g., automotive [102] or industrial manufacturing [103] can be met without use of these fundamental techniques.

8.2.1 Challenges

Several reports on the use of MDE in industry exist containing evidence that MDE can improve software quality and developer productivity [104–109]. Despite these successes and its thorough grounding in principles that have served computer science and other engineering disciplines so well, industrial adoption of MDE has been slow.

Selic explains this lack of adoption with cultural, social, and economic factors, and technical issues [98]. He classifies technical issues into capability, scalability, and usability challenges. Examples of capability challenges include (1) a lack of a sufficient theoretical foundation for modeling language design, specification, and semantics, as well as for model transformation and their languages, (2) model synchronization, and (3) model validation. His comments on usability challenges and that MDE tools "are far too complicated for most developers" are echoed by many studies on the use of MDE. But, perhaps somewhat surprisingly, Selic also identifies cultural and social impediments to a more widespread use of MDE such as an inadequate or flawed understanding fostering misconceptions [110], an overly technology-centric mindset which focuses more on the use of certain technologies than the end product and impedes significant technological change, and a lack of system perspective brought on by a lack of abstraction skills.

The extent to which tools affect the success of industrial use of MDE has been studied in [111]. The work concludes that factors to be considered range from technical (e.g., capabilities and properties of tool), to internal organizational (e.g., impact of tool on existing processes, availability of required skills), external organizational (e.g., impact of tool on standards, certification procedures, and interaction with suppliers), and social (e.g., reputation of tool vendor, availability of open user community).

Both papers highlight the need for MDE research on a broad set of topics that goes significantly beyond what is typically considered Computer Science.

8.2.2 Opportunities

Despite these challenges, some recent technological and cultural developments also present significant opportunities to advance the state of the art of MDE in theory and practice [25].

For instance, there has been significant progress in the specification and use of the formal semantics of highly complex artifacts including programming languages [112], operating system kernels [113], and compilers [114]. Combined with the recent advances in constraint solving [115] and synthesis [116, 117], there is some hope that these techniques will help address the capability challenges mentioned in [98] and make the use of descriptions of semantics as well-understood, widespread and useful as descriptions of syntax.

Also, there has been significant industrial backing for the development of open-source modeling tools [118, 119]. Open-source tools have the potential to facilitate the transfer of research problems and results between industry and academia, and the creation of an open community sharing materials and expertise.

As argued above, increasing levels of complexity in industrial software products and the way they are produced, together with novel uses of software will also create more opportunities for MDE. For instance, in its multiannual roadmap SPARC, the largest civilian-funded robotics research and innovation program in the world, has identified MDE as key technology for building robotic software and systems in a way that replaces the current lack of development processes and reliance on "craftmanship" by the adoption of "established engineering processes":

> *"Model based methods are needed at the core of all complex robot systems and through the lifecycle. To address increasing complexity, a shift from human-oriented document-driven approaches to computer-assisted tools and a computer processable model-driven approach is needed in order to gain from design support processes. [...] Model-driven software development and domain-specific languages are core technologies required in order to achieve a separation of roles in the robotics domain while also improving composability, system integration, and also addressing nonfunctional properties" [120].*

Examples of other domains undergoing significant changes many of which involve new uses of software include automotive [102], telecom, and industrial manufacturing [103]. The software systems industry is expecting to be able to build are highly complex and subject to stringent quality requirements. MDE has much to offer to meet these challenges. Indeed, as mentioned earlier, it is unclear how these systems can be built without increasing levels of abstraction, automation, and analysis.

Additional opportunities will be created by, e.g., tightening regulatory requirements for the certification of industrial software [121, 122], and increased use of provenance information to demonstrate the quality and trustworthiness of scientific, industrial, and business data and ensure the repeatability of scientific experiments [123].

Overall, we see significant potential for the continued successful use of MDE and increasing adoption, particularly if the challenges mentioned above can be addressed to a satisfactory extent.

APPENDIX
A.1 Ecore Meta-Model of CMLMs

In this Appendix, we present the Ecore meta-model of our CMLMs. For clarity, we show it in two parts. The first part highlights the elements used to represent individual MLMs. The second part highlights the elements used to represent communicating MLMs.

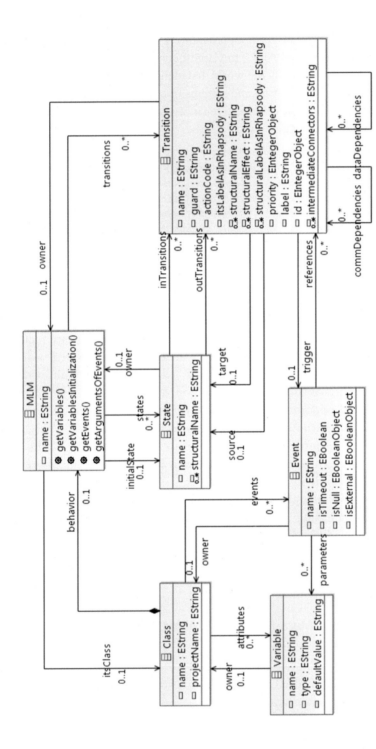

Fig. A.1 CMLMs Ecore meta-model—Part 1/2.

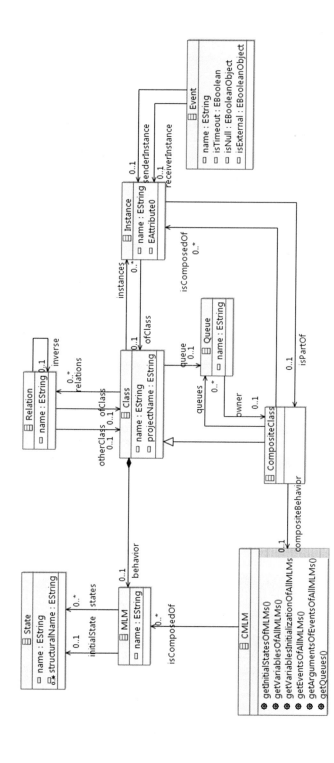

Fig. A.2 CMLMs Ecore meta-model—Part 2/2.

REFERENCES

[1] J.C. King, Symbolic execution and program testing, Commun. ACM 19 (7) (1976) 385–394.

[2] C. Gaston, P. Le Gall, N. Rapin, A. Touil, Symbolic execution techniques for test purpose definition, in: M.Ü. Uyar, A.Y. Duale, M.A. Fecko (Eds.), Testing of Communicating Systems, vol. 3964, Springer, 2006, pp. 1–18.

[3] K. Zurowska, J. Dingel, Symbolic execution of UML-RT state machines, in: 27th Annual ACM Symposium on Applied Computing (SAC'12)ACM, 2012, pp. 1292–1299.

[4] GRAMMATECH CodeSonar, GRAMMATECH CodeSonar, http://www.grammatech.com/codesonar/.

[5] IAR visualSTATE, IAR visualSTATE, https://www.iar.com/iar-embedded-workbench/add-ons-and-integrations/visualstate/.

[6] S. Person, G. Yang, N. Rungta, S. Khurshid, Directed incremental symbolic execution, ACM SIGPLAN Notices 46 (6) (2011) 504–515.

[7] G. Yang, C.S. Păsăreanu, S. Khurshid, Memoized symbolic execution, in: International Symposium on Software Testing and Analysis (ISSTA'12)ACM, 2012, pp. 144–154.

[8] C. Cadar, D. Dunbar, D.R. Engler, et al., KLEE: unassisted and automatic generation of high-coverage tests for complex systems programs, in: 8th USENIX Symposium on Operating Systems Design and Implementation (OSDI'08)8 2008, pp. 209–224.

[9] IBM Rational, Rational Rhapsody Developer, http://www-03.ibm.com/software/products/en/ratirhap/.

[10] D. Harel, Statecharts: a visual formalism for complex systems, Sci. Comput. Program. 8 (3) (1987) 231–274.

[11] L.A. Clarke, A system to generate test data and symbolically execute programs, IEEE Trans. Softw. Eng. 2 (3) (1976) 215–222.

[12] C. Cadar, K. Sen, Symbolic execution for software testing: three decades later, Commun. ACM 56 (2) (2013) 82–90.

[13] S. Khurshid, C.S. Păsăreanu, W. Visser, Generalized symbolic execution for model checking and testing, in: H. Garavel, J. Hatcliff (Eds.), Tools and Algorithms for the Construction and Analysis of SystemsSpringer, 2003, pp. 553–568.

[14] G. Yang, S. Person, N. Rungta, S. Khurshid, Directed incremental symbolic execution, ACM Trans. Softw. Eng. Methodol. 24 (1) (2014) 3.

[15] C. Cadar, D. Engler, Execution generated test cases: how to make systems code crash itself, in: P. Godefroid (Ed.), Model Checking SoftwareSpringer, 2005, pp. 2–23.

[16] P. Godefroid, N. Klarlund, K. Sen, DART: directed automated random testing, ACM SIGPLAN Notices 40 (6) (2005) 213–223.

[17] A. Thums, G. Schellhorn, F. Ortmeier, W. Reif, Interactive verification of statecharts, in: H. Ehrig, W. Damm, J. Desel, M. Große-Rhode, W. Reif, E. Schnieder, E. Westkämper (Eds.), Integration of Software Specification Techniques for Applications in EngineeringSpringer, 2004, pp. 355–373.

[18] M. Balser, S. Bäumler, A. Knapp, W. Reif, A. Thums, Interactive verification of UML state machines, in: J. Davies, W. Schulte, M. Barnett (Eds.), Formal Methods and Software EngineeringSpringer, 2004, pp. 434–448.

[19] K. Androutsopoulos, D. Clark, M. Harman, Z. Li, L. Tratt, Control dependence for extended finite state machines, in: M. Chechik, M. Wirsing (Eds.), Fundamental Approaches to Software Engineering, Springer, 2009, pp. 216–230.

[20] S. Labbé, J.-P. Gallois, Slicing communicating automata specifications: polynomial algorithms for model reduction, Form. Asp. Comput. 20 (6) (2008) 563–595.

[21] V.P. Ranganath, T. Amtoft, A. Banerjee, J. Hatcliff, M.B. Dwyer, A new foundation for control dependence and slicing for modern program structures, ACM Trans. Program. Lang. Syst. 29 (5) (2007) 27.

[22] V. Chimisliu, F. Wotawa, Improving test case generation from UML statecharts by using control, data and communication dependencies, in: 13th International Conference on Quality Software (QSIC), IEEE, 2013, pp. 125–134.

[23] M. Stephan, J.R. Cordy, A survey of model comparison approaches and applications, in: S. Hammoudi, L.F. Pires, J. Filipe, R.C. das Neves (Eds.), Modelsward 2013, pp. 265–277.

[24] S. Maoz, J.O. Ringert, B. Rumpe, A manifesto for semantic model differencing, in: J. Dingel, A. Solberg (Eds.), Models in Software Engineering, Springer, 2011, pp. 194–203.

[25] J. Dingel, Complexity is the only constant: trends in computing and their relevance to model driven engineering, in: R. Echahed, M. Minas (Eds.), International Conference on Graph Transformation, Springer, 2016, pp. 3–18.

[26] A.K. Kolawa, M. Kucharski, System and Method for Detecting Defects in a Computer Program Using Data and Control Flow Analysis, 2011, US Patent 7,900,193.

[27] S. Biswas, R. Mall, M. Satpathy, S. Sukumaran, Regression test selection techniques: a survey, Informatica 35 (3) (2011).

[28] P. Bhaduri, S. Ramesh, Model Checking of Statechart Models: Survey and Research Directions, 2004, arXiv:cs/0407038.

[29] C.S. Păsăreanu, W. Visser, A survey of new trends in symbolic execution for software testing and analysis, Int. J. Softw. Tools Technol. Transfer 11 (4) (2009) 339–353.

[30] S. Maoz, Using model-based traces as runtime models, IEEE Comput. 42 (2009) 28–36.

[31] D. Hearnden, M. Lawley, K. Raymond, Incremental model transformation for the evolution of model-driven systems, in: O. Nierstrasz, J. Whittle, D. Harel, G. Reggio (Eds.), Model Driven Engineering Languages and Systems, Springer, 2006, pp. 321–335.

[32] T.A. Henzinger, R. Jhala, R. Majumdar, M.A. Sanvido, Extreme model checking, in: N. Dershowitz (Ed.), Verification: Theory and Practice, Springer, 2003, pp. 332–358.

[33] G. Yang, M.B. Dwyer, G. Rothermel, Regression model checking, in: IEEE International Conference on Software Maintenance (ICSM 2009), IEEE, 2009, pp. 115–124.

[34] J. Backes, S. Person, N. Rungta, O. Tkachuk, Regression verification using impact summaries, in: E. Bartocci, C.R. Ramakrishnan (Eds.), Model Checking Software, Springer, 2013, pp. 99–116.

[35] D. Felsing, S. Grebing, V. Klebanov, P. Rümmer, M. Ulbrich, Automating regression verification, in: 29th ACM/IEEE International Conference on Automated Software Engineering (ASE'14), ACM, 2014, pp. 349–360.

[36] B. Godlin, O. Strichman, Regression verification, in: 46th Annual Design Automation Conference, ACM, 2009, pp. 466–471.

[37] B. Godlin, O. Strichman, Regression verification: proving the equivalence of similar programs, Softw. Test. Verif. Reliab. 23 (3) (2013) 241–258.

[38] S. Person, M.B. Dwyer, S. Elbaum, C.S. Pasareanu, Differential symbolic execution, in: 16th ACM SIGSOFT International Symposium on Foundations of Software Engineering (FSE'08), ACM, 2008, pp. 226–237.

[39] O. Strichman, B. Godlin, Regression verification—a practical way to verify programs, in: B. Meyer, J. Woodcock (Eds.), Verified Software: Theories, Tools, Experiments, Springer, 2005, pp. 496–501.

[40] G. Yang, S. Khurshid, S. Person, N. Rungta, Property differencing for incremental checking, in: 36th International Conference on Software Engineering, ACM, 2014, pp. 1059–1070.

[41] E. Mikk, Y. Lakhnech, M. Siegel, G.J. Holzmann, Implementing statecharts in PROMELA/SPIN, in: 2nd IEEE Workshop on Industrial Strength Formal Specification Techniques, IEEE, 1998, pp. 90–101.

[42] E. Clarke, W. Heinle, Modular Translation of Statecharts to SMV, Tech. Rep. CMU-CS-00-XXX, Carnegie Mellon University, Pittsburgh, PA 2000.

[43] P. Mehlitz, Trust your model-verifying aerospace system models with java pathfinder, in: Aerospace Conference, IEEE, 2008.

[44] K. Madhukar, R. Metta, P. Singh, R. Venkatesh, Reachability verification of rhapsody statecharts, in: IEEE Sixth International Conference on Software Testing, Verification and Validation Workshops (ICSTW), IEEE, 2013, pp. 96–101.

[45] H. Giese, M. Tichy, S. Burmester, W. Schäfer, S. Flake, ACM, Towards the compositional verification of real-time UML designs, ACM SIGSOFT Softw. Eng. Notes 28 (5) (2003) 38–47.

[46] I. Schinz, T. Toben, C. Mrugalla, B. Westphal, The rhapsody UML verification environment, in: Second International Conference on Software Engineering and Formal Methods (SEFM'04), IEEE, 2004, pp. 174–183.

[47] N.H. Lee, S.D. Cha, Generating test sequences using symbolic execution for event-driven real-time systems, Microprocess. Microsyst. 27 (10) (2003) 523–531.

[48] L. Frantzen, J. Tretmans, T. Willemse, Test generation based on symbolic specifications, in: Formal Approaches to Software Testing, Springer, 2005, pp. 1–15.

[49] T. Jéron, Symbolic model-based test selection, Electron. Notes Theor. Comput. Sci. 240 (2009) 167–184.

[50] C.S. Pasareanu, J. Schumann, P. Mehlitz, M. Lowry, G. Karsai, H. Nine, S. Neema, Model based analysis and test generation for flight software, in: Third IEEE International Conference on Space Mission Challenges for Information Technology, 2009. SMC-IT 2009, IEEE, 2009, pp. 83–90.

[51] S. Lauterburg, A. Sobeih, D. Marinov, M. Viswanathan, Incremental state-space exploration for programs with dynamically allocated data, in: 30th International Conference on Software Engineering (ICSE'08), ACM, 2008, pp. 291–300.

[52] C.L. Conway, K.S. Namjoshi, D. Dams, S.A. Edwards, Incremental algorithms for inter-procedural analysis of safety properties, in: Computer Aided Verification, Springer, 2005, pp. 449–461.

[53] J. Law, G. Rothermel, Incremental dynamic impact analysis for evolving software systems, in: 14th International Symposium on Software Reliability Engineering, 2003. ISSRE 2003, IEEE, 2003, pp. 430–441.

[54] W. Visser, J. Geldenhuys, M.B. Dwyer, Green: reducing, reusing and recycling constraints in program analysis, in: ACM SIGSOFT 20th International Symposium on the Foundations of Software Engineering, ACM, 2012.

[55] X. Jia, C. Ghezzi, S. Ying, Enhancing reuse of constraint solutions to improve symbolic execution, in: International Symposium on Software Testing and Analysis (ISSTA'15)ACM, 2015, pp. 177–187.

[56] A. Aquino, F.A. Bianchi, M. Chen, G. Denaro, M. Pezzè, Reusing constraint proofs in program analysis, in: International Symposium on Software Testing and Analysis (ISSTA'15), ACM, 2015, pp. 305–315.

[57] D. Beyer, P. Wendler, Reuse of verification results, in: E. Bartocci, C.R. Ramakrishnan (Eds.), Model Checking Software, Springer, 2013, pp. 1–17.

[58] D. Beyer, S. Löwe, E. Novikov, A. Stahlbauer, P. Wendler, Precision reuse for efficient regression verification, in: 9th Joint Meeting on Foundations of Software Engineering (ESEC/FSE'13), ACM, 2013, pp. 389–399.

[59] H. Kurshan, R.H. Hardin, R.P. Kurshan, K.L. Mcmillan, J.A. Reeds, N.J.A. Sloane, Efficient regression verification, in: IEE Proceedings of the International Workshop on Discrete Event Systems (WODES'96), 1996, pp. 147–150.

[60] O. Sery, G. Fedyukovich, N. Sharygina, Incremental upgrade checking by means of interpolation-based function summaries, in: Formal Methods in Computer-Aided Design (FMCAD), 2012, IEEE, 2012, pp. 114–121.

[61] J.R. Larus, Whole program paths, ACM SIGPLAN Notices 34 (5) (1999) 259–269.

[62] J. Law, G. Rothermel, Whole program path-based dynamic impact analysis, in: 25th International Conference on Software Engineering (ICSE'03), IEEE, 2003, pp. 308–318.

[63] N. Rungta, S. Person, J. Branchaud, A change impact analysis to characterize evolving program behaviors, in: 28th IEEE International Conference on Software Maintenance (ICSM), 2012, IEEE, 2012, pp. 109–118.

[64] S. Person, N. Rungta, Maintaining the health of software monitors, Innov. Syst. Softw. Eng. 9 (4) (2013) 257–269.

[65] M. Böhme, B.C. d. S. Oliveira, A. Roychoudhury, Partition-based regression verification., in: Proceedings of the 2013 International Conference on Software Engineering, IEEE Press, 2013, pp. 302–311.

[66] C. Ghezzi, Evolution, adaptation, and the quest for incrementality, in: R. Calinescu, D. Garlan (Eds.), Large-Scale Complex IT Systems. Development, Operation and Management, Springer, 2012, pp. 369–379.

[67] C.-M. Huang, J.-M. Hsu, An incremental protocol verification method, Comput. J. 37 (8) (1994) 698–710.

[68] B. Korel, L.H. Tahat, B. Vaysburg, Model based regression test reduction using dependence analysis, in: International Conference on Software Maintenance (ICSM'02), IEEE, 2002, pp. 214–223.

[69] K. El-Fakih, N. Yevtushenko, G. Bochmann, FSM-based incremental conformance testing methods, IEEE Trans. Softw. Eng. 30 (7) (2004) 425–436.

[70] Y. Chen, R.L. Probert, H. Ural, Model-based regression test suite generation using dependence analysis, in: 3rd International Workshop on Advances in Model-based Testing, ACM, 2007, pp. 54–62.

[71] E. Uzuncaova, S. Khurshid, D. Batory, Incremental test generation for software product lines, IEEE Trans. Softw. Eng. 36 (3) (2010) 309–322.

[72] E.J. Rapos, J. Dingel, Incremental test case generation for UML-RT models using symbolic execution, in: 2012 IEEE Fifth International Conference on Software Testing, Verification and Validation (ICST), IEEE, 2012, pp. 962–963.

[73] M. Lochau, I. Schaefer, J. Kamischke, S. Lity, Incremental model-based testing of delta-oriented software product lines, in: A.D. Brucker, J. Julliand (Eds.), Tests and Proofs, Springer, 2012, pp. 67–82.

[74] S. Lity, H. Baller, I. Schaefer, Towards incremental model slicing for delta-oriented software product lines, in: 2015 IEEE 22nd International Conference on Software Analysis, Evolution and Reengineering (SANER), IEEE, 2015, pp. 530–534.

[75] D. Clarke, M. Helvensteijn, I. Schaefer, Abstract delta modeling, ACM SIGPLAN Notices 46 (2) (2011) 13–22.

[76] A. Egyed, Automatically detecting and tracking inconsistencies in software design models, IEEE Trans. Softw. Eng. 37 (2) (2011) 188–204.

[77] X. Blanc, A. Mougenot, I. Mounier, T. Mens, Incremental detection of model inconsistencies based on model operations, in: International Conference on Advanced Information Systems Engineering, Springer, 2009, pp. 32–46.

[78] H. Zheng, Z. Zhang, C.J. Myers, E. Rodriguez, Y. Zhang, Compositional model checking of concurrent systems, IEEE Trans. Comput. 64 (6) (2015) 1607–1621.

[79] S. Burmester, H. Giese, M. Hirsch, D. Schilling, Incremental design and formal verification with UML/RT in the FUJABA real-time tool suite, in: International Workshop on Specification and Validation of UML Models for Real Time and Embedded Systems, SVERTS2004, Satellite Event of the 7th International Conference on the Unified Modeling Language, UML, Citeseer, 2004.

[80] M. Cordy, P.-Y. Schobbens, P. Heymans, A. Legay, Towards an incremental automata-based approach for software product-line model checking, in: 16th International Software Product Line Conference, 2 ACM, 2012, pp. 74–81.

[81] K. Johnson, R. Calinescu, S. Kikuchi, An incremental verification framework for component-based software systems, in: 16th International ACM Sigsoft Symposium on Component-based Software Engineering (CBSE'13), ACM, 2013, pp. 33–42.

[82] J. Hartmann, C. Imoberdorf, M. Meisinger, ACM, UML-based integration testing, ACM SIGSOFT Softw. Eng. Notes 25 (5) (2000) 60–70.

[83] C. Bourhfir, R. Dssouli, E. Aboulhamid, N. Rico, A guided incremental test case generation procedure for conformance testing for CEFSM specified protocols, in: A. Petrenko, N. Yevtushenko (Eds.), Testing of Communicating Systems, Springer, 1998, pp. 279–294.

[84] D.E. Long, Model checking, abstraction, and compositional verification (Ph.D. thesis), Carnegie Mellon University 1993.

[85] I. Kang, I. Lee, State minimization for concurrent system analysis based on state space exploration, in: Ninth Annual Conference on Computer Assurance, 1994. COM-PASS'94 Safety, Reliability, Fault Tolerance, Concurrency and Real Time, Security, IEEE, 1994, pp. 123–134.

[86] P.A. Abdulla, A. Annichini, S. Bensalem, A. Bouajjani, P. Habermehl, Y. Lakhnech, Verification of infinite-state systems by combining abstraction and reachability analysis, in: Computer Aided Verification, Springer, 1999, pp. 146–159.

[87] B. Boigelot, P. Godefroid, Symbolic verification of communication protocols with infinite state spaces using QDDs, Form. Methods Syst. Des. 14 (3) (1999) 237–255.

[88] S.C. Cheung, J. Kramer, Context constraints for compositional reachability analysis, ACM Trans. Softw. Eng. Methodol. 5 (4) (1996) 334–377.

[89] S.C. Cheung, D. Giannakopoulou, J. Kramer, Verification of Liveness Properties Using Compositional Reachability Analysis, in: 6th European Software Engineering Conference Held Jointly with the 5th ACM SIGSOFT Symposium on the Foundations of Software Engineering (ESEC-FSE '97), 1997, Springer, 1997, pp. 227–243.

[90] K. Zurowska, J. Dingel, Symbolic execution of communicating and hierarchically composed UML-RT state machines, in: A.E. Goodloe, S. Person (Eds.), NASA Formal Methods, Springer, 2012, pp. 39–53.

[91] E. Jöbstl, M. Weiglhofer, B.K. Aichernig, F. Wotawa, When BDDs fail: conformance testing with symbolic execution and SMT solving, in: Third International Conference on Software Testing, Verification and Validation (ICST), IEEE, 2010, pp. 479–488.

[92] S. Anand, C.S. Păsăreanu, W. Visser, Symbolic Execution with Abstract Subsumption Checking, in: International SPIN Workshop on Model Checking of Software, Springer, 2006, pp. 163–181.

[93] G. Luo, G.V. Bochmann, A. Petrenko, Test selection based on communicating nondeterministic finite-state machines using a generalized wp-method, IEEE Trans. Softw. Eng. 20 (2) (1994) 149–162.

[94] A. Khalil, J. Dingel, Incremental symbolic execution of evolving state machines, in: 2015 ACM/IEEE 18th International Conference on Model Driven Engineering Languages and Systems (MODELS), IEEE, 2015, pp. 14–23.

[95] A. Khalil, J. Dingel, Incremental Symbolic Execution of Evolving State Machines Using Memoization and Dependence Analysis, Queen's University, Tech. Rep. 2015–623, 2015, pp. 1–42.

[96] A.L.J. Dominguez, Detection of feature interactions in automotive active safety features (Ph.D. thesis), University of Waterloo, 2012.

[97] D. Lee, K.K. Sabnani, D.M. Kristol, S. Paul, Conformance testing of protocols specified as communicating finite state machines—a guided random walk based approach, IEEE Trans. Commun. 44 (5) (1996) 631–640.

[98] B. Selic, What will it take? A view on adoption of model-based methods, Softw. Syst. Model. 11 (2012) 513–526.

[99] J. Kramer, Is abstraction the key to computing? Commun. ACM 50 (4) (2007) 36–42.

[100] P. Denning, Virtual memory, ACM Comput. Surv. 2 (3) (1970) 153–189.
[101] D. Parnas, On the criteria to be used in decomposing systems into modules, Commun. ACM 15 (12) (1972) 1053–1058.
[102] P. Koopman, M. Wagner, Challenges in autonomous vehicle testing and validation, SAE Int. J. Trans. Saf. 4 (1) (2016) 15–24.
[103] S. Lohr, G.E., the 124-Year-Old Software Start-Up, 2016, The New York Times, August 27.
[104] T. Weigert, F. Weil, Practical experience in using model-driven engineering to develop trustworthy systems, in: IEEE International Conference on Sensor Networks, Ubiquitous, and Trustworthy Computing (SUTC'06), IEEE, 2006, pp. 208–217.
[105] S. Kirstan, J. Zimmermann, Evaluating costs and benefits of model-based development of embedded software systems in the car industryresults of a qualitative case study, in: Workshop "From Code Centric to Model Centric: Evaluating the Effectiveness of MDD (C2M: EEMDD)", 2010.
[106] M. Bone, R.R. Cloutier, The current state of model based system engineering: results from the OMGTM SysML request for information 2009, in: 8th Conference on Systems Engineering Research (CSER), 2010.
[107] J. Hutchinson, J. Whittle, M. Rouncefield, S. Kristofferson, Empirical assessment of MDE in industry, in: 33rd International Conference on Software Engineering (ICSE11), ACM, 2011, pp. 471–480.
[108] P. Mohagheghi, W. Gilani, A. Stefanescu, M. Fernandez, B. Nordmoen, M. Fritzsche, Where does model-driven engineering help? Experiences from three industrial cases, Softw. Syst. Model. 12 (3) (2013) 619–639.
[109] G. Liebel, N. Marko, M. Tichy, A. Leitner, J. Hansson, Assessing the state-of-practice of model-based engineering in the embedded systems domain, in: ACM/IEEE 17th International Conference on Model Driven Engineering Languages and Systems (MODELS'14), 2014, pp. 166–182.
[110] M. Petre, "No shit" or "Oh, shit!": responses to observations on the use of UML in professional practice, Softw. Syst. Model. 13 (4) (2014) 1225–1235.
[111] J. Whittle, J. Hutchinson, M. Rouncefield, H. Burden, R. Heldal, A taxonomy of tool-related issues affecting the adoption of model-driven engineering, Softw. Syst. Model. 16 (2) (2017) 313–331.
[112] D. Bogdănaş, G. Roșu, K-Java: a complete semantics of Java, in: ACM SIGPLAN/SIGACT Symposium on Principles of Programming Languages (POPL'15), ACM, 2015, pp. 445–456.
[113] G. Klein, K. Elphinstone, G. Heiser, J. Andronick, D. Cock, P. Derrin, D. Elkaduwe, K. Engelhardt, R. Kolanski, M. Norrish, T. Sewell, H. Tuch, S. Winwood, Formal verification of an OS kernel, in: ACM SIGOPS Symposium on Operating Systems Principles (SOSP'09), ACM, 2009, pp. 207–220.
[114] X. Leroy, Formal verification of a realistic compiler, Commun. ACM 52 (7) (2009).
[115] M. Vardi, The Automated-Reasoning Revolution: From Theory to Practice and Back, 2017, ICSE'17 keynote presentation.
[116] ExCAPE Annual Report of Activities, April 2014 to March 2015, 2015, Available from: https://excape.cis.upenn.edu/about.html.
[117] Syntax-Guided Synthesis Competition (SyGuS'17), July 22, 2017, Heidelberg, Germany.
[118] PolarSys: Open Source Solutions for Embedded Systems, Eclipse Working Group, https://www.polarsys.org.
[119] R. Barrett, F. Bordeleau, 5 Years of 'Papyrusing'—migrating industrial development from a proprietary commercial tool to Papyrus, in: Workshop on Open Source Software for Model Driven Engineering (OSS4MDE'15), 2015, pp. 3–12 (invited presentation).

[120] SPARC, Multi-Annual Roadmap (MAR) for Horizon 2020 Call ICT-2017 (ICT-25, 27 & 28). 2016, Available from: https://www.eu-robotics.net/sparc/newsroom/press/multi-annual-roadmap-mar-for-horizon-2020-call-ict-2017-ict-25-ict-27-ict-28-published.html.

[121] Radio Technical Commission for Aeronautics, DO-178C: Software Considerations in Airborne Systems and Equipment Certification., 2012.

[122] S. Kokaly, R. Salay, M. Sabetzadeh, M. Chechik, T. Maibaum, Model Management for Regulatory Compliance: a position paper, in: 8th Workshop on Models in Software Engineering (MiSE'16), Austin, USA, 2016.

[123] D. Monroe, When data is not enough, Commun. ACM 58 (12) (2015) 12–14.

ABOUT THE AUTHORS

Amal Khalil received her Ph.D. degree in computer science from Queen's University (2016) in the areas of model-driven engineering and model-based verification of behavioral software models in the context of automotive embedded systems. She is a former researcher in the Modeling and Analysis in Software Engineering Group at Queen's University. Amal Khalil was awarded a Silver Medal in the ACM Student Research Competition that was held in conjunction with the 2015 ACM/IEEE 18th International Conference on Model Driven Engineering Languages and Systems (MODELS). Before moving to Canada, Amal Khalil received her Bachelor and M.Sc. degrees in Electrical Engineering.

Juergen Dingel received an M.Sc. from Berlin University of Technology in Germany and a Ph.D. in Computer Science from Carnegie Mellon University (2000). He is Professor in the School of Computing at Queen's University where he leads the Modeling and Analysis in Software Engineering group. His research interests include software modeling, model-driven engineering, analysis of software artifacts, formal specification and verification, and software quality assurance. Juergen was PC co-chair of the ACM/IEEE 17th International Conference on Model Driven Engineering Languages and Systems (MODELS'14) and of the IFIP International Conference on Formal Techniques for Distributed Systems (FMOODS-FORTE'11). He is on the editorial boards of the Springer journals Software and Systems Modeling (SoSyM), and Software Tools for Technology Transfer (STTT) and currently serves as chair of the MODELS Steering Committee.

CHAPTER FIVE

A Tutorial on Software Obfuscation

Sebastian Banescu, Alexander Pretschner
Technische Universität München, München, Germany

Contents

Advances in Computers, Volume 108
ISSN 0065-2458
https://doi.org/10.1016/bs.adcom.2017.09.004

Abstract

Protecting a digital asset once it leaves the cyber trust boundary of its creator is a challenging security problem. The creator is an entity which can range from a single person to an entire organization. The trust boundary of an entity is represented by all the (virtual or physical) machines controlled by that entity. Digital assets range from media content to code and include items such as: music, movies, computer games, and premium software features. The business model of the creator implies sending digital assets to end-users—such that they can be consumed—in exchange for some form of compensation. A security threat in this context is represented by malicious end-users, who attack the confidentiality or integrity of digital assets, in detriment to digital asset creators and/or other end-users. Software obfuscation transformations have been proposed to protect digital assets against malicious end-users, also called Man-At-The-End (MATE) attackers. Obfuscation transforms a program into a functionally equivalent program which is harder for MATE to attack. However, obfuscation can be use both for benign and malicious purposes. Malware developers rely on obfuscation techniques to circumvent detection mechanisms and to prevent malware analysts from understanding the logic implemented by the malware. This chapter presents a tutorial of the most popular existing software obfuscation transformations and mentions published attacks against each transformation. We present a snapshot of the field of software obfuscation and indicate possible directions, which require more research.

1. INTRODUCTION

A common business model for commercial media content and software creators is to distribute digital assets (e.g., music, movies, proprietary algorithms in software executables) to end-users, in exchange for some form of compensation. Even with ubiquitous cloud-based services, digital asset creators still need to ship media content and client applications to end-users. For both performance and scalability reasons, software developers often choose to develop *thick client* applications, which contain sensitive code and/or data. For example, games or media players are often thick clients offering premium features or content, which should only be accessible if the end-user pays a license fee. Sometimes, the license is temporary and therefore the client software should somehow restrict access to these features and content once the license expires. Moreover, some commercial software developers also want to protect secret algorithms used in their client software, which give them an advantage over their competitors.

One open challenge in IT security is protecting digital assets once they leave the cyber trust boundary of their creator. The creator of digital assets can range from a single person to an organization. The security threat in this context is represented by malicious end-users, who among other things, may want to:

- Use digital assets without paying the license fees required by the creators of that digital asset.
- Redistribute illegal copies of digital assets to other end-users, sometimes in order to make a profit.
- Make changes to the digital assets (e.g., by tampering with its code), in order to modify its behavior.

Such malicious end-users are also called Man-At-The-End (MATE) attackers [1], and they have control of the (physical or virtual) machine where the digital asset is consumed. Practically, any device under the control of an end-user (e.g., PC, TV, game console, mobile device, smart meter) is exposed to MATE attacks. A model of the MATE attacker capabilities, akin to the degree of formalization of the Man-In-The-Middle (MITM) attacker introduced by Dolev-Yao [2], is still missing from the literature. However, MATE attackers are assumed to be extremely powerful. They can examine software both statically using manual or automatic static analysis, or dynamically using state-of-the-art software decompilers and debuggers [3]. Shamir et al. [4] present a MATE attack, which can retrieve a secret key used by a black-box cryptographic primitive to protect the system if it is stored somewhere in non/volatile memory. Moreover, the memory state can be inspected or modified during program execution and CPU or external library calls can be intercepted (forwarded or dropped) [5]. Software behavior modifications can also be performed by the MATE attacker by tampering with instructions (code) and data values directly on the program binary or after they are loaded in memory. The MATE attacker can even simulate the hardware platform on which software is running and alter or observe all information during software operation [6]. The only remaining line of defense in case of MATE attacks is to increase the complexity of an implementation to such an extent that it becomes economically unattractive to perform an attack [6].

Researchers, practitioners, and law makers have sought several solutions for this challenge, all of which have their advantages and disadvantages. Fig. 1 shows a classification of these solutions, proposed by Collberg et al. [7]. On the one hand, there are legal protection frameworks that apply to some geographic regions, such as the Digital Millennium Copyright Act [8] in the United States, the EU Directive 2009/24/EC [9], etc. On the other hand, there are technical protection techniques (complementing legal

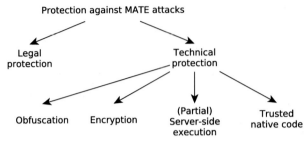

Fig. 1 Classification of protections against MATE attacks proposed in [7].

protection), which are divided into four subcategories, namely: (1) software based obfuscation, (2) encryption (via trusted hardware), (3) server-side execution, and (4) trusted (i.e., tamper-proof or tamper-evident) native code. The latter three subcategories will be briefly discussed in the related work section. The *obfuscation* subcategory is the main focus, i.e., software-only protection that does not rely on trusted entities.

An obfuscator is in essence a compiler that takes a program as input, and outputs a functionally equivalent program, which is harder to understand and analyze than the input program. The meaning of the phrases "functionally equivalent" and "harder to understand and analyze" will be discussed in this chapter. For instance, some classical compiler optimizations are also considered obfuscation transformations [7], because in order to make the code more efficient, such optimizations may replace control-flow abstractions that are easy to understand by developers (e.g., loops), with other constructs which are less straightforward (e.g., goto statements).

This chapter presents a tutorial of several popular obfuscation transformations together with illustrative examples. It also mentions the MATE attacks published in the literature, which have been proposed for defeating each obfuscation transformation. The rest of the chapter is structured as follows. Section 2 presents classification dimensions for obfuscation transformations. Section 3 presents classification dimensions for MATE attacks. Section 4 presents a survey of obfuscation transformations and state-of-the-art MATE attacks that claim to break each obfuscation. Section 5 discusses the current state of software obfuscation vs MATE attacks. Section 6 presents related work, and Section 7 concludes the chapter.

2. CLASSIFICATION OF CODE OBFUSCATION TRANSFORMATIONS

Several surveys and taxonomies for software obfuscation have been proposed in literature [7, 10–13]. This section describes the classification dimensions presented in those works and discusses their advantages, disadvantages, and overlaps. We present the classification dimensions in increasing order of importance, starting with the least important category.

2.1 Abstraction Level of Transformations

One common dimension of code transformations is the level of abstraction at which these transformations have a noticeable effect, i.e., *source code, intermediate representation*, and *binary machine code*. Such a distinction is relevant for usability purposes, e.g., a JavaScript developer will mostly be interested in source code-level transformations and a C developer will mainly be interested in binary level. However, none of the previously mentioned taxonomies and surveys classifies transformations according to the abstraction level. This is due to the fact that some obfuscation transformations have an effect at multiple abstraction levels. Moreover, it is common for papers to focus only on a specific abstraction level, disregarding transformations at other levels.

2.2 Unit of Transformations

Larsen et al. [12] proposed classifying transformations according to the of granularity at which they are applied. Therefore they propose the following levels of granularity:

- *Instruction-level* transformations are applied to individual instructions or sequences of instructions. This is due to the fact that at the source code level, a code statement can consist of one or more IR or Assembly instructions.
- *Basic block-level* transformations affect the position of one or more basic blocks. Basic blocks are a list of sequential instructions that have a single entry point and end in a branch instruction.
- *Loop-level* transformations alter the familiar loop constructs added by developers.
- *Function-level* transformations affect several instructions and basic blocks of a particular subroutine. Moreover, they may also affect the stack and heap memory corresponding to the function.
- *Program-level* transformations affect several functions inside an application. However, they also affect the data segments of the program and the memory allocated by that program.
- *System-level* transformations target the operating system or the runtime environment and they affect how other programs interact with them.

The unit of transformation is important in practice because developers can choose the appropriate level of granularity according to the asset they must protect. For example, loop-level transformations are not appropriate for hiding data, but they are appropriate for hiding algorithms. However, the same problem, as for the previous classification dimension, arises for the unit of transformation, namely, the same obfuscation transformation may be applicable to different units of transformation.

2.3 Dynamics of Transformations

The dynamics of transformation—used by Schrittwieser et al. [13]—indicate whether a transformation is applied to the program or its data *statically* or *dynamically*. Static transformations are applied once during: implementation, compilation, linking, installation, or update, i.e., the program and its data do not change during execution. Dynamic transformations are applied at the same times as static transformations, however, the program or its data also change during loading or execution, e.g., the program could be decoded at load time, because it was encoded on disk. Even though dynamic code transformations are generally considered stronger against MATE attacks than static ones, they require the code pages to be both writable and executable, because the code may modify itself during execution. This opens the door for remote attacks (e.g., code injection attacks [14]), which are more dangerous for end-users than MATE attacks. Moreover, dynamic transformations generally have a higher performance overhead than static transformations, because code has to first be written (generated or modified) and then executed. Therefore, on the one hand, many benign software developers avoid dynamic transformations entirely. On the other hand, dynamic transformations are heavily used by malware developers, because they are not generally concerned about high performance overhead.

2.4 Target of Transformations

The most common dimension for classifying obfuscation transformations is according to the target of transformations. This dimension was first proposed by Collberg et al. [7], who indicated four main categories: layout, data, control, and preventive transformations. In a later publication Collberg and Nagra [15] refined these categories into four broad classes: abstraction, data, control and dynamic transformations. Since the last class of Collberg and Nagra [15] (i.e., dynamic transformations), overlaps with the dynamics of transformation dimension, described in Section 2.3, we will use a simplification of these two proposals where we remove the dynamic transformations class and merge the abstraction, layout, and control classes. Therefore, the remaining transformation targets are:

- *Data transformations*, which change the representation and location of constant values (e.g., numbers, strings, keys) hard-coded in an application, as well as variable memory values used by the application.
- *Code transformations*, which transform the high-level abstractions (e.g., data structures, variable names, indentation) as well as the algorithm and control-flow of the application.

This dimension is important for practitioners, because it indicates the goal of the defender, i.e., whether the defender wants to protect data or code. Note that obfuscation transformations which target data may also affect the layout of the code and its control-flow, however, their *target* is hiding data, not code. In practice data transformations are often used in combination with code transformations, to improve the resilience of the program against MATE attacks.

2.4.1 Data Transformations

Data transformations can be divided into two subcategories:

1. *Constant data* transformations, which affect static (hard-coded) values. Abstractly, such transformations are encoding functions which take one or more constant data items i (e.g., byte arrays, integer variables), and convert them into one or more data items $i' = f(i)$. This means that any value assigned to, compared to and based on i is also changed according to the new encoding. There will be a trade-off between resilience on one hand, and cost on the other, because all operations performed on i require computing $f^{-1}(i)$, unless f is homomorphic w.r.t. those operations.

2. *Variable data* transformations, which modify the representation or structure of variable memory values. The goal of such transformations is to hamper the development of automated MATE attacks, which can assume that a certain variable memory value will always have a certain representation or a certain structure (e.g., an integer array of 100 contiguous elements). If such assumptions hold then automated attacks are easier to develop, because they mainly rely on pattern matching. Therefore, by employing the transformations presented in this section, one can raise the bar for these kinds of attacks.

Note that variable transformations are not the transformations that affect the names or the order of variables, but the representation of the data stored in those variables.

2.4.2 Code Transformations

Code transformations can also be divided into two subcategories, namely:

1. *Code logic* transformations, which affect the control-flow of the program, its ease of readability and maintainability. Such transformations are aimed at adding complexity to the code in order to prevent MATE attackers from understanding what the algorithm(s) implemented by the code are.
2. *Code abstraction* transformations, which destroy programming abstractions, are generally the first hints that MATE attackers use to start reverse engineering a program.

As opposed to *constant data transformations*, which in some cases perform one-way mappings of data to a completely different domain which requires no interpretation, code transformations must always perform a mapping to the same domain of executable code, because obfuscated code must be able to run on the underlying machine where it is installed.

2.5 Summary of Obfuscation Transformation Classification

Table 1 provides a summary of the classification dimensions described above along with the possible discrete values that each dimension can take. In the

Table 1 Classification Dimensions for Obfuscation Transformations

Dimension	Possible Values
Abstraction level	Source code
	Intermediate representation
	Binary machine code
Unit	Instruction
	Basic block
	Loop
	Function
	Program
	System
Dynamics	Static
	Dynamic
Target	Constant data
	Variable data
	Code logic
	Code abstraction

next section we choose to present a survey of obfuscation transformations classified according to their *target of transformation*, because it entails a clear partition of transformations.

3. CLASSIFICATION OF MATE ATTACKS

In contrast to the classification of software protection techniques (see Section 2), the classification of MATE attacks has been the topic of relatively few publications [1, 13, 16]. This section presents the most relevant classification dimensions of MATE attacks.

3.1 Attack Type Dimension

Basile et al. [17] argue that it is not feasible to consider every possible attacker goal since it represents the desired end-result for the attacker, e.g., see the position of other players in computer games, or play premium content without paying. Therefore, in this chapter we classify attacks according to their type, i.e., the means through which an attacker goal can be achieved. According to Collberg et al. [1, 18], there are four types of information a MATE attacker may be interested in recovering from an obfuscated program:

- The original or a simplified version of the *source code*. This is always the case for MATE attackers who are interested in intellectual property theft, i.e., stealing a competitor's algorithm.
- A statically embedded or dynamically generated *data item*. Common examples of such data items are decryption keys used by Digital Rights Management (DRM) technologies to play premium content only on authorized devices, for authorized users. However, data items may also include hard-coded passwords, IP addresses, etc.
- The sequence of obfuscation transformations and/or tools used to obfuscate (protect) the program, also called *metadata*. This kind of information is often used by antivirus engines to detect suspicious binaries, based on the fact that several previously seen malware have used the same obfuscation transformations and/or tools.
- The *location* (i.e., lines of code or bytes) of a particular function of the code. For instance, the attacker may be interested in the module which performs premium content decryption in order to copy it and reuse it in another program, without necessarily understanding how it works.

We compare these four information types proposed by Collberg et al. [1, 18], with the *analyst's aims* proposed by Schrittwieser et al. [13]:

- *Code understanding*, which according to its description in the paper maps to the *source code* information type described above. However, the name of this category suggests a more general attack type than a full recovery of the entire *source code*, because it could be sufficient to have a partial code understanding. For example, a malware analysis engine can decide that a software is malicious using its observed behavior (e.g., unsolicited calls to premium telephone numbers), which does not require full understanding of the source code. Moreover, *metadata* recovery also falls inside of this category of *code understanding*. Therefore, in this chapter we will use this more general category, i.e., *code understanding*.

- *Finding the location of data*, which maps perfectly onto the *data item* information type described above. However, the phrase *location of data* may be mistaken for the *location* information type. Therefore, this chapter will simply use *data item recovery*.
- *Finding the location of program functionality*, which maps onto the *location* information type described above. However, Schrittwieser et al. [13] also add that this type of information may be used to answer questions regarding if the program is malicious or not. In this chapter we move such questions to the *code understanding* category, because we believe an answer to such a question, requires some level of code understanding, but not necessarily recovering the entire *source code*.
- *Extraction of code fragments*, which does not directly map onto any of the information items described above. However, we believe that *finding the location of program functionality* is a prerequisite to this aim of the analyst, because extraction can only be done after the location of the code fragment has been recovered. Therefore, in this chapter we associate this aim of the analyst with the *location* information type.

In sum, we use the following three categories of attack types with the meanings discussed above: (1) *code understanding*, (2) *data item recovery*, and (3) *location recovery*.

3.2 Dynamics Dimension

Dynamics is one of the most commonly used classification criteria for automated MATE attacks and it refers to whether the attacked program is executed on a machine or not, i.e.:

- *Static analysis* attacks do not execute the program on the underlying (physical or virtual) machine. The subject of the analysis is the static code of the program.
- *Dynamic analysis* attacks run the program and record executed instructions, function calls and/or memory states during execution, which are the subject of analysis.

Static attacks are commonly faster than dynamic attacks. However, static analysis attacks do not handle many code obfuscation transformations as well as dynamic analysis attacks. This does not mean that dynamic analysis attacks can handle any kind of obfuscation easily. For instance, it is challenging to dynamically analyze programs employing code transformation techniques that achieve *temporal diversity*, i.e., the program has significantly different execution traces and/or memory states on every execution. Moreover, dynamic analysis is in general incomplete, i.e., it cannot explore or reason about all possible executions of a program, as opposed to static analysis.

3.3 Interpretation Dimension

Code interpretation refers to whether the program's code or the artifacts generated using it (e.g., static disassembly, dynamic traces) are treated as text or are interpreted according to a semantic meaning (e.g., operational semantics). Therefore the two types of code interpretation considered in this chapter are

- *Syntactic attacks* which treat the program's code or any other artifacts generated by executing or processing it, as a string of bytes (e.g., characters). For example, pattern matching on static code [19] and pattern recognition via machine learning traces of instructions [20], treat the code as a sequence of bytes.
- *Semantic attacks* which interpret the code according to some semantics, e.g., denotational semantics, operational semantics, axiomatic semantics, and variations thereof [21]. For example, abstract interpretation [22] uses denotational semantics, while fuzzing [23] uses operational semantics.

Syntactic attacks are generally faster than semantic attacks due to the missing layer of abstraction that interprets the code. Coincidentally, most syntactic attacks are performed via static analysis and most semantic attacks are performed via dynamic analysis. However, there are exceptions, e.g., the syntactic analysis of dynamically generated execution traces and semantic static analysis via abstract interpretation [22].

3.4 Alteration Dimension

Alteration refers to whether the automated MATE attack changes (alters) the code or not. This type of classification is analogous to the *message alteration* classification of MITM attacks on communication channels. Hence there are two types of code alteration:

- *Passive attacks* do not make any changes to the code or data of the program. For instance, extracting a secret key or password from a program does not require any code alterations.

Table 2 Classification Dimensions for Automated MATE Attacks

Dimension	Possible Values
Attack type	Code understanding
	Data recovery
	Location recovery
Dynamics	Static
	Dynamic
Code interpretation	Syntactic
	Semantic
Alteration	Passive
	Active

- *Active attacks* make changes to the code or data of a program. For example, removing data or code integrity checks (e.g., password checks) require modifying the code of the program. Also "disarming" malware may also involve tampering with its code.

If an attacker's goal can be achieved via either passive or active attacks, then the type of attack used depends on the complexity of the program under attack and the types of protection the program has in place. For instance, if a program is protected via dynamically verified checksums of the program input, then active attacks require finding and disabling these checksumming instructions, which could be more costly than a passive attack.

3.5 Summary of MATE Attack Classification

Table 2 provides a summary of the classification dimensions described above along with the possible discrete values that each dimension can take. In the remainder of this chapter we will refer to these dimensions when describing attack implementations.

4. SURVEY OF OBFUSCATION TRANSFORMATIONS

Obfuscation transformations can be implemented in different ways, i.e., the obfuscation transformation gives only a high-level description (e.g., pseudo-code) and it leaves it up to the obfuscation engine developer to take concrete implementation decisions. This section provides a description of the abstract idea behind common obfuscation transformations, it does not focus on any particular implementation of an obfuscation engine. This state-of-the-art survey groups obfuscation transformation techniques according to their target of transformation, namely, data and code. We also give examples using code snippets written in C, JavaScript, and Assembly language, to illustrate the transformations.

4.1 Constant Data Transformations

4.1.1 Opaque Predicates

Collberg et al. [7] introduce the notion of *opaque predicates*. The truth value of these opaque predicates is invariant w.r.t. the value of the variables which comprise it, i.e., opaque predicates have a value which is fixed by the obfuscator, e.g., the predicate $x^2 + x \equiv 0$ (mod 2) is always true. However, this property is hard for the attacker to deduce statically. Collberg et al. [7] also present an application of opaque predicates, which is called *extending loop condition*. This is done by adding an opaque predicate to loop conditions, which does not change the value of the loop condition, but makes it harder for an attacker to understand when the loop terminates.

Opaque predicates can be created based on mathematical formulas which are hard to solve statically, but they can also be built using any other problem which is difficult to compute statically, e.g., aliasing. Aliasing is represented by a state of a program where a certain memory location is referenced by multiple symbols (e.g., variables) in the program. Several works in literature show that pointer alias analysis (i.e., deciding at any given point during execution, which symbols may alias a certain memory location) is undecidable [24–26]. Therefore, Collberg et al. [7] propose to leverage this undecidability result to build opaque predicates using pointers in linked lists. For instance, consider the linked list illustrated in the top-left part of Fig. 2. This circular list consists of four elements and it has two pointers (i.e., q_1 and q_2) referencing its elements. After performing three list operations, i.e., inserting another list element (top-right part of Fig. 2), splitting the list in

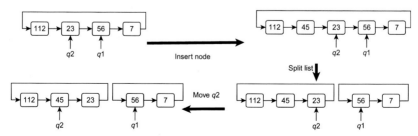

Fig. 2 Opaque expressions based on linked lists.

two parts (bottom-right) and then moving the pointer q_2 two elements forward (bottom-left), the obfuscator knows that the element referenced by q_1 is larger than the element referenced by q_2. However, this relation is hard to determine using static analysis techniques, therefore $q_1 > q_2$ represents an opaque predicate, which is always true. Wang et al. [27] employ such opaque expressions to hide code pointer values, hence, obfuscating control flow via data obfuscation.

One extension of opaque predicates was made by Palsberg et al. [28], who propose *dynamic opaque predicates* which change their truth values between different runs of the program. A further extension appeared in the work of Majumdar and Thomborson [29], who proposed *distributed opaque predicates* which change their truth values during the same execution of a program, depending on the location in code, where they are evaluated. This means that the values of the opaque predicate changes during execution of a program due to values being sent and received from other programs in a distributed system.

Attacks: Due to their popularity, opaque predicates have been the target of several MATE attacks published in the literature. Dalla Preda et al. [30] propose a location and data recovery attack which based on abstract interpretation [22], hence it is a static, semantic, and passive attack. Banescu et al. [31, 32] propose a data recovery and code understanding attack based on symbolic execution, which is dynamic, semantic and passive. This attack also aimed other obfuscation transformations which will be presented later in this section such as: converting static data to procedural data, encoding arithmetic, inserting dead code, virtualization, and control flow flattening. Salem and Banescu [33] propose a code understanding attack in order to identify which programs have been obfuscated using opaque predicates and which have been obfuscated using the same obfuscation transformations enumerated for the previous attacks, plus *program encoding*, which is presented later in this section.

Listing 1: Code Before Encode Literals

```
1 int main(int ac, char* av []) {
2   int a = 1;
3   // do stuff
4   return 0;
5 }
```

Listing 2: Code After Encode Literals

```
1 int main(int ac, char* av []) {
2     double s = sin(atof (av [1]));
3     double c = cos(atof (av [1]));
4     int a = (int) (s * s + c * c);
5     // do stuff
6     return 0;
7 }
```

4.1.2 Convert Static Data to Procedural Data (a.k.a. Encode Literals)

A simple way of obfuscating a hard-coded constant is to convert it into a function (program) that produces the constant at runtime [7]. This transformation implies choosing an invertible function (program) f, feeding the constant to f as input and storing the output. During runtime the inverse of that function, i.e., f^{-1} is applied to the output of f which was stored somewhere in the program. Obfuscating a hard-coded constant value (e.g., 5), by using simple encoding functions (e.g., $f(i) = a \cdot i + b$), leads to small execution overheads. However, since i is a constant, such functions can also be deobfuscated using compiler optimizations such as constant folding [34]. Therefore, another way of hiding constants is to build expressions dependent on external variables (e.g., user input). For instance, *opaque expressions*—similar to opaque predicates except that their value is non-Boolean—always have a certain fixed value during program execution, e.g., $cos^2(x) + sin^2(x)$ is always equal to 1, regardless of the value of x. Therefore, the constant value 1 from the C code from Listing 1 can be encoded using this opaque expression, which cannot be simplified away by the compiler. The resulting code after this obfuscation is shown in Listing 2. This transformation can also be applied to string constants, which can be split into substrings or even single characters, which can be interpreted as integers. At runtime these substrings or characters would be concatenated in the right order to form the original string.

Attacks: This obfuscation transformation has been successfully attacked by the two semantic attacks of Banescu et al. [31, 32] based on symbolic execution and the attack of Salem and Banescu [33] based on pattern recognition, mentioned in the attacks on opaque predicates.

4.1.3 Mixed Boolean-Arithmetic

Zhou et al. [35] propose a data encoding technique called Mixed Boolean-Arithmetic (MBA). MBA encodes data using linear identities involving Boolean and arithmetic operations, together with invertible polynomial functions. The resulting encoding is made dependent on external inputs such that it cannot be deobfuscated using compiler optimization techniques. The following example is taken from [35] and it aims to encode an integer value $k = 0x87654321$. The example gives k as an input to the following second degree polynomial with coefficients in $\mathbb{Z}/(2^{32})$:

$$f(x) = 727318528x^2 + 3506639707x + 6132886 \ (\text{mod } 2^{32}).$$

The output of computing $f(k)$ is 1704256593. This value can be inverted back to the value of k during runtime by using the following polynomial:

$$f^{-1}(x) = 1428291584x^2 + 1257694419x + 4129091678 \ (\text{mod } 2^{32}).$$

Listing 3: Hiding the Value of $k = 0x876554321$ Using Mixed Boolean-Arithmetic

```
1 int main(int argc, char* argv []) { // compiled on a 32-bit
    architecture
2 int x = atoi (argv [1]);
3 int x1 = atoi (argv [2]);
4 int x2 = atoi (argv [3]);
5
6 int a = x*(x1 | 3749240069);
7 int b = x*((-2*x1 - 1) | 3203512843);
8 int d = ((235810187*x + 281909696- x2) ^ (2424056794+x2));
9 int e = ((3823346922*x + 3731147903+2*x2)
    | (3741821003 + 4294967294*x2));
10
11 int k = 135832444*d + 4159134852*e + 272908530*a + 409362795*x
    + 136454265*b + 2284837645 + 415760384*a*b + 2816475136*a*d
    + 1478492160*a*e + 3325165568*b*b + 2771124224*b*x
    + 1247281152*a*x + 1408237568*b*d + 2886729728*b*e
    + 4156686336*x*x + 4224712704*x*d + 415760384*a*a
    + 70254592*x*e + 1428160512*d*d + 1438646272*d*e
    + 1428160512*e*e;
12 // do stuff
13 return 0;
14 }
```

Zhou et al. [35] describe how to pick such polynomials and how to compute their inverse. Since the polynomial $f^{-1}(x)$ does not depend on program inputs and the value of $f(k)$ is hard-coded in the program, an attacker can retrieve the value of k by using constant propagation. In order to create a dependency of $f^{-1}(k)$ on program inputs, the following Boolean-arithmetic identity is used

$$2y = -2(x \vee (-y-1)) - ((-2x-1) \vee (-2y-1)) - 3.$$

This identity makes the computation of a constant value (i.e., $2y$, which is the value of $f(k)$ in the running example), dependent on a program input value, i.e., x. Note that this relation can be applied multiple times for different program inputs. The resulting Boolean-arithmetic relation is further obfuscated by applying the following identity:

$$x + y = (x \oplus y) - ((-2x-1) \vee (-2y-1)) - 1.$$

Making the computation of $f^{-1}(k)$ dependent on three 32-bit integer input arguments of the program and applying the second Boolean-arithmetic relation multiple times gives the code in Listing 3, which dynamically computes the original value of $k = 0x87654321$. Note that in Listing 3, variables a, b, d, and e are input dependent, common sub-expressions of the MBA expression of k.

Attacks: Guinet et al. [36] propose a static analysis tool called Arybo, which is able to simplify MBA expressions. Their attack is meant for code understanding and data recovery and it uses code semantics to achieve this goal.

4.1.4 White-Box Cryptography

This transformation was pioneered by Chow et al. [37, 38], who proposed the first white-box data encryption standard (WB-DES) and white-box advanced encryption standard (WB-AES) ciphers in 2002. The goal of white-box cryptography (WBC) is the secure storage of secret keys (used by cryptographic ciphers), in software, without hardware keys or trusted entities. Instead of storing the secret key of a cryptographic cipher separately from the actual cipher logic, WBC embeds the key inside the cipher logic. For instance, for Advanced Encryption Standard (AES) ciphers, the key can be embedded by multiplication with the T-boxes of each encryption round [39]. However, simply embedding the key in the T-boxes of AES is prone to key extraction attacks since the specification of AES is publicly known. Therefore, WB-AES implementations use complex techniques to prevent key extraction attacks, e.g., wide linear encodings [40], perturbations to the cipher equations [41], and dual-ciphers [42].

Listing 4: 8-Bit XOR With MSB of Secret Key

```
1 char xor (char input) {
2   return input ^ 0x53;
3 }
```

Listing 5: LUT-Based 8-Bit XOR

```
1 char lut [256] = {
2   0x53, 0x52, 0x51, . . ., 0x5C,
3   0x43, 0x42, 0x41, . . ., 0x4C,
4   0x73, 0x72, 0x71, . . ., 0x7C,
5   |                        |
6   0xA3, 0xA2, 0xA1, . . ., 0xAC
7 };
8
9 char xor (char input) {
10   return lut [input];
11 }
```

The idea behind the white-box approach in [38] is to encode the internal AES cipher logic (functions) inside look-up tables (LUTs). One extreme and impractical instance of this idea is to encode all plaintext–ciphertext pairs corresponding to an AES cipher with a 128-bit key, as a LUT with 2^{128} entries, where each entry consists of 128-bits. Such a LUT would leak no information about the secret-key but exceed the storage capacity of currently available devices. However, this LUT-based approach also works for transforming internal AES functions (e.g., XOR functions, AddRoundKey, SubBytes, and MixColumns [39]) to table lookups, which can be divided such that they have a smaller input and output size. For instance, Listing 4 shows the implementation of a 8-bit XOR gate, which takes one byte value as an input argument and outputs the bitwise-XOR of this value and the most significant byte (MSB) of the secret key of the AES cipher instance, which is 0x53 in Listing 4. This function can easily be converted to a LUT-based implementation—as illustrated in Listing 5—by constructing a LUT containing all input-output combinations for the 8-bit XOR function from Listing 4. Therefore, this LUT contains 256 elements and the LUT-based version of the XOR function simply requires a look-up in this table, as shown on line 10 of Listing 5.

LUTs can also be used to encode random invertible bijective functions, which are used to further obfuscate the LUTs representing internal AES functions. This is necessary because an attacker could extract the MSB of the secret key from the LUT in Listing 5. In order to hide the key, WBC proposes to generate a random permutation of 256 bytes and apply it to the LUT. To be able to use the resulting LUT, the inverse permutation would be composed with the next operation following the XOR during AES encryption. Since the attacker would not know the randomly generated permutation value, she/he would no longer be able to extract the key directly from the LUT. Converting several steps of an AES cipher to LUT-based implementations and then applying random permutations to these LUTs leads to an implementation which is much more compact than the huge LUT with 2^{128} entries of 128-bits. Typically, the size of a WB-AES cipher is around a few megabytes. The same idea can also be applied to other ciphers as well.

Attacks: Several MATE attacks on WB-DES [5, 43] and WB-AES [44, 45] have been published in the literature, on the ground of algebraic attacks, which treat each cipher as an overdefined system of equations [46]. All of these attacks assume that the structure and purpose of the LUTs is known. Therefore, Banescu et al. [47] propose using data obfuscation techniques to hide the location and structure of the LUTs used by WBC ciphers.

4.1.5 One-Way Transformations

One-way transformations refer to mapping data values from one domain to another domain (e.g., $f(i) = i'$), without needing to perform the inverse mapping $f^{-1}(i')$ during runtime. This means that f must be homomorphic w.r.t. the operations performed using i. For instance, a cryptographic hash function such a secure hash algorithm (SHA), e.g., SHA-256 (denoted H) may be used as a one-way transformation. H can map a hard-coded string password s to a 256-bit value, i.e., $v = H(s)$. Generally, the only operation performed with a hard-coded password is an equality check with an external user input i. Hence, v does not need to be mapped back to s during runtime. Instead, the program can compute $v' = H(i)$ and verify the equality between the hard-coded value v and the dynamically computed v'. Since the implementation of H for cryptographic hash functions does not disclose the inverse mapping H^{-1}, the MATE attacker is forced to either guess s or identify the equality comparison and modify it such that it always indicates equality regardless of i. The latter tampering attack can be hampered if the code of the equality check is highly obfuscated, which can be achieved by applying code obfuscation transformations (see Section 4.3).

Attacks: More than a decade ago, Wang et al. [48] suggested how to find collisions in hash functions such as MD4, MD5, HAVAL 128, and RIPEMD. However, it was only recently that Stevens et al. [49] discovered the first collision for the SHA1 hash function. Both of these attacks are meant to recover input data for the hash function, which hashes to the same value as another input data. They are dynamic and semantic since they require executing the hash functions.

4.2 Variable Data Transformations

```
Listing 6: Split Variable
1 char b1, b2, b3, b4; // i = b1, b2, b3, b4
2
3 int add(int a) {
4     return a + ((b1 << 8 + b2)
5                       << 8 + b3)
6                       << 8 + b4;
7 }
```

```
Listing 7: Merged Variables
1 int i; // i = b1, b2, b3, b4
2 char add_b1(char a) {
3     return a + (i >> 24);
4 }
5 // ...
6 char add_b4(char a) {
7     return a + (i & 0xff);
8 }
```

4.2.1 Split Variables

The idea behind this transformation is to substitute one variable by two or more variables [7]. For instance, a 32-bit integer variable i can be split into four byte variables b_1, b_2, b_3, and b_4 that represent the integer i, i.e., $i = b_1 \cdot 256^3 + b_2 \cdot 256^2 + b_3 \cdot 256 + b_4$. This idea is similar to converting static data to procedural data, except that splitting variables does not apply to constant values, but to any value that a variable may hold at any moment during execution. Listing 6 shows a C-code snippet illustrating the previously mentioned example of replacing an integer by four bytes. This code snippet also shows that such an obfuscation transformation requires us to implement functions even for simple arithmetic operations such as addition (lines 3–7). We must reconstruct the 32-bit integer value before performing the actual arithmetic operation, i.e., addition. This kind of function must be implemented for all operations, which are needed in the original unobfuscated program.

Attacks: Slowinska et al. [50] propose a data recovery attack based on dynamic trace analysis of the so-called *temporal reuse intervals*, which indicate the usage patterns of certain memory locations. This attack is also applied to the merging variables transformation presented next.

4.2.2 Merge Variables

Two or more variables can be merged into a single variable, if the ranges of the combined variables fit within the precision of the compound variable [7]. For example, up to four 8-bit variables can be packed into a 32-bit variable, i.e., the inverse operation than *splitting variables*. Listing 7 shows an illustration of this example, where four byte variables, namely, b_1, b_2, b_3, and b_4 have been merged into an integer variable i. Any operations on the individual variables has to be carefully crafted in order not to affect the other variables. For instance, arithmetic addition functions—as shown on lines 2–8 of Listing 7—must be implemented for each of the four different variables. These functions could be further obfuscated using the code obfuscation transformations in order to raise the bar for attackers. A MATE attacker would first need to understand these functions in order to figure out that multiple variables are stored in a compound variable.

Attacks: The previously mentioned MATE attack proposed by Slowinska et al. [50] for splitting variables also applies to merging variables. In addition to this attack, Viticchi et al. [51] performed a user study where a group of students managed to successfully reverse engineer programs obfuscated using the merge variables transformation by employing a combination of both static and dynamic analysis tools and techniques.

Listing 8: Folded Array

```
1 int folded [10] [10] = {{1, 2, . . ., 10},
2                         {11, 12, . . ., 20},
3                         . . .
4                         {91, 92, . . ., 100}};
```

Listing 9: Flattened Array

```
1 int flattened [100] = {1, 2, . . ., 100};
```

4.2.3 Restructure Arrays

Similarly to variables, arrays can be split or merged [7]. However, in addition to that, arrays can be folded (increasing the number of dimensions) or flattened (decreasing the number of dimensions). For instance, if the original code contained an array of 100 integer elements, then this array could be folded into a 10 by 10 matrix as shown in Listing 8. We can also consider the dual, where the original code contained the 10 by 10 matrix and it would be flattened into a 1-dimensional array of 100 elements as shown in Listing 9. Folding and flattening break code abstractions put in by software developers (e.g., matrices are flattened into arrays), which force reverse engineers to first understand logic in the code before they can recover this useful abstraction.

Attacks: Slowinska et al. [52] propose a code understanding attack based on pattern recognition of access patterns on execution traces generated by symbolic execution. This attack mixes both static and dynamic techniques in order to recover data structures from obfuscated code. It is a passive attack which uses code semantics. This attack is not only aimed to break restructuring of arrays, but also loop transformations, presented later in this chapter.

Listing 10: Original Basic Block

```
1 mov eax, 0x10h
2 mul ebx, eax
3 cmp ebx, 0x10h
4 je 0x12345678
```

Listing 11: Basic Block After Reordering Variables
```
1 mov edx, 0x10h
2 mul ebx, edx
3 cmp ebx, 0x10h
4 je 0x12345678
```

4.2.4 Reorder Variables

This transformation changes the location or the name of variables in the program code, by permuting or substituting them [6]. Moreover, it can repurpose variables such that they are no longer used for the same (single) task. It has low cost and it improves resilience, because automated MATE attacks can no longer assume certain patterns, e.g., that variables are always laid-out in a certain order, or that one specific variable (name) is used for a certain purpose only. An example of such a transformation is illustrated in the Assembly-code snippets from Listings 10 and 11, where the register `eax` in the former listing is substituted by register `edx` in the latter listing. Pappas et al. [53] apply this transformation at the binary level by reassigning register operands at the basic block level. They show that this transformation is able to eliminate (on average) over 40% of return oriented programming (ROP) gadgets in different instances of the same program. This means that using this transformation breaks 40% of the patterns in the binary code.

Attacks: Griffin et al. [54] present an automated MATE attack, which is able to perform data recovery for the purpose of malware string signature extraction and is able to break this transformation of reordering variables. Their attack is static and semantic.

4.2.5 Dataflow Flattening

Dataflow flattening, proposed by Anckaert et al. [55], is an advanced version of variable reordering, inspired by the idea of oblivious RAM proposed by Goldreich and Ostrovsky [56]. It periodically reorders data stored on the heap via a memory management unit (MMU), such that the functionality of the program is not altered. It is not feasible to show a meaningful code snippet for the MMU that would fit in one page of this chapter, but the intuition behind the way in which it reorders data on the heap is simple.

1. The MMU allocates a new memory region on the heap, for a given variable.

2. It copies the value of that variable to the new memory region.
3. The MMU updates all pointers from the old to the new memory region of the variable.
4. Finally, it deallocates the old memory region.

In addition to reordering the data on the heap, dataflow flattening also proposes moving all local variables from the stack to the heap and scrambling pointers to hide the relation between the different pointers returned to the program. This transformation has a resilience against MATE attacks, nonetheless, its execution overhead is also high.

Attacks: At the time of writing this chapter, we are not aware of reported attacks on this obfuscation technique in the literature.

Listing 12: Original Function Prolog and Epilog

```
1 push ebp        ; save previous stack frame
2 mov ebp, esp    ; base of this stack frame
3 . . .           ; function body
4 mov esp, ebp    ; discard this stack frame
5 pop ebp         ; restore previous frame
6 ret
```

Listing 13: Prologue and Epilogue After Stack Randomization

```
1 sub esp, 0x43
2 push ebp
3 sub esp, 0x5f
4 mov ebp, esp
5 . . .
6 mov esp, ebp
7 add esp, 0x5f
8 pop ebp
9 add esp, 0x43
10 ret
```

4.2.6 Randomized Stack Frames

This transformation assigns each newly allocated stack frame a random position on the stack [57]. For this purpose it subtracts a random value from the stack pointer in the function prologue (simulating a push of multiple elements) and adds this value before the function returns in the function epilogue. Additionally, Fedler et al. [58] propose padding each stack frame internally by a random amount, such that return address and local variables are at random offsets. Note that, these random offsets could be generated at compile time or during runtime. This is again done by performing random subtractions from the stack pointer in the function prologue and undoing these subtractions in the function epilogue. A typical function prologue and epilogue are shown in the Assembly code snippet from Listing 12. The prologue starts by saving the address of the base of the previous stack frame (line 1). Then it sets the base of the new stack frame (line 2). After the body of the function is executed the contents of the new (current) stack frame are discarded (line 4) and the previous stack frame is restored (line 5). Listing 13 shows the prologue and epilogue after the stack randomization transformation is applied. Note the instructions inserted on lines 1, 3, 7, and 9, which subtract and add random values—0x43 and 0x5f in our example—from the stack pointer register (i.e., esp). Since the subtracting and adding instructions are stack operation, the value subtracted in the prologue, must be added in reverse order in the epilogue. These added instructions can be easily identified by the MATE attacker, however, they could be further obfuscated using both code and data transformations.

Attacks: Strackx et al. [59] propose a data recovery attack that is able to bypass the protection offered by randomized stack frames via a technique called a *buffer overread* (note that this is different from buffer overflow). This attack is dynamic, semantic and active because it changes the data of the program in process memory.

Listing 14: Original Addition Function

```
1 int a = 5;
2 int b = 10;
3 . . .
4 int sum = a + b;
```

Listing 15: Addition Function Obfuscated With DSR

```
1 int mask_a = rand ();
2 int a = 5 ^ mask_a;
3 int mask_b = rand ();
4 int b = 10 ^ mask_b;
5 . . .
6 int mask_sum = rand ();
7 int sum = ((a ^ mask_a) + (b ^ mask_b)) ^ mask_sum;
```

4.2.7 Data Space Randomization

Cadar et al. [60] and Bhatkar and Sekar [61] introduce a data transformation technique which they call Data Randomization and Data Space Randomization (DSR), respectively. The idea of these two techniques is to XOR (i.e., encrypt) data values stored in program memory (e.g., stack, heap) with randomly generated masks. The masks do not need to be fixed; they can be generated dynamically at runtime and used to encrypt the data values. Whenever a data value must be read by the program, it is first decrypted using the right mask. After an authorized modification of a decrypted data value occurs, the result is reencrypted with the same or with a different mask depending on the implementation. The technique is inspired by PointGuard [62], which encrypts code pointers. DSR offers protection against MATE attackers who want to extract or modify data values from/in process memory. One challenge of implementing DSR is that different pointers may point to the same encrypted memory value, therefore, they must use the same mask to decrypt a data value. This problem is solved in part by performing a static alias analysis of the code [63], before the transformation is applied. We have already mentioned that alias analysis is in general undecidable, therefore, an approximation is performed and a common mask is used for all pointers that cannot be statically determined.

Listing 14 shows a C-code snippet performing a simple summation operation. Listing 15 shows how the function from Listing 14, after it is obfuscated using DSR. Note that each value is XOR-ed with a random mask, which must be used for decryption when performing the addition operation on line 7 of Listing 15. Even though the actual values of the variables are now hidden by masking, the security issue is shifted to hiding the masks or the relation between values and masks from MATE attackers. This can be achieved by applying other obfuscation transformations on top of DSR. DSR is reported to introduce an average run-time overhead of 15% and it can protect against buffer- and heap-overflow attacks.

Attacks: The attack of Strackx et al. [59] which was presented as an attack for the *random stack frames* transformation is also able to bypass DSR.

4.3 Code Logic Transformations

Listing 16: Code Before Instruction Reordering

```
1 mov eax, ebx
2 add ecx, edx
3 add eax, ecx
4 . . .
```

Listing 17: Code After Instruction Reordering

```
1 add ecx, edx
2 mov eax, ebx
3 add eax, ecx
4 . . .
```

4.3.1 Instruction Reordering

This technique targets sequences of instructions, which when permuted, do not alter the original program execution [6]. Similarly to variable reordering, this transformation is meant to break MATE attacks based on pattern matching. However, it has very low resilience w.r.t. human-assisted MATE attacks, because it does not increase the difficulty of code understanding by much. An example of instruction reordering is shown in Listing 17, where the instructions on lines 1 and 2 have been reordered from their original positions in Listing 16. The candidate instruction sequences targeted by this technique are also candidates of parallel processing optimizations, because they can be independently performed by different execution threads, without any danger of race conditions. The reordered sequence of instructions must be equivalent to the original sequence. The cost of this transformation is low. Pappas et al. [53] have employed instruction reordering on binary basic block level and have shown that this transformation reduces the number of ROP gadgets by over 30%, hence, increasing the resilience against ROP attacks. Note that this transformation can be also performed at basic block level, however, this would have a lower increase in resilience compared to instruction reordering.

Attacks: Zhang et al. [64] propose a static code understanding attack in order to detect repackaged Android applications, which are suspected to be malicious. Their attack uses code semantics to build a so-called *view graph* of each application, which is compared to other applications in order to determine if they are repackaged versions of the same application. The authors indicate that this attack is resilient to multiple code obfuscation transformations presented in this section, namely: merging functions, opaque predicates, inserting dead code, removing functions, function argument randomization, and converting static data to procedural.

Listing 18: Assembly Code#1 Performing Swap

```
1 mov edx, eax
2 mov eax, ebx
3 mov ebx, edx
```

Listing 19: Assembly Code#2 Performing Swap

```
1 push eax
2 push ebx
3 pop eax
4 pop ebx
```

4.3.2 Instruction Substitution

This technique (first mentioned in [6]) is based on the fact that in some programming languages as well as in different instruction set architectures (ISAs), there exist several (sequences of) equivalent instructions. This means that substituting an instruction (sequence) with its equivalent will not change the semantic behavior of the program, nevertheless, it will result in a different binary representation. A concrete implementation and evaluation of this technique is presented by Jacob et al. [65] and it is also used in the *Hydan* tool [66]. Listing 18 shows an Assembly code snippet representing a swap function from register eax to register ebx, using register edx as an auxiliary variable. Listing 19 shows one of the many possible instruction sequences—equivalent to Listing 18—presented by Jacob et al. [65]. The transformation has a moderate cost, however, it offers low resilience against MATE attacks, due to the fact that the number of transformations available is limited. Regarding the resilience against remote attacks, Pappas et al. [53] measured the effect of this transformation at binary basic block level, against ROP attacks and discovered that it reduces less than 20% of ROP gadgets. Moreover, the use of uncommon instructions will decrease stealth, i.e., indicate to an attacker where the substitution occurred. In order to improve the stealth of this transform, De Sutter et al. [67] proposed a technique called *instruction set limitation*, which proposes candidates for substitution based on the statistical distribution of instruction types in the program. Mason et al. [68] also proposed a similar technique with the purpose of improving the stealth of shellcode by encoding it as text written in the English language.

Attacks: The code understanding and data recovery attack of Banescu et al. [31], presented as an attack for opaque predicates, are also applicable to bypass this transformation.

Listing 20: Code Before Encode Arithmetic

```
1 int main(int ac, char* av []) {
2   int x = atoi (av [1]);
3   int y = atoi (av [2]);
4   int w = atoi (av [3]);
5   int z = x + y + w;
6   // do stuff
7   return 0;
8 }
```

Listing 21: Code After Encode Arithmetic

```
1 int main(int ac, char* av []) {
2   int x = atoi (av [1]);
3   int y = atoi (av [2]);
4   int w = atoi (av [3]);
5   int z = (((x ^ y) + ((x & y) << 1)) | w) +
6           (((x ^ y) + ((x & y) << 1)) & w);
7   // do stuff
8   return 0;
9 }
```

4.3.3 Encode Arithmetic

This technique is proposed by Collberg [18] and it is a variant of *instruction substitution*, which substitutes Boolean or arithmetic expressions by expressions involving both Boolean and arithmetic operations, which are harder to understand. One example of such a transformation is illustrated by Listing 21, which shows a C code snippet after *encode arithmetic* has been applied to the right-hand side of the assignment to variable z from line 5 of Listing 20.

Attacks: The drawback of this approach is that there are a limited number of such Boolean-arithmetic identities available in the literature [69]. Eyrolles et al. [70] have proposed writing a reverse transformation for each of them, after identifying Mixed Boolean–Arithmetic (MBA) expressions via pattern matching.

Listing 22: Code Before Inserting Garbage Code

```
1 int sum = 0;
2 for (i = 0; i < arr_len; i++)
3   sum += arr [i];
4 int average = sum / arr_len;
```

Listing 23: Code After Inserting Garbage Code

```
1 int sum = 0;
2 int prod = 1;
3 for (i = 0; i < arr_len; i++) {
4   sum += arr [i];
5   prod *= arr [i];
6 }
7 int average = sqrt (prod);
8 average = sum / arr_len;
```

4.3.4 Garbage Insertion

This technique implies inserting arbitrary sequences of instructions that are independent of the data flow of the original program and do not affect its input–output (IO) behavior (functionality) [6]. The possible sequences that may be inserted are virtually infinite, nonetheless, the performance-cost grows proportionally to the number of inserted instructions. Listings 22 and 23 show a program that computes the average of all elements in an array arr, before and after garbage code has been inserted. Note that in Listing 23, lines 2, 5, and 7 represent garbage code that has no influence on the output value of the program. However, inserting garbage code changes the relative offset the original instructions of the program. It also raises the complexity of reverse-engineering by cluttering the original code. However, note that garbage code should be inserted only after performing compiler optimizations, because it can be identified and eliminated via taint analysis.

Attacks: Performing the generic attack based on taint analysis, proposed by Yadegari et al. [71], would remove lines 2, 5, and 7 in Listing 23, because the final value of average (line 8 in Listing 23) has no data dependency on prod. The transformation space for this technique is limited only by physical or practical run-time constraints such as time delays and memory consumption, because as opposed to dead code, garbage code is always executed.

Listing 24: Code Before Inserting Dead Code

```
1 int main(int ac, char* av []) {
2   int x = atoi (av [1]);
3   int y = sqrt (x);
4   // do stuff
5   return 0;
6 }
```

Listing 25: Code After Inserting Dead Code

```
1 int main(int ac, char* av []) {
2   int x = atoi (av [1]);
3   int y;
4   if (x*x + x % 2 == 0)
5       y = sqrt(x);
6   else
7       y = x*x;
8   // do stuff
9   return 0;
10 }
```

4.3.5 Insert Dead Code

This transformation was first proposed by Collberg et al. [7] and it modifies the control-flow of a program such that a dead branch is added, i.e., a branch that is never taken during runtime. Adding the dead branch is facilitated by opaque predicates, because one of the branches of a conditional statement that uses such an opaque predicate, will never be executed, while the other branch will always be executed. In order not to disclose the truth value of opaque predicates by leaving the dead branch empty, Collberg suggests inserting dummy code on the dead branch. To further confuse the attacker, the dead code can be a buggy version of the other branch, which is always chosen. For instance, consider the code snippet from Listing 24 which converts the first input argument to an integer and then sets the value of variable y to the square root of the input argument's value. Listing 25 shows that dead code can be inserted anywhere by first inserting a conditional statement with an opaque predicate that is always true (line 4) and then adding dead code in the branch that is never taken (line 7). Note that we can wrap as many lines of code as we want using the conditional statement. Moreover, the size of the dead code can be arbitrarily large.

Attacks: Along with the previously presented attacks by Salem and Banescu [33], Zhang et al. [64], Yadegari et al. [71], and Banescu et al. [31], there are numerous works in the filed of compilers presenting optimizations for dead code removal [72, 73].

Listing 26: Add Function Calls

```
1 int foo (int a, int b) {
2    return a + b;
3 }
4 . . .
5 int c = foo (a, b + 1);
```

Listing 27: Remove Function Calls

```
1 int c = a + b + 1;
```

4.3.6 Adding and Removing Function Calls

These two techniques were first proposed by Cohen [6] and can be applied at any unit of transformation. Adding a call to a subroutine implies

(1) selecting an arbitrary sequence of instructions,

(2) creating a subroutine using that sequence, and

(3) finally, substituting the original sequence with a call to that subroutine.

Removing a call to a subroutine implies

(1) substitute all calls to a subroutine with the body of that subroutine and

(2) delete the subroutine.

Listing 26 shows a code snippet that contains a function, while Listing 27 shows a code snippet where this function has been removed. The reverse transformation from Listings 27 to 26 is *adding function calls*. These transformations cause changes in the structure of a program, which creates more complexity for MATE attacks. The cost of this technique grows or decreases with the number of inserted and removed subroutine calls, respectively. This method has been extended by Banescu et al. [20], such that system calls are added or existing system calls are substituted with equivalent ones. They call this transformation *behavior obfuscation* because it hides the system call trace analyzed by behavioral malware analysis engines.

Attacks: We have already mentioned the attack of Zhang et al. [64], which also claims to bypass this obfuscation transformation. In addition to this work, many works on behavioral malware detection claim they are resilient to the addition or removal of system calls [20, 74]. Such machine learning-based malware detection approaches are classified as dynamic, passive, and syntactic, because they treat the function call traces generated by the obfuscated software as a sequence of bytes.

4.3.7 Loop Transformations

Several loop transformations have been proposed as compiler optimization passes by Bacon et al. [34]. Collberg et al. [7] argue that these loop transformations also increase software complexity metrics and can therefore be considered obfuscation transformations that increase resilience against attacks. *Loop tilling* or *blocking* is intended to improve cache locality, by dividing loop iteration lengths into parts that fit in the CPU cache. This increases the nesting level of loops and is therefore more potent. *Loop distribution* or *fission* breaks the independent instructions in a loop body into multiple loops with the same iteration length, which increases the number of loops in the code. *Loop unrolling* replicates the body of the loop a certain number of times and reduces the number of iterations correspondingly, which increases the number of lines of code in the program.

Attacks: The dynamic code understanding attack by Slowinska et al. [52], already presented as an attack on the *restructure arrays* transformation, is also applicable to loop transformations.

Listing 28: Code With Jumps Removed

```
1 mov edx, eax
2 mov eax, ebx
3 mov ebx, edx
4 . . .
```

Listing 29: Code With Jumps Added

```
1     mov edx, eax
2     jmp L1
3     . . .
4 L2:mov ebx, edx
5     . . .
6 L1:mov eax, ebx
7     jmp L2
8     . . .
```

4.3.8 Adding and Removing Jumps

These techniques change the control-flow of the program by adding spurious jumps or removing existing jumps [6]. Adding jumps can be done by substituting an arbitrary sequence of instructions I by:

(1) a jump to a random position,

(2) followed by I, and

(3) a jump to the instruction immediately following I in the original version of the program.

An example is illustrated in Listing 29, where the code from Listing 28 has been transformed by adding two unconditional jumps to labels L1 and L2. Removing jump instructions may also be done if it does not alter the original semantics of the program, e.g., unconditional jumps may be removed and the code from the address of the jump, merged with the code preceding the unconditional jump. An example is shown in Listing 28, where all jumps from the code in Listing 29 have been removed. However, in practice adding jumps is more frequently employed in order to increase the complexity of the MATE attack. The transformation space of this technique is bounded by the length of the program it is applied to. The cost of this method grows (decreases) with the number of inserted (removed) jump instructions. The resilience of adding jumps can be increased by further obfuscating the addresses of the jumps using data obfuscation techniques such as *opaque expressions* or *converting static data to procedural data*.

Attacks: If the target of the added jumps are not made dependent on input values using opaque expressions, then they are trivial to bypass using the already mentioned dynamic attack of Yadegari et al. [71]. On the other hand removing jumps can be bypassed by dynamic taint analysis on augmented control flow graphs (CFGs), proposed by Yadegari et al. [75], which is a dynamic, passive, and semantic attack.

4.3.9 Program Encoding

This technique keeps one or more instructions encoded (i.e., encrypted [76, 77] or compressed [78]), while the program is not executing and decodes the sequence(s) when the program is running [6]. The resilience of program encoding against attacks depends on the algorithm used for encoding, e.g., a compression algorithm can be undone without a secret key, while an encryption requires finding the key. However, the costs may also be relatively high compared to other obfuscation techniques, because the code has to be decoded before it can be executed. There is a trade-off between resilience and cost depending on the level of granularity at which this transformation is applied, i.e., if applied at instruction level, the cost, and resilience are high, because each instruction is decoded, executed and reencoded, hence the whole code is never stored in decoded form in memory, at one point in time. While if program encoding is applied at program level, the whole code is decoded and afterwards it starts executing, hence an attacker could perform a memory dump after decoding to have a copy the whole code. Additionally, this technique does not protect well against dynamic analysis attacks, e.g., during execution the code is decoded in memory and it can be read or modified directly in memory by the MATE attacker.

Attacks: Tang et al. [79] propose a static, passive, and syntactic attack against polymorphic malware, in order to extract signatures based on the position and distribution of byte values in obfuscated malware binaries. Qiu et al. [80] propose a dynamic, passive, and semantic attack based on taint analysis, in order to determine the location of integrity checks inside code.

4.3.10 Self-Modifying Code

This technique has been discussed in several works [6, 11, 81, 82]. It implies adding, modifying, and/or removing instructions of a program during its execution. Therefore, it creates a high complexity for static-analysis attacks. Kanzaki et al. [82] proposed replacing real instructions with bogus instructions that would get replaced by real instructions before they are executed and then replaced again by bogus instructions after execution. This transformation requires a sound analysis of all possible execution paths leading to and following the instruction(s) that are to be modified. Madou et al. [81] propose to apply self-modifying code at the function level by creating the so-called function templates, i.e., arrays of byte having a larger size than any single function in a chosen subset of functions. For instance, if the subset consists of functions f_1 and f_2, the code of function f_1 is 5 bytes long and the code of function f_2 is 4 bytes long, then the template T must be at least 5 bytes long as shown in Fig. 3. Note that function templates are generated by an intersection of the code bytes of all other functions in the subset, i.e., common code bytes are kept in place (e.g., 48 and a0 in Fig. 3) and other code bytes are randomly initialized. Each function f is associated to an edit script e, which indicates the locations (i.e., indices) of the function template that must be patched and the values they must be patched with, in order to reconstruct the code of f. Therefore, when any function f is called, the edit script e corresponding to f is first executed and then the resulting code in the function template is executed.

Attacks: Self-modifying code can be effective against dynamic-analysis attack, which aim to break the integrity of the code (e.g., via dynamic code patching), if the executed instructions are different in different runs of the program with the same inputs. However, Nguyen et al. [83] have shown

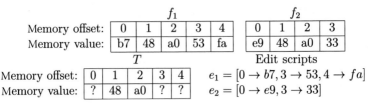

Fig. 3 Self-modifying code via function templates and edit scripts.

that applying such temporal diversity to instruction level encryption (i.e., instruction set randomization [84]) has negative effects on confidentiality of the encryption key because encrypting the same code with different keys leaks information about the code, similar to two-time pads [85]. In the case of self-modifying code, there is no dedicated decoder routine, as was the case for the *program encoding* transformation. Instead, different parts of the code are "responsible" for modifying other instructions and these are often spread throughout the entire program. The trade-off between resilience and cost is similar to that of *program encoding* and *virtualization*. Dalla Preda et al. [86] applied an abstract interpretation based attack to generate signatures for metamorphic malware samples, hence this attack is a code understanding, static, passive, and semantic attack.

```
Listing 30: Point Function Program
 1 int main(int argc, char* argv []) {    // Virtualization
                                                bytecode :
 2    char branch_cond = 1;                 // a5 00 07
 3
 4    branch_cond &= argv [1] [0] == '1';  // 87 00 00 02
 5    branch_cond &= argv [1] [1] == '2';  // 87 00 01 03
 6    branch_cond &= argv [1] [2] == '3';  // 87 00 02 04
 7    branch_cond &= argv [1] [3] == '4';  // 87 00 03 05
 8    branch_cond &= argv [1] [4] == '5';  // 87 00 04 06
 9
10    if (branch_cond)                      // 1f 00 02
11       printf ("You win !\n");            // 03 00
12    return 0;                             // 42 01
13 }
```

4.3.11 Virtualization Obfuscation

This technique is related to the *program encoding* technique, because it also implies an encoding of instructions [6]. Additionally, it also requires an interpretation engine (called "simulator" or "emulator"), which is able to decode the instructions and execute them on the underlying platform. The simulator may also be running on top of another simulator and so forth, giving an arbitrary nesting level. Normally, this creates complexity for static-analysis attacks, because the attacker has to first understand the custom interpreter logic and then the code running on top of it. The most significant difference of virtualization w.r.t. program encoding is that no code must be written to a memory location during decoding. However, the trade-off between resilience and cost for this transformation are the same as in the case of program encoding. For clarity, we provide all of the steps of the *Virtualize* transformation for the program in Listing 30. This program prints the message "You win!" on standard output if the first argument passed to this program is equal to "12345". The virtualization transformation is applied to this program using the following steps and the result is illustrated in Listing 31:

1. Map variables, function parameters, and constants to entries in a common `data` array, which represents the memory of the interpreter. This array is initialized on lines 3–4 in Listing 31. Its first position represents the `branch_cond` variable from Listing 30 and the following entries represent constants such as the return value, the ASCII codes of the characters from "1" to "5" and logical `true` encoded as 1.

2. Map all statements in a function to a new randomly chosen language, which represents the *instruction set architecture* (ISA) of the interpreter. In our example the ISA is defined by:

   - Variable assignment is encoded using 3 bytes, namely, the opcode (`0xa5`) and the index of the left- and right-hand operands inside the `data` array.

   - Equality comparison, followed by applying the logical *AND* operation to between the result and another variable. Examples of such instructions are shown on lines 4–5 in Listing 30. Such an instruction is encoded using 4 bytes, namely, the opcode (`0x87`), the variable to which the Boolean value is assigned and the two other byte values which are compared for equality.

   - Conditional branch statements are encoded using 3 bytes, namely, the opcode (`0x1f`), the Boolean variable which is tested and the number of bytes to jump over if the variable is false.

   - Printing a string on standard output is encoded using 2 bytes, namely, the opcode (`0x03`) and the index of the string to be printed in the list of hard-coded strings of the function. In our example the list of hard-coded strings contains only one string and is defined on line 2 of Listing 31.

- The return instruction is encoded using 2 bytes, namely, the opcode (0x42) and the value that should be returned by the program.

Now we can write virtualization bytecode corresponding to the C program in Listing 30, which is shown in the comments of the code from the same listing.

Listing 31: Point Function Program Obfuscated With Virtualization

```
1 int main(int argc, char* argv []) {
2   char const *strings = "You win !\0";
3   unsigned char data [8] = {0,  // branch_cond var
4     0, 49, 50, 51, 52, 53, 1}; // constants
5   unsigned char code [30] = {0xa5, 0x00, 0x07, 0x87, 0x00,
6     0x00, 0x02, 0x87, 0x00, 0x01, 0x03, 0x87,
7     0x00, 0x02, 0x04, 0x87, 0x00, 0x03, 0x05,
8     0x87, 0x00, 0x04, 0x06, 0x1f, 0x00, 0x02,
9     0x03, 0x00, 0x42, 0x01};
10  int vpc = 0;
11  while (1)
12  switch (code [vpc]) {
13    case 0xa5 : // variable assignment
14      data [code [vpc +1]] = data [code [vpc +2]];
15      vpc += 3;
16      break;
17    case 0x87 : // equality comparison plus and
18      data [code [vpc +1]] &=
19        (argv [1] [code [vpc +2]] == data [code [vpc +3]]);
20      vpc += 4;
21      break;
22    case 0x1f : // i f statement
23      vpc += (data [code [vpc +1]]) ? 0 : data [code [vpc +2]];
24      vpc += 3;
25      break;
26    case 0x03 : // printf string
27      printf ("% s\n", strings + code [vpc +1]);
28      vpc += 2;
29      break;
30    case 0x42 : // return
31      return data [code [vpc +1]];
32  }
33 }
```

3. Store the encoded bytecode inside the `code` array, which is initialized on lines 5–9 in Listing 31.

4. Create an interpreter for the previously generated ISA, which can execute the instructions in the `code` array using the `data` array as its memory. The input–output behavior of this execution must be the same as that of the original program. The interpreter can be seen on lines 11–32 of Listing 31. It consists of an infinite `while` loop, which has a `switch` statement inside. Each `case` section of the `switch` statement is an opcode handler, i.e., each possible opcode in the bytecode program is processed by a dedicated part of the interpreter. The current instruction to be processed by the interpreter is indicated by an integer variable of the interpreter called the *virtual program counter* (VPC). The VPC is used to index the instructions in the `code` array and it is initialized with the offset of the first instruction in that array (line 10). In every instruction handler the operands of the current instruction are used to perform the operation(s) corresponding to this instruction. Afterward, the VPC is set to the offset of the following bytecode instruction to be executed. This interpreter should be augmented with cases representing bogus opcodes for all possible byte values in order to increase the resilience against MATE attacks.

Together with virtualization a wide range of additional data and code obfuscation transformations can be applied. Banescu et al. [87] propose randomizing the layout of the bytecode instructions in the `code` array as well as the format of these instructions. Collberg [18] proposes encoding the value of the VPC using opaque expressions, changing the interpreter's dispatch method from a `switch`-statement to table lookups, and many more.

Attacks: Kinder [88] proposes a static, semantic attack based on abstract interpretation, and a technique called *VPC lifting*, in order to automatically recover memory values at any code location. Rolles [89] proposes a manual, static, semantic, and passive attack to understand the mapping between the bytecode and the instruction handlers of the interpreter. Banescu et al. [87] present an experiment where a dynamic, active, syntactic attack, which is able to remove a large portion of the code of the interpreter and recover the original logic of the program.

Listing 32: Code Before Control Flow Flattening (CFF)

```
1 int gcd(int a, int b){
2   while (a != b)
3     if (a > b)
4       a = a - b;
5     else
6       b = b - a;
7   return a;
8 }
```

Listing 33: Code After CFF

```
1 int gcd(int a, int b){
2 int next = 0;
3 while (1) {
4    switch (next) {
5       case 0:
6          if (a != b) next = 1;
7          else next = 2;
8          break;
9       case 1:
10         if (a > b) next = 3;
11         else next = 4;
12         break;
13      case 2: return a;
14         break;
15      case 3: a = a - b; next = 0;
16         break;
17      case 4: b = b - a; next = 0;
18         break;
19      default :
20         break;
21 }}}
```

4.3.12 Control Flow Flattening

Wang et al. [27] and Chow et al. [90] proposed CFF, which collapses all the basic blocks of a function into a flat CFG, which hides the original control flow of the program. A program transformed by CFF is similar to an interpreter (as we saw for *virtualization obfuscation*), which chooses the right basic block where to go on the second CFG level. An example is shown in Listing 33, where the function computing the greatest common divisor (GCD) of two integers a and b—shown in Listing 32—has been transformed by CFF. As opposed to virtualization obfuscation—which uses a virtual program counter (VPC)—the order in which the cases of the `switch` statement must be executed is indicated by a control variable (i.e., `next` in the example from Listing 33), which is updated by every case of the switch statement accordingly. Updating `next` can occur before, during or after part of the logic of the original program is executed. The infinite loop is exited when a case contains a return or break statement. Another difference w.r.t. *virtualization obfuscation* is that multiple cases of the switch statement of a program transformed by CFF may contain the same instructions. For virtualized programs each case represents a different instruction, which may appear multiple times in the original program. This means that in order to recover the original control flow of a program obfuscated using CFF, one must simply order the cases of the switch statement in the correct order, while this is not applicable to virtualized programs, because some cases may be needed multiple times.

Attacks: Udupa el al. [91] were the first to propose an automated MATE attack to recover the original control flow from a program obfuscated using CFF. Their method combined dynamic and static analysis techniques and was able to accurately recover the original control flow.

Listing 34: Code Before Branch Functions

```
1     mov edx, eax
2     jmp L1
3     . . .
4 L2:mov ebx, edx
5     jmp L3
6     . . .
7 L1:mov eax, ebx
8     jmp L2
9 L3:. . .
```

Listing 35: Code After Branch Functions

```
1     mov edx, eax
2     push 1
3     call f
4     . . .
5 L2:mov ebx, edx
6     push 3
7     call f
8     . . .
9 L1:mov eax, ebx
10    push 2
11    call f
12 L3:. . .
```

4.3.13 Branch Functions

Linn and Debray [92] propose hiding the control flow of calls, conditional, and unconditional jumps, from static disassembly algorithms, by replacing them with calls to a so-called *branch function*. An example is shown in Listing 35 where all jumps from the original code shown in Listing 34 have been replaced by a call to offset f, which represents the start address of the branch function. The branch function computes the actual target of the jump dynamically using a parameter passed by the callee—via push instructions in Listing 35—and a lookup-table. Instead of returning to the instruction immediately following the call instruction, the branch function either jumps to the address of the original jump instruction which it replaced, or to several "junk" bytes after the call instruction that it replaced. The structure of the control flow graph is also flat similar to CFF, however, the *switch*-statement is replaced by the branch function.

Schrittwieser and Katzenbeisser [93] present an extension of the idea from [92], which is explicitly aimed at defending against both static- and dynamic-analysis techniques. The targets of the branch functions are ROP gadgets (i.e., short instruction sequences ending in a return instruction) [94]. Additionally, they generate gadget graphs which add redundancy such that the one path in the original code can have multiple paths in the obfuscated code. This is meant to hamper dynamic analysis attacks by generating different traces for the same inputs. The disadvantage of this method is that its resilience increases inversely proportional to the size of the gadgets, while its cost decreases exponentially with this size.

Attacks: Kruegel et al. [95] present a static, passive, semantic approach to bypass branch functions in order to disassemble code obfuscated using this transformation to a large extent.

4.4 Code Abstraction Transformations

Listing 36: Splitting Functions

```
1 func1 (int a, int b) {
2    x = 4;
3    if (a < 3)
4        x = x + 6;
5    x *= b;
6 }
7
8 func2 (int a, int c) {
9    y = a + 12;
10   y = y/c;
11 }
```

Listing 37: Merging Functions

```
1 func3 (int a, int b, int c) {
2    if (c % 2 == 0) {
3        x = 4;
4        if (a < 3)
5            x = x + 6;
6        x *= b;
7    } else {
8        y = a + 12;
9        y = y/b;
10   }
11 }
```

4.4.1 Merging and Splitting Functions

These two techniques are the code correspondents to the data obfuscation transformations of merging and splitting variables [6]. Merging is done by creating larger functions with more inputs and outputs, some of which are independent. An example is shown in Listing 37, where func3 is the result of merging func1 and func2 from Listing 36. Note that func3 uses a flag variable c. The body of func1 is executed when c is even and the body of func2 is executed with c is odd. Splitting is done by dividing large functions into smaller functions, e.g., the function in Listing 37 can be split into two functions as shown in Listing 36. Similarly to the *adding and removing calls* transformations, this technique changes the structure of the program, breaking abstractions added by developers, which makes the code more difficult to understand. Moreover, the cost of this transformation is increased or decreased with the number of split, respectively, merged functions.

Attacks: Rugaber et al. [96] present ideas for implementing a static, semantic and active attack for code obfuscated by merging functions. The goal of their attack is to both detect the location of the lines of code belonging to different functions and to extract this code into separate functions.

Listing 38: Original JavaScript Code

```
1 function NewObject(prefix) {
2    var count = 0;
3    // This function generates a pop—up
4    this . SayHello = function (msg) {
5       count++;
6       alert (prefix+ msg);
7    }
8 }
9 var obj = new NewObject("Message : ");
10 obj . SayHello ("You are welcome . ");
```

Listing 39: JavaScript code after removing comments and formatting

```
1 function NewObject(prefix){var count=0;
      this ["SayHello"] = function (msg){count++;
      alert (prefix + msg)}} var obj =
      new NewObject("Message : ");
      obj . SayHello ("You are welcome. ")
```

4.4.2 Remove Comments and Change Formatting

This transformation is only applicable to programs which are delivered as source code (e.g., JavaScript). Comments are removed if they exist and all space, tab, and newline characters are also removed, which results in a continuous string of code which is more potent against human attackers, than the original code. An example is shown in Listing 39, which is the result of removing comments and all formatting of the code from Listing 38. The original formatting and comments cannot be recovered [7]; however, the code can be easily reformatted using automated tools. The cost of this transformation is low and in many cases it even improves memory costs and execution speed. Therefore, such transformations have found their way into commercial products, such as Stunnix [97], DashO [98], Dotfuscator [99], Thicket [100], ProGuard [101], and yGuard [102]. However, the resilience against MATE attacks of these obfuscation transformations is very low because a similar alignment can be automatically generated even by free and open source integrated development environments (IDEs).

Attacks: Bichsel et al. [103] propose a code understanding attack against this transformation, based on probabilistic learning of large code bases. This attack is implemented as a service called DeGuard and it is applied statically, syntactically, and passively to android applications.

Listing 40: Code Before Scrambling Identifiers

```
1 sum = 0;
2 for (i = 0; i < arr_len; i++)
3   sum += arr [i];
4 average /= arr_len;
```

Listing 41: Code After Scrambling Identifiers

```
1 za82b547bcb = 0;
2 for(z1c0ab7cf0c = 0; z1c0ab7cf0c < za862d19cbc; z1c0ab7cf0c++)
3   za82b547bcb += zc1c28ca67f [z1c0ab7cf0c];
4 z8c8f7c7867 /= za862d19cbc
```

4.4.3 Scrambling Identifier Names

This transformation implies changing all symbol names (e.g., variables, constants, functions, classes) into random strings [7]. One example is shown in Listing 41 where all the variable names of the code from Listing 40, have been changed to random identifiers. This is a one-way transformation, because the names of the symbols cannot be automatically recovered by a deobfuscator. Therefore, the MATE is forced to understand what a symbol is from a semantics point of view. It has a much higher resilience than formatting removal since identifiers contain useful abstractions added by software developers. Similarly to removing comments and changing formatting, this transformation has a very low cost and it is also used as an optimization that reduces code size, because long symbol names can be replaced by shorter ones.

Attacks: Ceccato et al. [104] investigate the relative strength of two different obfuscation transformations, namely, *scrambling identifier names* and *opaque predicates.* They find that *scrambling identifier names* poses more challenges for human-assisted attacks than *opaque predicates.* This holds in the context where identifiers in the original program have a proper semantic meaning (in English). On the other hand the machine learning based attack of Bichsel et al. [103], which was already mentioned as an attack for removing comments and changing formatting, is able to automatically recover meaningful identifier names with high accuracy.

4.4.4 Removing Library Calls and Programming Idioms

Most programs perform calls to external libraries providing useful data structures (e.g., lists, maps) and algorithms (e.g., sorting, searching). MATE attackers often start by inspecting calls made to external libraries to give a high-level indication of what the program is doing. This transformation implies replacing such dependencies on external libraries with own implementations where possible [7]. Note that such a transformation is stronger than static linking, which only copies the code of the library routines in the executable. Static linking can be easily reverse engineered by pattern matching attacks [19]. Techniques from the field automatic program recognition [105] can be used to identify common programming patterns and replace them with less obvious ones. For example, consider iterating over a linked list; the standard list data structure can be replaced with a less common one, such as cursors into an array of elements.

Attacks: The machine learning based attack of Bichset et al. [103], which was already mentioned as an attack for removing comments and changing formatting, is also applicable for bypassing this transformation.

4.4.5 Modify Inheritance Relations

Programs written in some object-oriented programming languages are distributed in some intermediate format to end-users (e.g., C#, Java). These intermediate formats are only compiled to native code on the client's machine and contain useful object-oriented programming abstractions. In such programs it is important to break the useful abstractions offered by classes, their structure and their relations (e.g., aggregation, inheritance). According to Collberg et al. [7], the complexity of a class grows with its depth in the inheritance hierarchy and the number of its direct descendants. This can be done by splitting classes and inserting dummy classes. One variant of class insertion is called *false refactoring* [7]. False refactoring is performed on two or more classes that have no common behavior. All instance variables of these classes having the same type are moved into the new parent class. Methods of the parent class can be buggy versions of methods from its child classes. This approach has been further extended by Foket et al. [106], who propose a technique called *class hierarchy flattening*. In this approach a common interface that contains all methods of all classes is created. All classes implement this common interface and they have no other relationship between each other. This effectively destroys class hierarchies and forces the attacker to analyze the code.

Attacks: At the time of writing this chapter there have been no reported attacks on this obfuscation technique in the literature.

4.4.6 Function Argument Randomization

Randomizing the order of formal parameters of methods and inserting bogus arguments is a technique implemented by tools such as Tigress [107]. The purpose of this transformation is to hide common function signatures across a large diverse set of instances. This transformation is straightforward to perform for programs which do not offer an external interface (e.g., libraries). However, if this obfuscation is applied to a library, then it changes the interface (of that library), and all the corresponding programs using that library will have to be updated as well. The resilience and cost of this transformation are low. However, the resilience can be improved by combining this transformation with the *encode arithmetic* transformation such that computations inside the function are made dependent on the randomly added arguments similarly to how we did for MBA.

Attacks: The static code understanding attack of Zhang et al. [64] is also able to bypass this obfuscation transformations.

4.5 Summary of Survey

In the survey presented above, we have enumerated several practical data and code obfuscation transformations. Practical obfuscation does not offer provable security guarantees like cryptographic obfuscation [108, 109] does. Nevertheless, many contexts mandate the use of practical obfuscation transformations to protect digital software assets, e.g., secret keys, premium content, intellectual property of code. In such contexts, the goal is to raise the bar against the majority of MATE attackers, not all possible attackers, e.g., developers are concerned about malicious end–users, not governmental organizations, which are highly funded.

Table 3 shows an overview and classification of all the obfuscation transformation presented in this section. We note that most of the presented transformations are applicable for all levels of abstraction (i.e., source code, IR, and binary). Most of the presented transformations are applicable at the function unit of transformation, followed closely by the program unit and then by instruction and basic block, both in third place. Only randomized stack frame have an effect on the system unit of transformation, because they change the layout of the stack. Two of the presented techniques are dynamic, meaning that they must allow code pages to be writable during

Table 3 Overview of the Classification of Obfuscation Transformations

Obfuscation Transformation	Abstraction	Unit	Dynamics	Target
Opaque predicates	All	Function	Static	Data constant
Convert static data to procedural data	All	Instruction	Static	Data constant
Mixed Boolean Arithmetic	All	Basic block	Static	Data constant
White-box cryptography	All	Function	Static	Data constant
One-way transformations	All	Instruction	Static	Data constant
Split variables	All	Function	Static	Data variable
Merge variables	All	Function	Static	Data variable
Restructure arrays	Source	Program	Static	Data variable
Reorder variables	All	Basic block	Static	Data variable
Dataflow flattening	Binary	Program	Static	Data variable

Table 3 Overview of the Classification of Obfuscation Transformations—cont'd

Obfuscation Transformation	Abstraction	Unit	Dynamics	Target
Randomized stack frames	Binary	System	Static	Data variable
Data space randomization	All	Program	Static	Data variable
Instruction reordering	All	Basic block	Static	Code logic
Instruction substitution	All	Instruction	Static	Code logic
Encode Arithmetic	All	Instruction	Static	Code logic
Garbage insertion	All	Basic block	Static	Code logic
Insert dead code	All	Function	Static	Code logic
Adding and removing calls	All	Program	Static	Code logic
Loop transformations	Source, IR	Loop	Static	Code logic
Adding and removing jumps	Binary	Function	Static	Code logic
Program encoding	All	All buy system	Dynamic	Code logic
Self-modifying code	All	Program	Dynamic	Code logic
Virtualization obfuscation	All	Function	Static	Code logic
Control flow flattening	All	Function	Static	Code logic
Branch functions	Binary	Instruction	Static	Code logic
Merging and splitting functions	All	Program	Static	Code abstraction
Remove comments and change formatting	Source	Program	Static	Code abstraction
Scrambling identifier names	Source	Program	Static	Code abstraction
Removing library calls and programming idioms	All	Function	Static	Code abstraction
Modify inheritance relations	Source, IR	Program	Static	Code abstraction
Function argument randomization	All	Function	Static	Code abstraction

program execution. The number of presented transformations are relatively balanced between the two different targets of transformation and their sub-categories, however, code transformations are slightly more numerous.

5. DISCUSSION

In this section we present an overview of the previous survey and discuss the observations stemming from it. Based on this discussion we identify gaps in the field of software obfuscation which require further investigation/research.

Table 4 provides an overview of the various MATE attacks (already mentioned in the previous section), which have been successfully applied in order to defeat the obfuscation transformations presented in Section 4, w.r.t. the attack dimensions presented in Section 3. These attacks are placed in the cells of Table 4 to indicate the obfuscation transformation(s) they claim to break and where they stand w.r.t. the attack dimensions presented in Section 3. The following observations stem from Table 4:

- Some works present attacks which are applicable to multiple obfuscation transformations and multiple attack dimensions. For instance, the attacks by Salem and Banescu [33], Zhang et al. [64], Yadegari et al. [71], and Banescu et al. [31] have a wide range of application. However, not that this does not imply that these attacks are equally effective against all the obfuscation transformation they apply to. For instance symbolic execution-based attacks have a much harder task when dealing with *virtualization obfuscation* or *encode arithmetic*, than when dealing with *converting static data to procedural data*.

- Except for comparing different MATE attacks w.r.t. the classification dimensions presented in Section 3, there is no set of standard benchmarks used in the field of software obfuscation, which would allow comparing different attacks to one another. This indicates that more research is necessary in order to be able to compare these attack techniques to each other. Such research is needed for evaluating the strength of different obfuscation transformations. Therefore, the development of: (1) standard obfuscation benchmarks for MATE attacks and (2) standard attack benchmarks for obfuscation transformations is still an open problem in the field of software protection.

- It is not shown in Table 4 due to lack of horizontal space, however, there are fewer active attacks than passive attacks. This may be due to the fact that active attacks are often associated with malicious exploitation

Table 4 Overview of Obfuscation Resilience Against MATE Attacks

Obfuscation Transformation	Code Understanding				Data item recovery				Location recovery			
	Static		Dynamic		Static		Dynamic		Static		Dynamic	
	Syn	Sem	Syn	Sem	Syn	Sem	Syn	Sem	Syn	Sem	Syn	Sem
Opaque predicates	[33]	[64]	[33]	[31]		[30]		[31]		[30]		
Convert static data to procedural data	[33]	[64]	[33]	[31]				[31]				
Mixed Boolean Arithmetic						[36]						
White-box cryptography								[43, 45]				
One-way transformations								[48, 49]				
Split variables								[50]				
Merge variables								[50, 51]				[51]
Restructure arrays				[52]								
Reorder variables						[54]						
Dataflow flattening												
Randomized stack frames								[59]				
Data space randomization								[59]				
Instruction reordering		[64]										
Instruction substitution				[31]				[31]				
Encode Arithmetic	[33]	[33]		[31]		[70]		[31]				
Garbage insertion				[71]								

Continued

Table 4 Overview of Obfuscation Resilience Against MATE Attacks—cont'd

Obfuscation Transformation	Code Understanding				Data item recovery				Location recovery			
	Static		Dynamic		Static		Dynamic		Static		Dynamic	
	Syn	Sem	Syn	Sem	Syn	Sem	Syn	Sem	Syn	Sem	Syn	Sem
Insert dead code	[33]	[64]	[33]	[31, 71]				[31]				
Adding and removing calls		[64]	[20, 74]									
Loop transformations				[52]								
Adding and removing jumps				[71, 75]								
Program encoding	[33, 79]		[33]								[80]	
Self-modifying code		[86]					[83]				[80]	
Virtualization obfuscation	[33]	[89]	[33, 87]	[31, 71]	[88]			[31]				
Control flow flattening	[33]		[33]	[31, 91]				[31]				
Branch functions		[95]										
Merging and splitting functions		[96]							[96]			
Remove comments and change formatting	[103]											
Scrambling identifier names	[103]											
Removing library calls and prog. idioms	[103]				[19]							
Modify inheritance relations												
Function argument randomization		[64]										

(e.g., cracking software), which is not ethical for most researchers. However, more research is required in the direction of active attacks in order to enable the development of stronger protection against such attacks.

- Also there are fewer attacks which aim to recover the location of data and/or code, in comparison to attacks on code understanding or data recovery. Location attacks are often associated with active attacks; therefore the previous argument applies to this observation as well.

- As expected, the majority of data recovery attacks target data obfuscation transformations and the majority of code understanding attacks target code obfuscation transformations. However, there are also some attacks which target both types of transformations, when they are used in combination with each other. Nevertheless, much more research is needed for investigating the strength of combinations of multiple software obfuscation transformations.

- Most attacks on obfuscated code are dynamic, because the majority of the obfuscation transformations presented in this paper aim to break static analysis techniques. Similarly, most attacks are semantic because most dynamic attacks are also semantic. However, note that there are also exceptions. Nevertheless, there are relatively few works about static semantic and dynamic syntactic attacks, which indicates a topic that needs further exploration.

- Obfuscation transformations which have few or no attacks against them are not an indication that they are more secure. For instance, in the case of *dataflow flattening*, the most probable reason why there are no attacks against it is because it is too expensive to implement this obfuscation transformation in most practical scenarios. One of the strongest obfuscation transformation categories against MATE attackers are dynamic transformations, however, these transformations generally require making executable code pages in process memory also writable, which opens the door to remote code injection attacks [110]. Since most software today is connected to the Internet, the risk of a remote attacker performing code injection in the software of an end-user is higher than the risk of that end-user being a MATE attacker. Therefore, such self-modifying code techniques are avoided by commercial software developers.

- Most attacks published in literature are dynamic, because even if code is statically encrypted/encoded, it must be decrypted/decoded during execution. This is the "Achilles heel" of obfuscation, i.e., code must still be executable, no matter how intricate the code is obfuscated when inspected statically. To prevent dynamic MATE attacks in practice obfuscation is used in combination with antidynamic–analysis techniques such as antidebugging [111] and antiemulation [112]. Such methods aim to prevent dynamic analysis attacks, i.e., they force attackers to take a static analysis approach, because obfuscation is assumed to be harder to break using static attacks.

• In some contexts, the MATE attacker is not forced to perform static analysis due to antidynamic-analysis techniques, but due to resource constraints. For instance, some malware analysis systems have to analyze millions of instances per day [113]. Performing dynamic analysis (e.g., execution trance analysis) is in general more costly than static analysis (e.g., code pattern recognition). Therefore, benign MATE attackers (i.e., malware analysts) choose to perform static analysis attacks as a way of quickly filtering previously analyzed malware instances, which employ simple obfuscation transformations such as *instruction reordering*. Generally, any software which is detected as highly obfuscated and not seen before is subjected to a more thorough (dynamic) analysis.

Given the publications in the field of software obfuscation presented in this chapter, we can see that the field of software protection research is divided into two camps:

1. *The software protection camp*, which aims to create defenses against malicious MATE attackers who want steal intellectual property and bypass license or integrity checks. The main disadvantage of this camp is that it is subject to significant performance constraints, because commercial software developers generally do not want to compromise the speed or responsiveness of their products by adding security.

2. *The malware analysis camp*, which are benign MATE attackers, who aim to develop attacks against obfuscation techniques employed by malware. The main disadvantage of this camp is again related to performance constraints of their attacks, which should be automatic and scalable to millions of programs per day.

These constraints are due to the high competitiveness on the commercial software markets. In today's software market the end-user experience is one of the main selling points of software and this implies having transparent but effective security solutions.

Probably the main beneficiaries of the publications from this field are malicious MATE attackers and malware developers (shown in the top–right and bottom–left of Fig. 4) who can borrow any techniques proposed by the second and first camps, respectively, because they do not impose such high performance constraints on themselves, as those imposed on the two camps

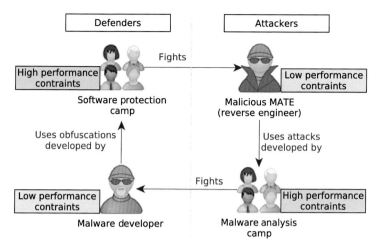

Fig. 4 Overview of obfuscation related research camps and their connection via malicious entities.

of researchers. For instance, a malware developer is likely much more interested in the security added by combining multiple obfuscation transformations, because even if the performance of the malware is decreased by three orders of magnitude, for the malware developer it is crucial that malware analysis engines have a hard-time disarming and reverse engineering the malware. This is very different for game developers, who must have highly responsive graphical user interfaces. Similarly to malware developers, malicious reverse engineers may afford to leave their laptops to run an attack for "a few more hours/days," because the return on investment (e.g., a free, cracked game) is likely higher than the power bill.

Given this observation, one may jump to the conclusion that it may be beneficial for benign software developers and antimalware developers, if less research in this field was published, because then malicious MATE attackers and malware developers would not have as many techniques and tools at their disposal. This would not stop malicious parties from developing their own techniques, but it would keep many benign parties "in the dark" about advances in this field. This was the case in the 1980s, before this research community was born and malicious parties were already using software obfuscation to protect malware and MATE attacks to break licenses. Therefore, having such a two camp research community, where each side challenges the weaknesses of the other side leads to continuous improvements, which accelerate progress and practical applicability of such research.

6. RELATED WORK

As indicated in Fig. 1, there are several ways in which one could protect against MATE attacks. Since in this chapter we focus on technical protection via obfuscation, in this section we present the other types of protections. This is followed by related work from the field of cryptographic obfuscation and by other surveys of software obfuscation.

6.1 Encryption via Trusted Hardware

Software protection via encryption is usually enabled by trusted hardware, also called *trusted computing*. Intel has released a hardware based technology [114], known as *Software Guard eXtension* (SGX), which enables software developers to protect the confidentiality of their applications' code via protected execution areas called *enclaves*. Dewan et al. [115] also use a trusted hypervisor to protect the sensitive memory of programs against unauthorized access by leveraging trusted hardware. Feng et al. [116] propose performing randomly timed stealthy measurements which can be validated locally, using Intel's Active Management Technology [117]. These approaches provide high security guarantees. However, they require trusted hardware to be available and the installation of a hypervisor. Software developers of popular software (e.g., web browsers) generally do not want to restrict their user base by imposing such requirements.

6.2 Server-Side Execution

Tamper protection via communication with trusted servers is employed in massive multiplayer online games (MMOGs) to detect cheating. Anticheat software such as PunkBuster [118], Valve Anti-Cheat (VAC) [119], Fides [120], and Warden [121] perform client-side computation, which are validated by a trusted server.

Martignoni et al. [122] and Seshandri et al. [123] propose establishing a *trusted computing base* to achieve verifiable code execution on a remote untrusted system. The trusted computing base in the two methods is established using a verification function. The verification function is composed of three components: (i) a checksum function, (ii) a send function, and (iii) a checksum function. However, the main difference between the two methods is the checksum function. In the work of Martignoni et al. [122] generates a new checksum function each time and sends it encrypted to the untrusted system. In the work of Seshandri et al. [123], the checksum function is known a priori and the challenge issued by the dispatcher consists in a seed that initializes this function. Since the remote component in both methods knows precisely in which execution environment the function must be executed and knows the hardware characteristics of the untrusted system, it can compute the expected checksum value and can estimate the amount of time that will be required by the untrusted system to decrypt and execute the function, and to send back the result. Since Intel x86 architecture, the architecture for which the approach of Seshandri et al. [123], was developed, is full of subtle details, researchers have found ways to circumvent the remote component. Also, a limitation of the approach of Martignoni et al. [122], is the impossibility to bootstrap a tamper-proof environment on simultaneous multi threading (SMT) or simultaneous multi processing (SMP) systems. On such systems, the attacker can use the secondary computational resources (parallel threads, for example) to forge checksums or to regain control of the execution after attestation.

Jakobsson and Johansson [124] propose a similar technique for detecting malware on mobile devices. Collberg et al. [107] propose tamper protection by pushing continuous updates from a trusted server to the client, which force the attacker to repeat reverse engineering and patching on each update. One disadvantage of these protection techniques is their dependence on external trusted servers. This dependence may cause a denial-of-service to end-users of the protected software applications which are also meant to be used offline, in case Internet connectivity is unavailable. Therefore in this chapter we focus on solutions that operate locally, i.e., without dependence on a trusted server.

6.3 Code Tamper-detection and Tamper-proofing

Code tamper-detection and tamper-proofing are complementary techniques to software diversity and obfuscation and they aim to detect, respectively, prevent unauthorized modifications of a program's code. However, these techniques are not generally stealthy, and hence they should be combined with diverse obfuscation in order to hamper MATE attackers from disabling such mechanisms.

Chang and Atallah [125] propose building a network of code regions, where a region can be a block of user code, a checker, or a responder. In this method checkers check each other in addition to user code by comparing a known checksum of piece of code to runtime checksum of the same code. If the checker has discovered that a region has been tampered with, a responder will replace the tampered region with a copy stored elsewhere. An important aspect of this algorithm is that it is not enough for checkers to check just the code, they must check each other as well. If checkers are not checked, they are easy to remove. Horne et al. [126] build on top of [125], by hiding the expected (precomputed checksum) value which is easy to identify, because of its randomness. The idea is to construct the checksum function such that unless the code has been tampered with, the function always checksums to a known number (usually zero). Having this function allows to insert an empty slot within the region under protection, and later give this slot a value that makes the region checksum to zero. The technique of Horne et al. [126] randomly places large numbers of checkers all over the program, but makes sure that every region of code is covered by multiple checkers. To minimize pattern-matching attacks, this method describes how to generate a large number of variants of lightweight checksum functions. The disadvantage of the *code introspection* approach used by both [125] and [126] is its stealthiness, because code that reads itself is seldom used for other purposes.

Chen et al. [127] propose an idea called *oblivious hashing*, where the checksum value is computed over the *execution trace* rather than the static code. The checksum can be computed by inserting instructions that monitor changes to variables and the execution of instructions. A problem with automating this technique is that it is hard to predict what side effects a function might have. It might destroy valuable global data or allocate extraneous dynamic memory that will never be properly freed. Furthermore, there is a problem with nondeterministic functions that depend on: user input, the time of day, network traffic, thread scheduling, and so on, because they do not have a fixed output that can be checked. This technique also faces the issue of automatically generating challenge data (test inputs) that most of the code of a function. The approach by Ibrahim and Banescu [128] implements a variant of oblivious hashing and therefore it also suffers from the same disadvantages. However, we address the last issue by proposing the use of symbolic execution in order to generate the challenge data.

Jacob et al. [129] propose an approach which depends on a unique property of the x86 instruction set architecture (ISA). The x86 ISA has a *variable instruction length* (1–15 bytes) with no alignment, this means instructions can start at any offset in the code. This results in the possibility of having overlapping or even nested instructions. So the basic idea will be that when a block is executed it computes a checksum of another block. For the purpose of protecting the code, we need two blocks to share instruction bytes. Having two blocks to share instruction bytes, can be achieved by interleaving the instructions and inserting jumps to maintain semantics. The advantage in this technique is that the code checksumming computations will not require reading the code explicitly. The disadvantage is mainly the performance overhead of the added instructions. Jacob et al.[129] report that the protected binary can be up to three times slower than the original. Even though this overhead may be acceptable in many circumstances, this technique cannot be applied to programs that execute on the Common Language Runtime such as programs written in C#.

Cappaert et al. [77] propose a technique that hinders both code analysis and tampering attacks simultaneously through code encryption. During run-time, code decryption can be done at a chosen granularity (e.g., one function at a time), when that part of code is needed at run-time. This technique performs integrity-checking of the code by using it to compute the keys for decryption and encryption. The basic idea is using the checksum value of a function, as the decryption key of another function. The advantage of this technique is that the encryption key is computed at run time, which means the key is not hard-coded in the binary and therefore hard to find through static analysis. The disadvantage of this technique is the run-time overhead as well as the its stealth.

6.4 Cryptographic Obfuscation

In addition to the practical software protection techniques presented so far, there is also an entire subfield of cryptography dedicated to obfuscation. The first formal study of obfuscation was published in 2001 by Barak et al. [108]. They propose that an ideal obfuscator should be able to take any program and transform it into a *virtual black box*, i.e., a MATE attacker would be able to interact with it in the same manner as with a program running on a remote server, however, the attacker would not be able to learn anything from the program in addition to what can be learned from its input–output behavior. Barak et al. [108] formally prove that such an obfuscator cannot exist for all programs. However, they do not exclude that such an obfuscator may exist for particular programs.

Over a decade later, Garg et al. [109] proposed a construction for *indistinguishability obfuscation*, a different obfuscation notion than black-box obfuscation, which guarantees that the obfuscations of two programs implementing the same functionality are computationally indistinguishable.

This was a major breakthrough in cryptography, since a few years earlier it was proven by Goldwasser and Rothblum [130] that indistinguishability obfuscation is the best possible type of obfuscation that can be achieved for all programs. Therefore, we are currently seeing a revival of interest in obfuscation from the cryptographic community, because the construction of Garg et al. [109] may be employed to construct functional encryption, public key encryption, digital signatures, etc. However, multiple works have signaled that the current constructions of indistinguishability obfuscation are still far from being practical, i.e., applicable to real-world software applications [131–133].

6.5 Other Surveys of Software Obfuscation

The seminal work of Collberg et al. [7] provides one of the first taxonomies of software obfuscation. Their work also provides an illustration of how different obfuscation transformations work, however, since this paper was published 20 years before this chapter, they do not include recently developed obfuscation transformations and attacks against them, as presented in this chapter. Mavrogiannopoulos et al. [11] provide a taxonomy of self-modifying code techniques and implementations, which we have compressed into one single transformation.

Larsen et al. [12] provide a survey of software diversification transformations, which are closely related to obfuscation, i.e., all software obfuscation transformations can be used to achieve diversity, but not vice-versa. However, as opposed to our work which focuses on MATE attacks, the work of Larsen et al. [12] focuses on remote attacks such as memory corruption, code injection, and code reuse.

Schrittwiesser et al. [13] recently published a survey complementary to this work. Their focus is on providing an overview of MATE attacks and how they affect existing obfuscation transformations. However, they do not provide many details or illustrations of how the different obfuscation transformations look like, as we do in this chapter. Their classification of obfuscation transformations and MATE attacks differ from the classification used in this chapter to some extent. Therefore, they do not touch on the same points we have discussed in Section 5. For instance, as opposed to our work, they do not discuss obfuscation transformations that target the readability of the source code, hence, they also do not cite MATE attacks applicable to such transformations. Regarding the MATE attack classification, we emphasize the difference between syntactic and semantic attacks and claim that this dimension of classification is orthogonal to the dynamics dimension. This means that syntactic attacks such as pattern recognition can also be applied to dynamically generated execution traces, which is often done by malware analysis engines. Moreover, we also propose using the *alternation* classification dimension, which differentiates between active and passive attacks. This dimension allowed us to identify the fact that there are far less active attacks published in the literature than passive attacks, which indicates a gap in this field. On the other hand, they propose a separation between automated attacks and human-assisted attacks. They claim human-assisted attacks are the hardest attacks to defeat using obfuscation.

7. CONCLUSION

This paper presents a tutorial on software obfuscation. This tutorial includes a state-of-the-art survey of obfuscation transformations and their classifications. Each obfuscation transformation is accompanied by an illustrative example (often in the form of code snippets), as well as an enumeration of existing associated attacks published in the literature. We discuss how the current landscape of software obfuscation research is split into two complementary camps, namely, software protection and malware analysis, and why this division is important for accelerating progress. An interesting observation is that malicious entities may have higher benefits from techniques proposed in the field of research, due to the fact that they do not have high performance constraints as benign entities (e.g., commercial software vendors) have. However, the only way to defend against malicious entities is to push these areas of research forward. One way to increase the performance constraints of malicious MATE attackers is by employing frequent software updates, which shrinks the window of opportunity of such attackers. Therefore, future research in the field of software obfuscation should also investigate obfuscation techniques which allow incremental updates.

Our outlook on the field of obfuscation is positive. We believe that even with advances in hardware-based software protection techniques such as Intel SGX [134], software obfuscation will still be used for applications where such hardware is not available (e.g., mobile devices), or where such hardware is too costly too incorporate (e.g., resource constrained embedded devices). Moreover, Intel SGX may be able to defend software assets against a software-based attacks, but MATE attackers may also resort to side-channel attacks [135] or reverse engineering attacks at the hardware level [136]. There is an entire field of research, parallel to software obfuscation, namely, *hardware obfuscation*, where the goal is to mangle the structure or layout of logical gates such that MATE attackers cannot steal the intellectual property directly from hardware, e.g., an application-specific integrated circuit (ASIC) [137]. Note that it is straightforward to map many of the obfuscation transformations presented in this chapter to the field of hardware obfuscation, and vice-versa. Furthermore, the field of cryptograpic obfuscation is popular nowadays due to the indistinguishability obfuscation candidate proposed by Garg el al. [109], which makes us optimistic in believing that obfuscation will still be a highly dynamic and innovative field of research in the following decade.

The main challenge in the field of software obfuscation is that there is no standard methodology or benchmarks for evaluating the strength of different obfuscation transformations or combinations thereof. The first steps in the direction of obfuscation evaluation have been made by Banescu et al. [31, 32, 138], who use automated attack effort as a way of comparing the strength of different obfuscation transformations. However, they mainly focus on attacks based on symbolic execution, hence more research is needed in this direction to cover other types of attacks. This observation is on par with the conclusions of other surveys [13]. We also note that more research is required for topics such as the development of active attacks, static semantic attacks, and self-modifying code techniques that do not increase the attack surface of the software they protect.

We envision that side-channel attacks against obfuscation will become increasingly popular. This premonition is due to the recent *differential computation analysis* (DCA) presented by Bos et al. [139]. DCA is the software counterpart of *differential power analysis* (DPA), which has been successfully applied to recover secret keys from ASICs, e.g., smart-cards with little effort and prerequisites [140]. Similar to DPA, DCA able to recover a symmetric cryptographic key from a white-box cryptographic cipher binary, in a matter of seconds, without needing to disassemble the binary or to know anything about its structure. However, these side-channel attacks also make some assumptions, which could be broken by cleaver obfuscation transformations. Therefore, we urge researchers to invest more resources in designing obfuscation transformations, which are able to block such side-channel attacks.

REFERENCES

[1] C. Collberg, J. Davidson, R. Giacobazzi, Y.X. Gu, A. Herzberg, F. Wang, Toward digital asset protection, IEEE Intell. Syst. 26 (6) (2011) 8–13.

[2] D. Dolev, A.C. Yao, On the security of public key protocols, in: Proceedings of the 22nd Annual Symposium on Foundations of Computer ScienceIEEE Computer Society, Washington, DC, USA, 1981, pp. 350–357, https://doi.org/10.1109/SFCS. 1981.32.

[3] A. Main, P.C. van Oorschot, Software Protection and Application Security: Understanding the Battleground. 2003, http://citeseerx.ist.psu.edu/viewdoc/download? doi=10.1.1.216.3498&rep=rep1&type=pdf. International Course on State of the Art and Evolution of Computer Security and Industrial Cryptography.

[4] A. Shamir, N. Van Someren, Playing 'hide and seek' with stored keys, in: Financial Cryptography, Springer-Verlag, London, UK, 1999, pp. 118–124. https://dl.acm. org/citation.cfm?id=728464.

[5] B. Wyseur, White-box cryptography (Ph.D. thesis), Katholieke Universiteit Leuven, Leuven-Heverlee, 2009, https://repository.libis.kuleuven.be/dspace/handle/1979/ 2098.

[6] F.B. Cohen, Operating system protection through program evolution, Comput. Secur. 12 (6) (1993) 565–584, ISSN: 0167-4048, https://doi.org/10.1016/0167-4048 (93)90054-9. http://www.sciencedirect.com/science/article/pii/0167404893900549.

[7] C. Collberg, C. Thomborson, D. Low, A Taxonomy of Obfuscating Transformations, Department of Computer Science, The University of Auckland, New Zealand, 1997, https://researchspace.auckland.ac.nz/handle/2292/3491.

[8] U. Congress, Digital Millennium Copyright Act, Public Law 105 (304) (1998) 112.

[9] A. Kur, M. Planck, T. Dreier, European intellectual property law: text, cases and materials, Edward Elgar Publishing, Cheltenham, 2013.

[10] L. Badger, L. D'Anna, D. Kilpatrick, B. Matt, A. Reisse, T. Van Vleck, Self-Protecting Mobile Agents Obfuscation Techniques Evaluation, Network Associates Laboratories, Report 01036, 2002.

[11] N. Mavrogiannopoulos, N. Kisserli, B. Preneel, A taxonomy of self-modifying code for obfuscation, Comput. Secur. 30 (8) (2011) 679–691, ISSN: 01674048, https://doi. org/10.1016/j.cose.2011.08.007. http://linkinghub.elsevier.com/retrieve/pii/S0167 404811001076.

[12] P. Larsen, A. Homescu, S. Brunthaler, M. Franz, SoK: automated software diversity, in: IEEE Symposium on Security and Privacy, IEEE, 2014, pp. 276–291.

[13] S. Schrittwieser, S. Katzenbeisser, J. Kinder, G. Merzdovnik, E. Weippl, Protecting software through obfuscation: can it keep pace with progress in code analysis? ACM Comput. Surv. 49 (1) (2016) 4.

[14] P. Team, PaX Non-Executable Pages Design & Implementation, 2003. Available from: http://pax.grsecurity.net.

[15] C. Collberg, J. Nagra, Surreptitious Software, Addison-Wesley Professional, Upper Saddle River, NJ, 2010.

[16] A. Akhunzada, M. Sookhak, N.B. Anuar, A. Gani, E. Ahmed, M. Shiraz, S. Furnell, A. Hayat, M.K. Khan, Man-at-the-end attacks: analysis, taxonomy, human aspects, motivation and future directions, J. Netw. Comput. Appl. 48 (2015) 44–57.

[17] C. Basile, S. Di Carlo, T. Herlea, V. Business, J. Nagra, B. Wyseur, Towards a formal model for software tamper resistance, in: Second International Workshop on Remote Entrusting (ReTtust 2009), vol. 16.

[18] C. Collberg, The Tigress C Diversifier/Obfuscator. http://tigress.cs.arizona.edu/, n.d. (accessed 29.11.16).

[19] I. Skochinsky, IDA F.L.I.R.T. technology: in-depth, https://www.hex-rays.com/products/ida/tech/flirt/in_depth.shtml, 2013 (accessed: 03.03.17).

[20] S. Banescu, T. Wuchner, A. Salem, M. Guggenmos, A. Pretschner, et al., A framework for empirical evaluation of malware detection resilience against behavior obfuscation, in: 2015 10th International Conference on Malicious and Unwanted Software (MALWARE), IEEE, 2015, pp. 40–47.

[21] R.W. Floyd, Assigning meanings to programs, in: Mathematical Aspects of Computer Science, vol. 19, 1967, pp. 19–32.

[22] P. Cousot, R. Cousot, Abstract interpretation: a unified lattice model for static analysis of programs by construction or approximation of fixpoints, in: Proceedings of the 4th ACM SIGACT-SIGPLAN Symposium on Principles of Programming Languages, ACM, 1977, pp. 238–252.

[23] M. Sutton, A. Greene, P. Amini, Fuzzing: Brute Force Vulnerability Discovery, Pearson Education, Crowfordsville, IN, USA, 2007.

[24] W. Landi, Undecidability of static analysis, ACM Lett. Program. Lang. Syst. 1 (4) (1992) 323–337.

[25] G. Ramalingam, The undecidability of aliasing, ACM Trans. Program. Lang. Syst. 16 (5) (1994) 1467–1471.

[26] S. Horwitz, Precise flow-insensitive may-alias analysis is NP-hard, ACM Trans. Program. Lang. Syst. 19 (1) (1997) 1–6.

[27] C. Wang, J. Davidson, J. Hill, J. Knight, Protection of software-based survivability mechanisms, in: International Conference on Dependable Systems and Networks (DSN 2001), 2001, pp. 193–202, https://doi.org/10.1109/DSN.2001.941405.

[28] J. Palsberg, S. Krishnaswamy, M. Kwon, D. Ma, Q. Shao, Y. Zhang, Experience with software watermarking, in: Computer Security Applications, 2000. ACSAC'00. 16th Annual Conference, IEEE, 2000, pp. 308–316.

[29] A. Majumdar, C. Thomborson, Manufacturing opaque predicates in distributed systems for code obfuscation, in: Proceedings of the 29th Australasian Computer Science Conference, vol. 48, Australian Computer Society, Inc., 2006, pp. 187–196.

[30] M. Dalla Preda, M. Madou, K. De Bosschere, R. Giacobazzi, Opaque predicates detection by abstract interpretation, in: M. Johnson, V. Vene (Eds.), Algebraic Methodology and Software Technology, Lecture Notes in Computer Science, vol. 4019, Springer, Berlin, Heidelberg, ISBN 978-3-540-35633-2, 2006, pp. 81–95, https://doi.org/10.1007/11784180_9.

[31] S. Banescu, C. Collberg, V. Ganesh, Z. Newsham, A. Pretschner, Code obfuscation against symbolic execution attacks, in: Proc. of 2016 Annual Computer Security Applications Conference, ACM, 2016.

[32] S. Banescu, M. Ochoa, A. Pretschner, A framework for measuring software obfuscation resilience against automated attacks, in: 2015 IEEE/ACM 1st International Workshop on Software Protection (SPRO), IEEE, 2015, pp. 45–51.

[33] A. Salem, S. Banescu, Metadata recovery from obfuscated programs using machine learning, in: Proceedings of the 6th Software Security, Protection and Reverse Engineering Workshop, ACM, 2016, p. 8.

[34] D.F. Bacon, S.L. Graham, O.J. Sharp, Compiler transformations for high-performance computing, ACM Comput. Surv. 26 (4) (1994) 345–420.

[35] Y. Zhou, A. Main, Y.X. Gu, H. Johnson, Information hiding in software with mixed Boolean-arithmetic transforms, in: International Workshop on Information Security Applications, Springer, 2007, pp. 61–75.

[36] A. Guinet, N. Eyrolles, M. Videau, Arybo: Manipulation, canonicalization and identification of mixed boolean-arithmetic symbolic expressions, Proceedings of GreHack 2016, Grenoble, France, 2016.

[37] S. Chow, P. Eisen, H. Johnson, P.C. van Oorschot, A white-box DES implementation for DRM applications, in: J. Feigenbaum (Ed.), Digital Rights Management. DRM 2002, Lecture Notes in Computer Science, vol. 2696, Springer, Berlin, Heidelberg, 2003.

[38] S. Chow, P. Eisen, H. Johnson, P.C.V. Oorschot, White-box cryptography and an AES implementation, in: Selected Areas in Cryptography, no. 2595 in LNCS, Springer, Berlin, Heidelberg, ISBN 978-3-540-00622-0, 978-3-540-36492-4, 2003, pp. 250–270.

[39] P.U.B.S. FIPS, 197: Advanced Encryption Standard (AES), 2001, National Institute of Standards and Technology.

[40] Y. Xiao, X. Lai, A secure implementation of white-box AES, in: 2nd International Conference on Computer Science and its Applications, 2009. CSA '092009, pp. 1–6, https://doi.org/10.1109/CSA.2009.5404239.

[41] J. Bringer, H. Chabanne, E. Dottax, White box cryptography: another attempt, 2006, Available from (accessed 22.06.11). http://citeseerx.ist.psu.edu/viewdoc/download?doi=10.1.1.99.661&rep=rep1&type=pdf.

[42] M. Karroumi, Protecting White-Box AES with Dual Ciphers, in: K.-H. Rhee, D. Nyang (Eds.), Information Security and Cryptology—ICISC 2010, no. 6829 in Lecture Notes in Computer Science, Springer, Berlin, Heidelberg, ISBN 978-3-642-24208-3, 978-3-642-24209-0, 2011, pp. 278–291, http://link.springer.com/chapter/10.1007/978-3-642-24209-0_19.

[43] B. Wyseur, W. Michiels, P. Gorissen, B. Preneel, Cryptanalysis of white-box DES implementations with arbitrary external encodings, in: Selected Areas in Cryptography, no. 4876 in LNCS, Springer, Berlin, Heidelberg, ISBN 978-3-540-77359-7, 978-3-540-77360-3, 2007, pp. 264–277, http://link.springer.com/chapter/10.1007/978-3-540-77360-3_17.

[44] O. Billet, H. Gilbert, C.E. Chatbi, Cryptanalysis of a White Box AES Implementation, Springer-Verlag, Waterloo, Canada, 2005, pp. 227–240. ISBN 3-540-24327-5, 978-3-540-24327-4.

[45] Y. De Mulder, P. Roelse, B. Preneel, Cryptanalysis of the Xiao-Lai white-box AES implementation, in: L.R. Knudsen, H. Wu (Eds.), Selected Areas in CryptographySpringer, Berlin, Heidelberg, 2013, pp. 34–49. http://link.springer.com/chapter/10.1007/978-3-642-35999-6_3.

[46] N.T. Courtois, J. Pieprzyk, Cryptanalysis of block ciphers with overdefined systems of equations, in: Y. Zheng (Ed.), International Conference on the Theory and Application of Cryptology and Information SecuritySpringer, Berlin, Heidelberg, 2002, pp. 267–287.

[47] S. Banescu, A. Pretschner, D. Battré, S. Cazzulani, R. Shield, G. Thompson, Software-based protection against changeware, in: Proceedings of the Conference on Data and Application Security and Privacy, CODASPY '15, 2015, pp. 231–242.

[48] X. Wang, D. Feng, X. Lai, H. Yu, Collisions for hash functions MD4, MD5, HAVAL-128 and RIPEMD, IACR Cryptology ePrint Archive. Report 2004:199, (2004).

[49] M. Stevens, E. Bursztein, P. Karpman, A. Albertini, Y. Markov, The First Collision for Full SHA-1, 2017, https://shattered.it/static/shattered.pdf.

[50] A. Slowinska, I. Haller, A. Bacs, S. Baranga, H. Bos, Data structure archaeology: scrape away the dirt and glue back the pieces! in: International Conference on Detection of Intrusions and Malware, and Vulnerability AssessmentSpringer, Cham, 2014, pp. 1–20.

[51] A. Viticchié, L. Regano, M. Torchiano, C. Basile, M. Ceccato, P. Tonella, R. Tiella, Assessment of source code obfuscation techniques. in: 2016 IEEE 16th International Working Conference on Source Code Analysis and Manipulation (SCAM), Raleigh, NC, 2016, pp. 11–20, https://doi.org/10.1109/SCAM.2016.17.

[52] A. Slowinska, T. Stancescu, H. Bos, Howard: a dynamic excavator for reverse engineering data structures, in: NDSS, 2011.

[53] V. Pappas, M. Polychronakis, A.D. Keromytis, Smashing the gadgets: hindering return-oriented programming using in-place code randomization, in: IEEE Symposium on Security and Privacy (SP), IEEE, ISBN 978-1-4673-1244-8, 978-0-7695-4681-0. 2012, pp. 601–615, https://doi.org/10.1109/SP.2012.41. http://ieeexplore.ieee.org/lpdocs/epic03/wrapper.htm?arnumber=6234439.

[54] K. Griffin, S. Schneider, X. Hu, T.-C. Chiueh, Automatic generation of string signatures for malware detection, in: International Workshop on Recent Advances in Intrusion Detection, Springer, 2009, pp. 101–120.

[55] B. Anckaert, M.H. Jakubowski, R. Venkatesan, C.W. Saw, Runtime protection via dataflow flattening, in: 2009 Third International Conference on Emerging Security Information, Systems and Technologies, IEEE, 2009, pp. 242–248.

[56] O. Goldreich, R. Ostrovsky, Software protection and simulation on oblivious RAMs, J. ACM 43 (3) (1996) 431–473.

[57] S. Forrest, A. Somayaji, D.H. Ackley, Building diverse computer systems, in: The Sixth Workshop on Hot Topics in Operating Systems, IEEE, 1997, pp. 67–72.

[58] R. Fedler, S. Banescu, A. Pretschner, ISA2R: improving software attack and analysis resilience via compiler-level software diversity, in: International Conference on Computer Safety, Reliability, and Security, Springer, 2015, pp. 362–371.

[59] R. Strackx, Y. Younan, P. Philippaerts, F. Piessens, S. Lachmund, T. Walter, Breaking the memory secrecy assumption, in: Proceedings of the Second European Workshop on System Security, ACM, 2009, pp. 1–8.

[60] C. Cadar, P. Akritidis, M. Costa, J.-P. Martin, M. Castro, Data Randomization, Tech. rep. TR-2008-120, Microsoft Research, 2008.

[61] S. Bhatkar, R. Sekar, Data space randomization, in: D. Zamboni (Ed.), Detection of Intrusions and Malware, and Vulnerability Assessment, no. 5137 in Lecture Notes in Computer Science, Springer, Berlin, Heidelberg, 2008, pp. 1–22.

[62] C. Cowan, S. Beattie, J. Johansen, P. Wagle, Pointguard TM: protecting pointers from buffer overflow vulnerabilities, in: Proceedings of the 12th Conference on USENIX Security Symposium, vol. 12, 2003, pp. 91–104.

[63] L.O. Andersen, Program analysis and specialization for the C programming language (Ph.D. thesis), University of Copenhagen, 1994.

[64] F. Zhang, H. Huang, S. Zhu, D. Wu, P. Liu, ViewDroid: towards obfuscation-resilient mobile application repackaging detection, in: Proceedings of the 2014 ACM Conference on Security and Privacy in Wireless & Mobile Networks, ACM, 2014, pp. 25–36.

[65] M. Jacob, M.H. Jakubowski, P. Naldurg, C.W. Saw, R. Venkatesan, The superdiversifier: peephole individualization for software protection, in: K. Matsuura, E. Fujisaki (Eds.), Advances in Information and Computer Security. IWSEC 2008, Lecture Notes in Computer Science, vol. 5312, Springer, Berlin, Heidelberg, 2008.

[66] R. El-Khalil, A.D. Keromytis, Hydan: hiding information in program binaries, in: International Conference on Information and Communications Security. Springer, 2004, pp. 187–199.

[67] B. De Sutter, B. Anckaert, J. Geiregat, D. Chanet, K. De Bosschere, Instruction set limitation in support of software diversity, in: International Conference on Information Security and Cryptology, Springer, 2008, pp. 152–165.

[68] J. Mason, S. Small, F. Monrose, G. MacManus, English shellcode, in: Proceedings of the 16th ACM Conference on Computer and Communications Security, ACM, 2009, pp. 524–533.

[69] H.S. Warren, Hacker's Delight, Addison-Wesley Longman Publishing Co., Inc., Boston, MA, 2002, ISBN 0201914654.

[70] N. Eyrolles, L. Goubin, M. Videau, Defeating MBA-based Obfuscation, in: Proceedings of the 2016 ACM Workshop on Software PROtection, ACM, 2016, pp. 27–38.

[71] B. Yadegari, B. Johannesmeyer, B. Whitely, S. Debray, A generic approach to automatic deobfuscation of executable code, in: 2015 IEEE Symposium on Security and Privacy (SP), IEEE, 2015, pp. 674–691.

[72] J. Knoop, O. Rüthing, B. Steffen, Partial dead code elimination, ACM SIGPLAN Notices 29 (6) (1994) 147–158.

[73] R. Gupta, D. Benson, J.Z. Fang, Path profile guided partial dead code elimination using predication, in: Proceedings International Conference on Parallel Architectures and Compilation Techniques, IEEE, 1997, pp. 102–113.

[74] T. Wüchner, M. Ochoa, A. Pretschner, Robust and effective malware detection through quantitative data flow graph metrics, in: International Conference on Detection of Intrusions and Malware, and Vulnerability Assessment, Springer, 2015, pp. 98–118.

[75] B. Yadegari, J. Stephens, S. Debray, Analysis of exception-based control transfers, in: Proceedings of the Seventh ACM on Conference on Data and Application Security and Privacy, ACM, 2017, pp. 205–216.

[76] Vrba, Z. cryptexec: Next-generation runtime binary encryption using on-demand function extraction. Phrak 0x0b (0x3f). #0x0d of 0x14.[online] (2003), http://www.phrack.org/archives/issues/63/13.txt.

[77] J. Cappaert, B. Preneel, B. Anckaert, M. Madou, K. De Bosschere, Towards tamper resistant code encryption: practice and experience, in: Information Security Practice and Experience, Springer, 2008, pp. 86–100.

[78] M. Oberhumer, L. Molnár J.F. Reiser, UPX: The Ultimate Packer for eXecutables, 2004, Published online at http://upx.sourceforge.net/.

[79] Y. Tang, S. Chen, An automated signature-based approach against polymorphic internet worms, IEEE Trans. Parallel Distrib. Syst. 18 (7) (2007).

[80] J. Qiu, B. Yadegari, B. Johannesmeyer, S. Debray, X. Su, Identifying and understanding self-checksumming defenses in software, in: Proceedings of the 5th ACM Conference on Data and Application Security and Privacy, ACM, 2015, pp. 207–218.

[81] M. Madou, B. Anckaert, P. Moseley, S. Debray, B. De Sutter, K. De Bosschere, Software protection through dynamic code mutation, in: Information Security Applications, Springer, Berlin, Heidelberg, 2006, pp. 194–206.

[82] Y. Kanzaki, A. Monden, M. Nakamura, K.-I. Matsumoto, Exploiting self-modification mechanism for program protection, in: Proceedings. 27th Annual International Computer Software and Applications Conference. COMPSAC, 2003, pp. 170–179.

[83] A. Nguyen-Tuong, A. Wang, J.D. Hiser, J.C. Knight, J.W. Davidson, On the effectiveness of the metamorphic shield, in: Proceedings of the Fourth European Conference on Software Architecture: Companion Volume, ACM, 2010, pp. 170–174.

[84] E.G. Barrantes, D.H. Ackley, T.S. Palmer, D. Stefanovic, D.D. Zovi, Randomized instruction set emulation to disrupt binary code injection attacks, in: Proceedings of the 10th ACM Conference on Computer and Communications Security, ACM, 2003, pp. 281–289.

[85] J. Mason, K. Watkins, J. Eisner, A. Stubblefield, A natural language approach to automated cryptanalysis of two-time pads, in: Proceedings of the 13th ACM Conference on Computer and Communications Security, ACM, 2006, pp. 235–244.

[86] M. Dalla Preda, R. Giacobbazzi, S. Debray, K. Coogan, G.M. Townsend, Modelling metamorphism by abstract interpretation, in: International Static Analysis Symposium, Springer, 2010, pp. 218–235.

[87] S. Banescu, C. Lucaci, B. Krämer, A. Pretschner, VOT4CS: a virtualization obfuscation tool for C, in: Proceedings of the 2016 ACM Workshop on Software PROtection, ACM, 2016, pp. 39–49.

[88] J. Kinder, Towards static analysis of virtualization-obfuscated binaries, in: 19th Working Conference on Reverse Engineering (WCRE), ISSN 1095-1350, 2012, pp. 61–70, https://doi.org/10.1109/WCRE.2012.16.

[89] R. Rolles, Unpacking Virtualization Obfuscators, in: Proceedings of the 3rd USENIX Conference on Offensive Technologies, Montreal, Canada, WOOT'09, USENIX Association, Berkeley, CA, USA, 2009, pp. 1. http://dl.acm.org/citation.cfm?id=1855876.1855877.

[90] S. Chow, Y. Gu, H. Johnson, V.A. Zakharov, An approach to the obfuscation of control-flow of sequential computer programs, in: International Conference on Information Security, Springer, 2001, pp. 144–155.

[91] S. Udupa, S. Debray, M. Madou, Deobfuscation: reverse engineering obfuscated code. in: 12th Working Conference on Reverse Engineering, ISSN 1095-1350, 2005. https://doi.org/10.1109/WCRE.2005.13.

[92] C. Linn, S. Debray, Obfuscation of executable code to improve resistance to static disassembly, in: Proceedings of the 10th ACM Conference on Computer and Communications Security, ACM, 2003, pp. 290–299.

[93] S. Schrittwieser, S. Katzenbeisser, Code obfuscation against static and dynamic reverse engineering, in: T. Filler, T. Pevný, S. Craver, A. Ker (Eds.), Information Hiding. IH 2011, Lecture Notes in Computer Science, vol. 6958, Springer, Berlin, Heidelberg, 2011.

[94] H. Shacham, The geometry of innocent flesh on the bone: return-into-libc without function calls (on the x86), in: Proceedings of the 14th ACM Conference on Computer and Communications Security, 2007, pp. 552–561. http://dl.acm.org/citation.cfm?id=1315313.

[95] C. Kruegel, W. Robertson, F. Valeur, G. Vigna, Static disassembly of obfuscated binaries, in: USENIX Security Symposium13 2004, p. 18.

[96] S. Rugaber, K. Stirewalt, L.M. Wills, The interleaving problem in program understanding, in: Proceedings of 2nd Working Conference on Reverse Engineering, 1995, IEEE, 1995, pp. 166–175.

[97] Stunnix, C/C++ Obfuscator. http://stunnix.com/prod/cxxo/ (accessed 03.03.17).

[98] PreEmptiveSolutions, DashO: Java & Android Obfuscator & Runtime Protection. https://www.preemptive.com/products/dasho (accessed 03.03.17).

[99] PreEmptiveSolutions, Dotfuscator: .NET App Self Protection and Obfuscation. https://www.preemptive.com/products/dotfuscator (accessed 03.03.17).

[100] SemanticDesigns, Thicket Family of Source Code Obfuscators. http://www.semanticdesigns.com/Products/Obfuscators/ (accessed 03.03.17).

[101] GuardSquare, ProGuard: The Open Source Optimizer for Java Bytecode. https://www.guardsquare.com/en/proguard (accessed 03.03.17).

[102] yWorks, yGuard Java Bytecode Obfuscator and Shrinker. https://www.yworks.com/products/yguard (accessed 03.03.17).

[103] B. Bichsel, V. Raychev, P. Tsankov, M. Vechev, Statistical deobfuscation of android applications, in: Proceedings of the 2016 ACM SIGSAC Conference on Computer and Communications Security, ACM, 2016, pp. 343–355.

[104] M. Ceccato, M.D. Penta, J. Falcarin, F. Ricca, M. Torchiano, P. Tonella, A family of experiments to assess the effectiveness and efficiency of source code obfuscation techniques. Empir. Softw. Eng. 19 (4) (2013) 1040–1074. ISSN: 1382-3256, 1573-7616, https://doi.org/10.1007/s10664-013-9248-x.

[105] L.M. Wills, Automated program recognition: a feasibility demonstration, Artif. Intell. 45 (1–2) (1990) 113–171. ISSN: 0004-3702, https://doi.org/10.1016/0004-3702(90)90039-3.

[106] C. Foket, B. De Sutter, B. Coppens, K. De Bosschere, A novel obfuscation: class hierarchy flattening, in: J. Garcia-Alfaro, F. Cuppens, N. Cuppens-Boulahia, A. Miri, N. Tawbi (Eds.), Foundations and Practice of Security. FPS 2012, Lecture Notes in Computer Science, vol. 7743, Springer, Berlin, Heidelberg, 2013.

[107] C. Collberg, S. Martin, J. Myers, J. Nagra, Distributed application tamper detection via continuous software updates, in: Proceedings of the 28th Annual Computer Security Applications Conference, Orlando, Florida, USA, ACSAC '12, ACM, New York, NY, USA, ISBN 978-1-4503-1312-4, 2012, pp. 319–328, https://doi.org/10.1145/2420950.2420997. http://doi.acm.org/10.1145/2420950.2420997.

[108] B. Barak, O. Goldreich, R. Impagliazzo, S. Rudich, A. Sahai, S. Vadhan, K. Yang, On the (im) possibility of obfuscating programs, in: Advances in Cryptology CRYPTO 2001, Springer, 2001, pp. 1–18.

[109] S. Garg, C. Gentry, S. Halevi, M. Raykova, A. Sahai, B. Waters, Candidate indistinguishability obfuscation and functional encryption for all circuits. in: Proc. of the 54th Annual Symp. on Foundations of Computer Science, 2013, pp. 40–49, https://doi.org/10.1109/FOCS.2013.13.

[110] Molnar, I., Method and Apparatus for Creating an Execution Shield. U.S. Patent Application No. 10/420,253.

[111] B. Abrath, B. Coppens, S. Volckaert, J. Wijnant, B. De Sutter, Tightly-coupled self-debugging software protection, in: Proceedings of the 6th Workshop on Software Security, Protection, and Reverse Engineering, ACM, 2016, p. 7.

[112] P. Ferrie, Attacks on more virtual machine emulators, Symantec Technology Exchange 55 (2007).

[113] McAfee, McAfee Labs Threats Report, tech. rep. 2016, http://www.mcafee.com/us/resources/reports/rp-quarterly-threats-mar-2016.pdf.

[114] I. Anati, S. Gueron, S. Johnson, V. Scarlata, Innovative technology for CPU based attestation and sealing, in: Proceedings of the 2nd International Workshop on Hardware and Architectural Support for Security and Privacy, vol. 13, 2013.

[115] P. Dewan, D. Durham, H. Khosravi, M. Long, G. Nagabhushan, A hypervisor-based system for protecting software runtime memory and persistent storage, in: Proceedings of the 2008 Spring Simulation Multiconference, Society for Computer Simulation International, 2008, pp. 828–835.

[116] W.-C. Feng, E. Kaiser, T. Schluessler, Stealth measurements for cheat detection in on-line games, in: Proceedings of the 7th ACM SIGCOMM Workshop on Network and System Support for Games, 2008, pp. 15–20.

[117] Intel, Intel Active Management Technology—Query, Restore, Upgrade, and Protect Devices Remotely (Online), 2016. http://www.intel.com/content/www/us/en/architecture-and-technology/intel-active-management-technology.html (accessed 20.09.16).

[118] Evenbalance, PunkBuster—Online Countermeasures (Online), 2015. http://www.evenbalance.com/pbsetup.php (accessed 20.09.16).

[119] Valve, Valve Anti-Cheat System (VAC) (Online), 2015. https://support.steampowered.com/kb_article.php?p_faqid=370 (accessed 20.09.16).

[120] E. Kaiser, W.-C. Feng, T. Schluessler, Fides: remote anomaly-based cheat detection using client emulation, in: Proceedings of the 16th ACM Conference on Computer and Communications Security, ACM, New York, NY, USA, 2009, pp. 269–279. ISBN 978-1-60558-894-0.

[121] G. Hoglund, Hacking World of Warcraft: An Exercise in Advanced Rootkit Design, 2006. Black Hat.

[122] L. Martignoni, R. Paleari, D. Bruschi, Conqueror: tamper-proof code execution on legacy systems, in: International Conference on Detection of Intrusions and Malware, and Vulnerability Assessment, Springer, 2010, pp. 21–40.

[123] A. Seshadri, M. Luk, E. Shi, A. Perrig, L. van Doorn, P. Khosla, Pioneer: verifying code integrity and enforcing untampered code execution on legacy systems, ACM SIGOPS Operating Systems Review, 39 (5) (2005) 1–16.

[124] M. Jakobsson, K.-A. Johansson, Retroactive detection of malware with applications to mobile platforms, in: Proceedings of the 5th USENIX Conference on Hot Topics in Security, USENIX Association, Washinton, DC, 2010, pp. 1–13.

[125] H. Chang, M.J. Atallah, Protecting software code by guards, in: Security and Privacy in Digital Rights Management, Springer, 2001, pp. 160–175.

[126] B. Horne, L. Matheson, C. Sheehan, R.E. Tarjan, Dynamic self-checking techniques for improved tamper resistance, in: T. Sander (Ed.), Security and Privacy in Digital Rights Management. DRM 2001, Lecture Notes in Computer Science, vol. 2320, Springer, Berlin, Heidelberg, 2002.

[127] Y. Chen, R. Venkatesan, M. Cary, R. Pang, S. Sinha, M.H. Jakubowski, Oblivious hashing: a stealthy software integrity verification primitive, in: International Workshop on Information Hiding, Springer, 2002, pp. 400–414.

[128] A. Ibrahim, S. Banescu, StIns4CS: a state inspection tool for C, in: Proceedings of the 2016 ACM Workshop on Software PROtection, ACM, 2016, pp. 61–71.

[129] M. Jacob, M.H. Jakubowski, R. Venkatesan, Towards integral binary execution: implementing oblivious hashing using overlapped instruction encodings, in: Proceedings of the 9th Workshop on Multimedia & Security, ACM, 2007, pp. 129–140.

[130] S. Goldwasser, G.N. Rothblum, On best-possible obfuscation, in: S.P. Vadhan (Ed.), Theory of Cryptography. TCC 2007, Lecture Notes in Computer Science, vol. 4392, Springer, Berlin, Heidelberg, 2007.

[131] S. Banescu, M. Ochoa, A. Pretschner, N. Kunze, Benchmarking indistinguishability obfuscation—a candidate implementation, in: LNCS, Proc. of 7th International Symposium on ESSoS, 2015. 8978.

[132] D. Apon, Y. Huang, J. Katz, A.J. Malozemoff, Implementing cryptographic program obfuscation. IACR Cryptology ePrint Archive, Report 2014/779, 2014.

[133] B. Barak, Hopes, fears, and software obfuscation, Commun. ACM 59 (3) (2016) 88–96.

[134] V. Costan, S. Devadas, Intel SGX Explained. IACR Cryptology ePrint Archive, Report 2016/86, 2016.

[135] F. Brasser, U. Müller, A. Dmitrienko, K. Kostiainen, S. Capkun, A.-R. Sadeghi, Software Grand Exposure: SGX Cache Attacks are Practical, 2017, arXiv preprint arXiv:1702.07521.

[136] K. Nohl, D. Evans, S. Starbug, H. Plötz, Reverse-engineering a cryptographic RFID tag, in: USENIX Security Symposium, vol. 28, 2008.

[137] G.T. Becker, M. Fyrbiak, C. Kison, Hardware obfuscation: techniques and open challenges, in: Foundations of Hardware IP Protection, Springer, 2017, pp. 105–123.

[138] S. Banescu, C. Collberg, A. Pretschner, Predicting the resilience of obfuscated code against symbolic execution attacks via machine learning, in: 26th USENIX Security Symposium (USENIX Security 17), USENIX Association, Vancouver, BC, 2017, https://www.usenix.org/conference/usenixsecurity17/technical-sessions/presentation/banescu.

[139] J.W. Bos, C. Hubain, W. Michiels, P. Teuwen, Differential computation analysis: hiding your white-box designs is not enough, in: International Conference on Cryptographic Hardware and Embedded Systems, Springer, 2016, pp. 215–236.

[140] P. Kocher, J. Jaffe, B. Jun, Differential power analysis, in: M. Wiener (Ed.), Advances in Cryptology—CRYPTO' 99. CRYPTO 1999, Lecture Notes in Computer Science, vol. 1666, Springer, Berlin, Heidelberg, 1999.

ABOUT THE AUTHORS

Sebastian Banescu is an IT Security Specialist at BMW AG in Munich, where he is involved in various projects regarding the security of the connected car against remote attackers, tuning garages, and malicious car owners. In July 2017, he received his PhD, with distinction, at the Technical University of Munich under the supervision of Prof. Alexander Pretschner. The topic of his PhD thesis was to characterize the strength of software obfuscation against automated man-at-the-end attackers. Before that, he received a MSc in Computer Science and Engineering, from Eindhoven University of Technology in the Netherlands, and a BSc in Computer Science and Engineering, from the Technical University of Cluj-Napoca in Romania.

Alexander Pretschner is full professor of software and systems engineering at the Technical University of Munich, scientific director at the Fortiss Research and Technology Transfer Institute in Munich, and speaker of the board of the Munich Center for Internet Research. Research focus on all aspects of systems quality, specifically testing and security. Diploma and MS degrees in computer science from RWTH Aachen University and the University of Kansas; PhD from Technical University of Munich. Prior positions include those of a senior researcher at ETH Zurich; of a group manager at the Fraunhofer Institute for Experimental Software Engineering and of an adjunct associate professor at the Technical University of Kaiserslautern; and of a full professor at Karlsruhe Institute of Technology. Program or general (co) chair of ICST, MODELS, ESSOS, and CODASPY; associate editor of the Journal of Software Testing, Verification, and Reliability (Wiley); the Journal of Computer and System Sciences (Elsevier); and the Journal of Software Systems Modeling (Springer); former associate editor of the IEEE Transactions on Dependable and Secure Computing.